LIAL
VIDEO WORKBOOK
WITH INTEGRATED REVIEW WORKSHEETS
CHRISTINE VERITY

INTRODUCTORY &
INTERMEDIATE ALGEBRA
SIXTH EDITION

Margaret L. Lial
American River College

John Hornsby
University of New Orleans

Terry McGinnis

 Pearson

ISBN-13: 978-0-13-477550-0
ISBN-10: 0-13-477550-3

CONTENTS for INTEGRATED REVIEW WORKSHEETS

CONTENTS for LIAL VIDEO WORKBOOK

Chapter 1 THE REAL NUMBER SYSTEM

Learning Objectives
1.R.1   Multiply and divide fractions and mixed numbers.
1.R.2   Add and subtract fractions and mixed numbers.
1.R.3   Add and subtract decimals.
1.R.4   Multiply and divide decimals.

Key Terms

Use the vocabulary terms listed below to complete each statement in exercises 1–12.

numerator	denominator	proper fraction
improper fraction	equivalent fractions	lowest terms
prime number	composite number	prime factorization
decimals	place value	percent

1.   Two fractions are _____ when they represent the same portion of a whole.

2.   A fraction whose numerator is larger than its denominator is called an _____.

3.   In the fraction $\frac{2}{9}$, the 2 is the _____.

4.   A fraction whose denominator is larger than its numerator is called a _____.

5.   The _____ of a fraction shows the number of equal parts in a whole.

6.   A _____ has at least one factor other than itself and 1.

7.   In a _____ every factor is a prime number.

8.   The factors of a _____ are itself and 1.

9.   A fraction is written in _____ when its numerator and denominator have no common factor other than 1.

10.   We use _____ to show parts of a whole.

11.   A _____ is assigned to each place to the left or right of the decimal point.

12.   _____ means per one hundred.

Objective 1.R.1 Multiply and divide fractions and mixed numbers.

Video Examples

Review these examples:

Find each quotient, and write it in lowest terms.

$$\frac{2}{5} \div \frac{8}{7}$$

$\frac{2}{5} \div \frac{8}{7} = \frac{2}{5} \cdot \frac{7}{8}$ Multiply by the reciprocal.

$\qquad = \frac{2 \cdot 7}{5 \cdot 4 \cdot 2}$ Multiply and factor.

$\qquad = \frac{7}{20}$

$$\frac{7}{9} \div 14$$

$\frac{7}{9} \div 14 = \frac{7}{9} \cdot \frac{1}{14}$ Multiply by the reciprocal.

$\qquad = \frac{7 \cdot 1}{9 \cdot 2 \cdot 7}$ Multiply and factor.

$\qquad = \frac{1}{18}$

Now Try:

Find each quotient, and write it in lowest terms.

$$\frac{6}{7} \div \frac{9}{8}$$

_____

$$\frac{4}{5} \div 8$$

_____

Practice Exercises

Find each product or quotient, and write it in lowest terms.

1. $\frac{25}{11} \cdot \frac{33}{10}$

1. _____

2. $\frac{5}{4} \div \frac{25}{28}$

2. _____

3. $4\frac{3}{8} \cdot 2\frac{4}{7}$

3. _____

Objective 1.R.2   Add and subtract fractions and mixed numbers.

Video Examples

Review these examples:

Add. Write sum in lowest terms.

$$\frac{5}{21} + \frac{3}{14}$$

Step 1 To find the LCD, factor the denominators to prime factored form.

$21 = 3 \cdot 7$ and $14 = 2 \cdot 7$

7 is a factor of both denominators.

$$21 \quad 14$$
$$\wedge \quad \wedge$$

Step 2 $LCD = 3 \cdot 7 \cdot 2 = 42$

In this example, the LCD needs one factor of 3, one factor of 7 and one factor of 2.

Step 3 Now we can use the second property of 1 to write each fraction with 42 as the denominator.

$$\frac{5}{21} = \frac{5}{21} \cdot \frac{2}{2} = \frac{10}{42} \quad \text{and} \quad \frac{3}{14} = \frac{3}{14} \cdot \frac{3}{3} = \frac{9}{42}$$

Now add the two equivalent fractions to get the sum.

$$\frac{5}{21} + \frac{3}{14} = \frac{10}{42} + \frac{9}{42}$$
$$= \frac{19}{42}$$

Subtract. Write difference in lowest terms.

$$\frac{14}{15} - \frac{5}{8}$$

Since 15 and 8 have no common factors greater than 1, the LCD is $15 \cdot 8 = 120$.

$$\frac{14}{15} - \frac{5}{8} = \frac{14}{15} \cdot \frac{8}{8} - \frac{5}{8} \cdot \frac{15}{15}$$
$$= \frac{112}{120} - \frac{75}{120}$$
$$= \frac{37}{120}$$

Now Try:

Add. Write sum in lowest terms.

$$\frac{7}{12} + \frac{3}{8}$$

_____

Subtract. Write difference in lowest terms.

$$\frac{17}{6} - \frac{13}{7}$$

_____

Practice Exercises

Find each sum or difference, and write it in lowest terms.

4. $\dfrac{23}{45} + \dfrac{47}{75}$

4. _____

5. $2\dfrac{3}{4} + 7\dfrac{2}{3}$

5. _____

6. $12\dfrac{5}{6} - 7\dfrac{7}{8}$

6. _____

Objective 1.R.3   Add and subtract decimals.

Video Examples

Review this example:

Add or subtract as indicated.

$5 - 0.832$

A whole number is assumed to have the decimal point at the right of the number. Write 5 as 5.000.

$$
\begin{array}{r}
5.000 \\
- \ 0.832 \\
\hline
4.168
\end{array}
$$

Now Try:

Add or subtract as indicated.

$8 - 0.976$

_____

Practice Exercises

Add or subtract as indicated.

7. $45.83 + 20.923 + 5.7$

7.   _____

8. $768.5 - 13.402$

8.   _____

9. $689 - 79.832$

9.   _____

Objective 1.R.4   Multiply and divide decimals.

Video Examples

Review these examples:

Multiply.

37.4×5.26

There is 1 decimal place in the first number and
2 decimal places in the second number.
Therefore, there are $1 + 2 = 3$ decimal places in
the answer.

$$
\begin{array}{r}
37.4 \\
\times\ 5.26 \\
\hline
2244 \\
748 \\
1870 \\
\hline
196.724
\end{array}
$$

Divide.

$1191.45 \div 23.5$

Write the problem as follows.

$$23.5\overline{)1191.45}$$

To change 23.5 into a whole number, move the
decimal point one place to the right. Move the
decimal point in 1191.45 the same number of
places to the right, to get 11,914.5.

$$235\overline{)11{,}914.5}$$

Move the decimal point straight up and divide as
with whole numbers.

$$
\begin{array}{r}
50.7 \\
235\overline{)11{,}914.5} \\
1175 \\
\hline
1645 \\
1645 \\
\hline
0
\end{array}
$$

Now Try:

Multiply.

26.8×9.37

_____

Divide.

$2472.12 \div 76.3$

_____

Multiply or divide as indicated.

97.648×100

Move the decimal point 2 places to the right because 100 has two zeros.
$97.648 \times 100 = 9764.8$

$85.1 \div 1000$

Move the decimal point three places to the left because 1000 has three zeros. Insert a zero in front of the 8 to do this.
$85.1 \div 1000 = 0.0851$

Multiply or divide as indicated.

51.302×100

_____

$98.6 \div 1000$

_____

Practice Exercises

Multiply or divide as indicated.

10.　　14.64×0.16

10. _____

11.　　$498.624 \div 21.2$

11. _____

12.　　$429.2 \div 1000$

12. _____

Chapter 2 EQUATIONS, INEQUALITIES, AND APPLICATIONS

Learning Objectives
2.R.1   Complete a table of fractions, decimals, and percents.
2.R.2   Write statements that change the direction of inequality symbols.
2.R.3   Translate word phrases to algebraic expressions.
2.R.4   Evaluate algebraic expressions, given values for the variables.
2.R.5   Identify solutions of equations.
2.R.6   Translate sentences to equations.
2.R.7   Classify numbers and graph them on number lines.
2.R.8   Identify properties of real numbers.
2.R.9   Simplify expressions.

Key Terms

Use the vocabulary terms listed below to complete each statement in exercises 1–20.

decimals	percent	inequality	variable	constant
algebraic expression		equation	solution	integers
natural numbers		whole numbers	number line	
negative number		positive number	signed numbers	
rational number		irrational number	real numbers	
identity element for addition			identity element for multiplication	

1. We use _____ to show parts of a whole.

2. _____ means per one hundred.

3. An _____ is a statement that two expressions may not be equal.

4. A(n) _____ is a statement that says two expressions are equal.

5. A _____ is a symbol, usually a letter, used to represent an unknown number.

6. A collection of numbers, variables, operation symbols, and grouping symbols is an_____.

7. Any value of a variable that makes an equation true is a(n) _____ of the equation.

8. A _____ is a fixed, unchanging number.

9. The set {0, 1, 2, 3, …} is called the set of _____.

10. The whole numbers together with their opposites and 0 are called _____.

11.   The set { 1, 2, 3, …} is called the set of _____.

12.   A _____ shows the ordering of the real numbers on a line.

13.   A real number that is not a rational number is called a(n) _____.

14.   A number located to the left of 0 on a number line is a _____.

15.   A number located to the right of 0 on a number line is a _____.

16.   Numbers that can be represented by points on the number line are
       _____.

17.   Positive numbers and negative numbers are _____.

18.   A number that can be written as the quotient of two integers is a
       _____.

19.   When the _____, which is 0, is added to a
       number, the number is unchanged.

20.   When a number is multiplied by the _____,
       which is 1, the number is unchanged.

Objective 2.R.1 Complete a table of fractions, decimals, and percents.

Video Examples

Review these examples:

Write each fraction as a decimal. For repeating decimals, write the answer by first using bar notation and then rounding to the nearest thousandth.

$$\frac{27}{8}$$

Divide 27 by 8. Add a decimal point and as many 0s as necessary.

```
      3.375
  8)27.000
    24
    ──
     30
     24
     ──
      60
      56
      ──
       40
       40
       ──
        0
```

$$\frac{27}{8} = 3.375$$

$$\frac{26}{9}$$

```
      2.888...
  9)26.000...
    18
    ──
     80
     72
     ──
      80
      72
      ──
       80
       72
       ──
        8
```

$$\frac{26}{9} = 2.888...$$

The remainder is never 0. Because 8 is always left after the subtraction, this quotient is a repeating decimal. A convenient notation for a repeating decimal is a bar over the digit (or digits) that repeats.

$$\frac{26}{9} = 2.\overline{8} \quad \text{or} \quad \frac{26}{9} \approx 2.889 \quad \text{rounded}$$

Now Try:

Write each fraction as a decimal. For repeating decimals, write the answer by first using bar notation and then rounding to the nearest thousandth.

$$\frac{7}{20}$$

_____

$$\frac{32}{9}$$

_____

Convert each percent to a decimal and each decimal to a percent.

0.29%

0.29% = 0.0029

0.54

0.54 = 54%

Convert each percent to a decimal and each decimal to a percent.

0.43%

_____

0.91

_____

Write the percent as a fraction. Give answers in lowest terms as needed.

65%

We use the fact that, $1\% = \dfrac{1}{100}$, and convert as follows.

$$65\% = 65 \cdot 1\% = 65 \cdot \dfrac{1}{100} = \dfrac{65}{100}$$

In lowest terms,

$$\dfrac{65}{100} = \dfrac{13 \cdot 5}{20 \cdot 5} = \dfrac{13}{20}.$$

Thus,

$$65\% = \dfrac{13}{20}.$$

Write the percent as a fraction. Give answers in lowest terms as needed.

52%

_____

Practice Exercises

Write the fraction as a decimal and as a percent.

1. $\dfrac{3}{100}$

1. _____

Write the decimal as a fraction and as a percent.

2. 0.875

2. _____

Write the percent as a decimal and as a fraction.

3. 87%

3. _____

Objective 2.R.2 Write statements that change the direction of inequality symbols.

Video Examples

Review this example:

Write the statement as another true statement with the inequality symbol reversed.

$9 > 7$

$7 < 9$

Now Try:

Write the statement as another true statement with the inequality symbol reversed.

$15 > 11$

_____

Practice Exercises

Write each statement with the inequality symbol reversed.

4. $\dfrac{3}{4} > \dfrac{2}{3}$

4. _____

5. $12 \geq 8$

5. _____

6. $0.002 > 0.0002$

6. _____

Objective 2.R.3 Translate word phrases to algebraic expressions.

Video Examples

Review these examples:

Write each word phrase as an algebraic expression, using x as the variable.

A number subtracted from 13

$13 - x$

The product of 3 and the difference between a number and 5

$3 \cdot (x - 5)$, or $3(x - 5)$

Now Try:

Write each word phrase as an algebraic expression, using x as the variable.

A number subtracted from 12

_____

The product of 5 and the difference between a number and 6

_____

Practice Exercises

Write each word phrase as an algebraic expression. Use x as the variable.

7. Ten times a number, added to 21

7. _____

8. 11 fewer than eight times a number

8. _____

9. Half a number subtracted from two-thirds of the number

9. _____

Objective 2.R.4 Evaluate algebraic expressions, given values for the variables.

Video Examples

Review these examples:	**Now Try:**
Find the value of each algebraic expression for $p = 4$ and then $p = 7$.	Find the value of each algebraic expression for $k = 6$ and then $k = 9$.

$$5p^2$$

For $p = 4$,

$$5p^2 = 5 \cdot 4^2 \quad \text{Let } p = 4.$$
$$= 5 \cdot 16 \quad \text{Square 4.}$$
$$= 80 \quad \text{Multiply.}$$

For $p = 7$,

$$5p^2 = 5 \cdot 7^2 \quad \text{Let } p = 7.$$
$$= 5 \cdot 49 \quad \text{Square 7.}$$
$$= 245 \quad \text{Multiply.}$$

Now Try:

$$7k^2$$

_____

Find the value of each expression for $x = 7$ and $y = 6$.

$$3x + 4y$$

Replace x with 7 and y with 6.
$$3x + 4y = 3 \cdot 7 + 4 \cdot 6$$
$$= 21 + 24 \quad \text{Multiply.}$$
$$= 45 \quad \text{Add.}$$

$$\frac{8x - 6y}{4x - 3y}$$

Replace x with 7 and y with 6.
$$\frac{8x - 6y}{4x - 3y} = \frac{8 \cdot 7 - 6 \cdot 6}{4 \cdot 7 - 3 \cdot 6}$$
$$= \frac{56 - 36}{28 - 18} \quad \text{Multiply.}$$
$$= \frac{20}{10} \quad \text{Subtract.}$$
$$= 2 \quad \text{Divide.}$$

Find the value of each expression for
$x = 8$ and $y = 4$.
$$5x + 6y$$

_____

$$\frac{9x + 2y}{3x - 5y}$$

_____

Practice Exercises

Find the value of each expression if $x = 2$ and $y = 4$.

10. $9x - 3y + 2$

10. _____

11. $\dfrac{2x + 3y}{3x - y + 2}$

11. _____

12. $\dfrac{3y^2 + 2x^2}{5x + y^2}$

12. _____

Objective 2.R.5 Identify solutions of equations.

Video Examples

Review these examples:

Decide whether the given number is a solution of the equation.

$$6p+5=29; \quad 4$$

$$6p+5=29$$

$$6\cdot 4+5\overset{?}{=}29$$

$$24+5\overset{?}{=}29$$

$$29=29 \quad \text{True – the left side of the equation equals the right side.}$$

The number 4 is a solution of the equation.

$$8n-7=41; \quad \frac{9}{4}$$

$$8n-7=41$$

$$8\cdot\frac{9}{4}-7\overset{?}{=}41$$

$$18-7\overset{?}{=}41$$

$$11=41 \quad \text{False – the left side does not equal the right side.}$$

The number $\frac{9}{4}$ is not a solution of the equation.

Now Try:

Decide whether the given number is a solution of the equation.

$$7k+5=26; \quad 3$$

_____

$$9m-8=41; \quad 5$$

_____

Practice Exercises

Decide whether the given number is a solution of the equation.

13. $5+3x^2=19; \quad 2$

13. _____

14. $\dfrac{m+2}{3m-10}=1; \quad 8$

14. _____

15. $3y+5(y-5)=7; \quad 4$

15. _____

Objective 2.R.6 Translate sentences to equations.

Video Examples

Review this example :

Write the word sentence as an equation.
Use x as the variable.

Nine less than five times a number is equal to twenty.

Five times
a number   less   nine   is equal to   twenty.
$\downarrow$ $\downarrow$ $\downarrow$ $\downarrow$ $\downarrow$
$5x$ $-$ 9 $=$ 20

$$5x - 9 = 20$$

Now Try:

Write the word sentence as an equation. Use x as the variable.

Seven less than three times a number is equal to twelve.

_____

Practice Exercises

Write each word sentence as an equation. Use x as the variable.

16.   Ten divided by a number is two more than the number.

16. _____

17.   The product of six and five more than a number is nineteen.

17. _____

18.   Seven times a number subtracted from 61 is 13 plus the number.

18. _____

Objective 2.R.7 Classify numbers and graph them on number lines.

Video Examples

Review these examples:

Graph each number on a number line.

$$-3\frac{1}{2},\ -\frac{3}{2},\ 0,\ \frac{7}{2},\ 1$$

To locate the improper fractions on the number line, write them as mixed numbers or decimals.

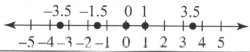

Now Try:

Graph each number on a number line.

$$\frac{1}{2},\ 0,\ -3,\ -\frac{5}{2}$$

$$\xleftarrow{\hspace{0.3cm}}\underset{-5\,-4\,-3\,-2\,-1\ \ 0\ \ 1\ \ 2\ \ 3\ \ 4\ \ 5}{+\!+\!+\!+\!+\!+\!+\!+\!+\!+\!+}\xrightarrow{\hspace{0.3cm}}$$

List the numbers in the following set that belong to each set of numbers.

$$\left\{-6,\ -\frac{5}{6},\ 0,\ 0.\overline{3},\ \sqrt{3},\ 4\frac{1}{5},\ 6,\ 6.7\right\}$$

Natural numbers

Answer: 6

Whole numbers

Answer: 0 and 6

Integers

Answer: −6, 0, and 6

Rational numbers

Answer: $-6,\ -\frac{5}{6},\ 0,\ 0.\overline{3},\ 4\frac{1}{5},\ 6,\ 6.7$

Irrational numbers

Answer: $\sqrt{3}$

Real numbers

Answer: all the numbers in the set

List the numbers in the following set that belong to each set of numbers.

$$\left\{-10, -\frac{5}{8},\ 0,\ 0.\overline{4},\ \sqrt{5},\ 5\frac{1}{2},\ 7,\ 9.9\right\}$$

Natural numbers

_____

Whole numbers

_____

Integers

_____

Rational numbers

_____

Irrational numbers

_____

Real numbers

_____

Practice Exercises

List the numbers set that belong to each set of numbers. $\left\{-8, -\frac{3}{5}, \sqrt{2}, 1.\overline{6},\ 3\frac{1}{4},\ 5,\ 7.9\right\}$

19. Whole numbers

20. Irrational numbers

21. Rational numbers

19. _____

20. _____

21. _____

Graph the group of rational numbers on a number line.

22. −4.5, −2.3, 1.7, 4.2

22.

$$\xleftarrow{\hspace{0.3cm}}\underset{-5\,-4\,-3\,-2\,-1\ \ 0\ \ 1\ \ 2\ \ 3\ \ 4\ \ 5}{+\!+\!+\!+\!+\!+\!+\!+\!+\!+\!+}\xrightarrow{\hspace{0.3cm}}$$

Objective 2.R.8 Identify properties of real numbers.

Practice Exercises

Decide whether each statement is an example of a commutative, *an* associative, *an* identity, *an* inverse, *or the* distributive *property.*

23. $12 \cdot 1 = 12$ **23.** _____

24. $4(ab) = (4a)b$ **24.** _____

25. $-2(5y) + 2(9z) = -2(5y - 9z)$ **25.** _____

26. $-\dfrac{3}{5} \cdot \left(-\dfrac{5}{3}\right) = 1$ **26.** _____

27. $-4(4 + z) = (4 + z)(-4)$ **27.** _____

28. $-4 + 4 = 0$ **28.** _____

29. $y + 4 = 4 + y$ **29.** _____

30. $4 + 0 = 4$ **30.** _____

31. $4r + (3s + 14t) = (4r + 3s) + 14t$ **31.** _____

Objective 2.R.9 Simplify expressions.

Video Examples

Review these examples:

Simplify each expression.

$$7 + 5(9k + 6)$$

$$7 + 5(9k + 6) = 7 + 5(9k) + 5(6)$$
$$= 7 + 45k + 30$$
$$= 37 + 45k$$

$$9 - (4y - 6)$$

$$9 - (4y - 6) = 9 - 1(4y - 6)$$
$$= 9 - 4y + 6$$
$$= 15 - 4y$$

Now Try:

Simplify each expression.

$$8 + 9(2x + 7)$$

_____

$$8 - (7x - 3)$$

_____

Practice Exercises

Simplify each expression.

32. $4(2x + 5) + 7$

32. _____

33. $-4 + s - (12 - 21)$

33. _____

34. $-2(-5x + 2) + 7$

34. _____

Chapter 3 GRAPHS OF LINEAR EQUATIONS AND INEQUALITIES IN TWO VARIABLES

Learning Objectives
3.R.1   Solve linear equations using more than one property of equality.
3.R.2   Solve a formula for one variable, given the values of the other variables.
3.R.3   Solve linear inequalities by using both properties of inequality.

Key Terms

Use the vocabulary terms listed below to complete each statement in exercises 1 8.

conditional equation	**identity**	**contradiction**
formula	**inequalities**	**interval**
interval notation	**linear inequality**	

1.   An equation with no solution is called a(n) _____ .

2.   A(n) _____ is an equation that is true for some values of the variable and false for other values.

3.   An equation that is true for all values of the variable is called a(n) _____ .

4.   An equation in which variables are used to describe a relationship is called a(n) _____ .

5.   A portion of a number line is called a(n) _____ .

6.   A(n) _____ can be written in the form $Ax + B < C$, $Ax + B \leq C$, $Ax + B > C$, or $Ax + B \geq C$, where A, B, and C are real numbers with $A \neq 0$.

7.   Algebraic expressions related by $<$, $\leq$, $>$, or $\geq$ are called _____ .

8.   The _____ for $a \leq x < b$ is $[a, b)$.

Objective 3.R.1 Solve linear equations using more than one property of equality.

Video Examples

Review these examples:

Solve $5(k-4)-k=k-2$.

Step 1 Clear parentheses using the distributive property.

$$5(k-4)-k=k-2$$
$$5(k)+5(-4)-k=k-2$$
$$5k-20-k=k-2$$
$$4k-20=k-2$$

Step 2 $4k-20-k=k-2-k$
$$3k-20=-2$$
$$3k-20+20=-2+20$$
$$3k=18$$

Step 3 $$\frac{3k}{3}=\frac{18}{3}$$
$$k=6$$

Step 4 Check by substituting 6 for k in the original equation.

$$5(k-4)-k=k-2$$
$$5(6-4)-6\overset{?}{=}6-2$$
$$5(2)-6\overset{?}{=}4$$
$$10-6\overset{?}{=}4$$
$$4=4 \quad \text{True}$$

The solution, 6, checks, so the solution set is {6}.

Solve $6x-18=6(x-3)$.

$$6x-18=6(x-3)$$
$$6x-18=6x-18$$
$$6x-18-6x=6x-18-6x$$
$$-18=-18$$
$$-18+18=-18+18$$
$$0=0$$

The solution set is {all real numbers}.

Now Try:

Solve $9(k-2)-k=k+10$.

_____

Solve $3x+4(x-5)=7x-20$.

_____

Solve $3x+4(x-5)=7x+5$.

$$3x+4(x-5)=7x+5$$
$$3x+4x-20=7x+5$$
$$7x-20=7x+5$$
$$7x-20-7x=7x+5-7x$$
$$-20=5 \quad \text{False}$$

There is no solution. The solution set is $\varnothing$.

Solve $-5x+17=x-6(x+3)$.

_____

Practice Exercises

Solve each equation and check your solution.

1. $7t+6=11t-4$

1. _____

2. $3a-6a+4(a-4)=-2(a+2)$

2. _____

3. $3(t+5)=6-2(t-4)$

3. _____

Solve each equation and check your solution.

4. $3(6x-7)=2(9x-6)$

4. _____

5. $6y-3(y+2)=3(y-2)$

5. _____

6. $3(r-2)-r+4=2r+6$

6. _____

Objective 3.R.2 Solve a formula for one variable, given the values of the other variables.

Video Examples

Review this example:

Find the value of the remaining variable in the formula.

$A = LW$; $A = 54, L = 8$

Substitute the given values for A and L into the formula.

$$A = LW$$
$$54 = 8W$$
$$\frac{54}{8} = \frac{8W}{8}$$
$$6.75 = W$$

The width is 6.75. Since $8(6.75) = 54$, the answer checks.

Now Try:

Find the value of the remaining variable in the formula.

$A = LW$; $A = 88, L = 16$

_____

Practice Exercises

In the following exercises, a formula is given, along with the values of all but one of the variables in the formula. Find the value of the variable that is not given.

7. $S = \dfrac{a}{1-r}$; $S = 60, r = 0.4$

7. _____

8. $I = prt$; $I = 288, r = 0.04, t = 3$

8. _____

9. $A = \dfrac{1}{2}(b + B)h$; $b = 6, B = 16, A = 132$

9. _____

Objective 3.R.3 Solve linear inequalities by using both properties of inequality.

Video Examples

Review this example:

Solve $4x + 3 - 7 > -2x + 8 + 3x$. Graph the solution set.

Step 1 Combine like terms and simplify.

$$4x + 3 - 7 > -2x + 8 + 3x$$

$$4x - 4 > x + 8$$

Step 2 Use the addition property of inequality.

$$4x - 4 - x > x + 8 - x$$

$$3x - 4 > 8$$

$$3x - 4 + 4 > 8 + 4$$

$$3x > 12$$

Step 3 Use the multiplication property of inequality.

$$\frac{3x}{3} > \frac{12}{3}$$

$$x > 4$$

The solution set is $(4, \infty)$. The graph is shown below.

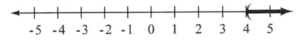

Now Try:

Solve $8x - 5 + 4 \geq 6x - 3x + 9$. Graph the solution set.

Practice Exercises

Solve each inequality. Write the solution set in interval notation and then graph it.

10. $4(y-3)+2>3(y-2)$

10. _____

11. $-3(m+2)+3\le-4(m-2)-6$

11. _____

12. $7(2-x)\le-2(x-3)-x$

12. _____

Chapter 4 SYSTEMS OF LINEAR EQUATIONS AND INEQUALITIES

Learning Objectives
4.R.1 Decide whether a given ordered pair is a solution of a given equation.
4.R.2 Graph linear equations by plotting ordered pairs.
4.R.3 Use a linear equation to model data.
4.R.4 Graph linear inequalities in two variables.

Key Terms

Use the vocabulary terms listed below to complete each statement in exercises 1–19.

line graph	linear equation in two variables	ordered pair
table of values	rectangular (Cartesian) coordinate system	
x-axis	y-axis	origin	quadrants
plane	coordinates	plot	scatter diagram
graph	graphing	y intercept	x-intercept
linear inequality in two variables	boundary line	

1. A _____ uses dots connected by lines to show trends.

2. An equation that can be written in the form $Ax + By = C$, where A, B, and C are real numbers and A, $B \neq 0$, is called a _____ _____.

3. _____ are the numbers in the ordered pair that specify the location of a point on a rectangular coordinate system.

4. In a coordinate system, the horizontal axis is called the _____.

5. In a coordinate system, the vertical axis is called the _____.

6. A pair of numbers written between parentheses in which order is important is called a(n) _____.

7. Together, the x-axis and the y-axis form a _____.

8. A coordinate system divides the plane into four regions called _____.

9. The axis lines in a coordinate system intersect at the _____.

10. To _____ an ordered pair is to find the corresponding point on a coordinate system.

11. A graph of ordered pairs is called a _____.

12. A table showing selected ordered pairs of numbers that satisfy an equation is called a
_____.

13. A flat surface determined by two intersecting lines is a _____.

14. If a graph intersects the y-axis at k, then the _____ is (0, k).

15. If a graph intersects the x-axis at k, then the _____ is (k, 0).

16. The process of plotting the ordered pairs that satisfy a linear equation and drawing a
line through them is called _____.

17. The set of all points that correspond to the ordered pairs that satisfy the equation is
called the _____ of the equation.

18. In the graph of a linear inequality, the _____ separates
the region that satisfies the inequality from the region that does not satisfy the
inequality.

19. An inequality that can be written in the form $Ax + By < C$, $Ax + By > C$,
$Ax + By \leq C$, or $Ax + By \geq C$ is called a _____.

Objective 4.R.1 Decide whether a given ordered pair is a solution of a given equation.

Video Examples

Review these examples:

Decide whether each ordered pair is a solution of the equation $4x + 5y = 40$.

(5, 4)

Substitute 5 for x and 4 for y in the given equation.

$$4x + 5y = 40$$
$$4(5) + 5(4) \overset{?}{=} 40$$
$$20 + 20 \overset{?}{=} 40$$
$$40 = 40 \quad \text{True}$$

This result is true, so $(5, 4)$ is a solution of $4x + 5y = 40$.

(−3, 6)

Substitute −3 for x and 6 for y in the given equation.

$$4x + 5y = 40$$
$$4(-3) + 5(6) \overset{?}{=} 40$$
$$-12 + 30 \overset{?}{=} 40$$
$$18 = 40 \quad \text{False}$$

This result is false, so $(-3, 6)$ is not a solution of $4x + 5y = 40$.

Now Try:

Decide whether each ordered pair is a solution of the equation $3x - 4y = 12$.

(8, 3)

_____

(5, −4)

_____

Practice Exercises

Decide whether the given ordered pair is a solution of the given equation.

1. $4x - 3y = 10;\ (1,\ 2)$ 1. _____

2. $2x - 3y = 1;\ \left(0,\ \frac{1}{3}\right)$ 2. _____

3. $x = -7;\ (-7,\ 9)$ 3. _____

Objective 4.R.2 Graph linear equations by plotting ordered pairs.

Video Examples

Review this example:

Graph $2x + 3y = 6$.

First let $x = 0$ and then let $y = 0$ to determine two ordered pairs.

$$
\begin{array}{c|c}
2(0) + 3y = 6 & 2x + 3(0) = 6 \\
0 + 3y = 6 & 2x + 0 = 6 \\
3y = 6 & 2x = 6 \\
y = 2 & x = 3
\end{array}
$$

The ordered pairs are $(0, 2)$ and $(3, 0)$. Find a third ordered pair by choosing a number other than 0 for x or y. We choose $y = 4$.

$$2x + 3(4) = 6$$
$$2x + 12 = 6$$
$$2x = -6$$
$$x = -3$$

This gives the ordered pair $(-3, 4)$. We plot the three ordered pairs $(0, 2)$, $(3, 0)$, and $(-3, 4)$ and draw a line through them.

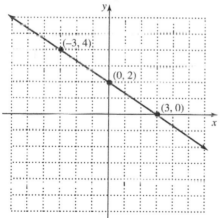

Now Try:

Graph $x + y = 3$.

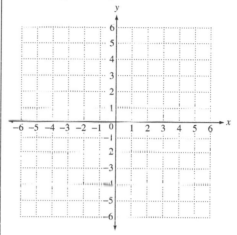

Practice Exercises

Complete the ordered pairs for each equation. Then graph the equation by plotting the points and drawing a line through them.

4. $y = 3x - 2$

$(0,\ \ )$

$(\ \ , 0)$

$(2,\ \ )$

4.

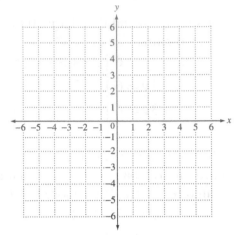

5. $x - y = 4$

$(0,\ \ )$

$(\ \ , 0)$

$(-2,\ \ )$

5.

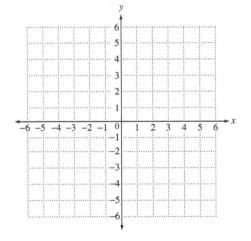

6. $x = 2y + 1$

$(0,\ \ )$

$(\ \ , 0)$

$(\ \ , -2)$

6.

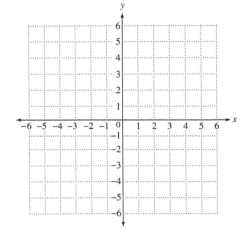

Objective 4.R.3 Use a linear equation to model data.

Video Examples

Review these examples:

Every year sea turtles return to a certain group of islands to lay eggs. The number of turtle eggs that hatch can be approximated by the equation $y = -70x + 3260$, where y is the number of eggs that hatch and $x = 0$ representing 1990.

a. Use this equation to find the number of eggs that hatched in 1995, 2000, and 2005, and 2015.

Substitute the appropriate value for each year x to find the number of eggs hatched in that year.

For 1995:

$y = -70(5) + 3260$ 1995 1990 − 5

$y = 2910$ eggs   Replace x with 5.

For 2000:

$y = -70(10) + 3260$ $2000 - 1990 = 10$

$y = 2560$ eggs   Replace x with 10.

For 2005:

$y = -70(15) + 3260$ $2005 - 1990 = 15$

$y = 2210$ eggs   Replace x with 15.

For 2015:

$y = -70(25) + 3260$ $2015 - 1990 = 25$

$y = 1510$ eggs   Replace x with 25.

b. Write the information from part (a) as four ordered pairs, and use them to graph the given linear equation.

Since x represents the year and y represents the number of eggs, the ordered pairs are (5, 2910), (10, 2560), (15, 2210), and (25, 1510).

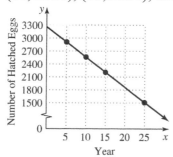

Now Try:

Suppose that the demand and price for a certain model of calculator are related by the equation $y = 45 - \frac{3}{5}x$, where y is the price (in dollars) and x is the demand (in thousands of calculators).

a. Assuming that this model is valid for a demand up to 50,000 calculators, use this equation to find the price of calculators at each level of demand.

0 calculators _____

5000 calculators _____

20,000 calculators _____

45,000 calculators _____

b. Write the information from part (a) as four ordered pairs, and use them to graph the given linear equation.

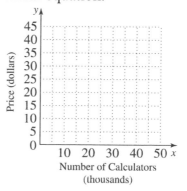

c. Use this graph and the equation to estimate the number of eggs that will hatch in 2010.

For 2010, $x = 20$. On the graph, find 20 on the horizontal axis, move up to the graphed line and then across to the vertical axis. It appears that in 2010, there were about 1900 eggs.

To use the equation, substitute 20 for x.

$$y = -70(20) + 3260$$

$$y = 1860 \text{ eggs}$$

This result for 2020 is close to our estimate of 1900 eggs from the graph.

c. Use this graph and the equation to estimate the price of 30,000 calculators.

_____

Practice Exercises

Solve each problem. Then graph the equation.

7. The profit y in millions of dollars earned by a small computer company can be approximated by the linear equation $y = 0.63x + 4.9$, where $x = 0$ corresponds to 2014, $x = 1$ corresponds to 2015, and so on. Use this equation to approximate the profit in each year from 2014 through 2017.

7. 2014 _____

2015 _____

2016 _____

2017 _____

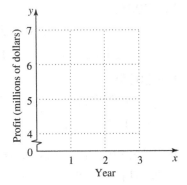

8. The number of band instruments sold by Elmer's
    Music Shop can be approximated by the equation
    $y = 325 + 42x$, where y is the number of
    instruments sold and x is the time in years, with
    $x = 0$ representing 2013. Use this equation to
    approximate the number of instruments sold in each
    year from 2013 through 2016.

8. 2013 _____

   2014 _____

   2015 _____

   2016 _____

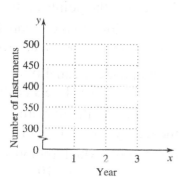

9. According to *The Old Farmer's Almanac*, the
    temperature in degrees Celsius can be determined by
    the equation $y = \frac{1}{3}x + 4$, where x is the number of
    cricket chirps in 25 seconds and y is the temperature
    in degrees Celsius. Use this equation to find the
    temperature when there are 48 chirps, 54 chirps, 60
    chirps, and 66 chirps.

9. 48 _____

   54 _____

   60 _____

   66 _____

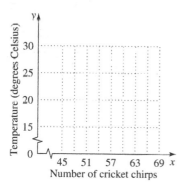

Objective 4.R.4 Graph linear inequalities in two variables.

Video Examples

Review these examples:

Graph $2x + 5y > -10$.

This equation does not include equality. Therefore, the points on the line $2x + 5y = -10$ do not belong to the graph. However, the line still serves as a boundary for two regions.

To graph the inequality, first graph the equation $2x + 5y > -10$. Use a dashed line to show that the points on the line are not solutions of the inequality $2x + 5y > -10$. Then choose a test point to see which region satisfies the inequality.

$$2x + 5y > -10$$

$$2(0) + 5(0) \overset{?}{>} -10$$

$$0 + 0 \overset{?}{>} -10$$

$$0 > -10 \quad \text{True}$$

Since $0 > -10$ is true, the graph of the inequality is the region that contains (0, 0). Shade that region.

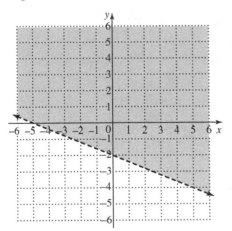

Now Try:

Graph $5x + 4y > 20$.

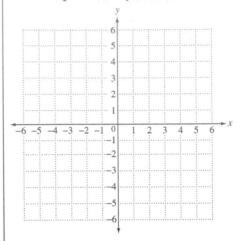

Graph $x - 4 \leq -1$.

First, solve the inequality for x.

$x \leq 3$

Now graph the line $x = 3$, a vertical line through the point $(3, 0)$. Use a solid line, and choose $(0, 0)$ as a test point.

$0 \leq 3$ True

Because $0 \leq 3$ is true, we shade the region containing $(0, 0)$.

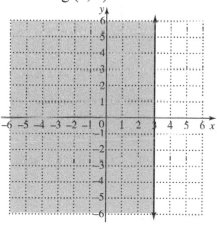

Graph $y \geq -1$.

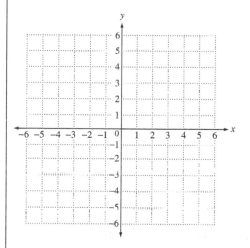

Practice Exercises

Graph each linear inequality.

10. $y \geq x - 1$

10.

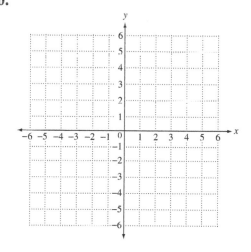

11. $y > -x + 2$

11.

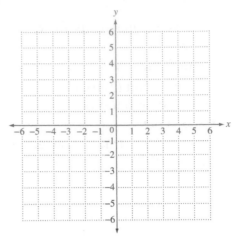

12. $3x - 4y - 12 > 0$

12.

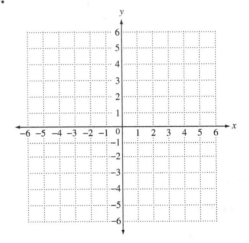

Chapter 5 EXPONENTS AND POLYNOMIALS

Learning Objectives
5.R.1   Evaluate algebraic expressions, given values for the variables.
5.R.2   Solve applied problems that involve real number operations.
5.R.3   Simplify expressions.
5.R.4   Solve a formula for one variable, given the values of the other variables.

Key Terms

Use the vocabulary terms listed below to complete each statement in exercises 1–14.

variable	constant	algebraic expression	equation
solution	product	quotient	dividend
divisor	reciprocals	term	numerical coefficient
like terms	formula		

1.   A(n) _____ is a statement that says two expressions are equal.

2.   A _____ is a symbol, usually a letter, used to represent an unknown number.

3.   A collection of numbers, variables, operation symbols, and grouping symbols is an_____.

4.   Any value of a variable that makes an equation true is a(n) _____ of the equation.

5.   A _____ is a fixed, unchanging number.

6.   The answer to a division problem is called the _____.

7.   Pairs of numbers whose product is 1 are called _____.

8.   The answer to a multiplication problem is called the _____.

9.   In the division $x \div y$, x is called the _____.

10.   In the division $x \div y$, y is called the _____.

11.   In the term $4x^2$, "4" is the_____.

12.   A number, a variable, or a product or quotient of a number and one or more variables raised to powers is called a _____.

13.   Terms with exactly the same variables, including the same exponents, are called _____.

14.   An equation in which variables are used to describe a relationship is called a(n) _____.

Objective 5.R.1 Evaluate algebraic expressions, given values for the variables.

Video Examples

Review these examples:

Find the value of each algebraic expression for $p = 4$ and then $p = 7$.

$$5p^2$$

For $p = 4$,

$$5p^2 = 5 \cdot 4^2 \quad \text{Let } p = 4.$$
$$= 5 \cdot 16 \quad \text{Square 4.}$$
$$= 80 \quad \text{Multiply.}$$

For $p = 7$,

$$5p^2 = 5 \cdot 7^2 \quad \text{Let } p = 7.$$
$$= 5 \cdot 49 \quad \text{Square 7.}$$
$$= 245 \quad \text{Multiply.}$$

Now Try:

Find the value of each algebraic expression for $k = 6$ and then $k = 9$.

$$7k^2$$

_____

Find the value of each expression for $x = 7$ and $y = 6$.

$$3x + 4y$$

Replace x with 7 and y with 6.

$$3x + 4y = 3 \cdot 7 + 4 \cdot 6$$
$$= 21 + 24 \quad \text{Multiply.}$$
$$= 45 \quad \text{Add.}$$

$$\frac{8x - 6y}{4x - 3y}$$

Replace x with 7 and y with 6.

$$\frac{8x - 6y}{4x - 3y} = \frac{8 \cdot 7 - 6 \cdot 6}{4 \cdot 7 - 3 \cdot 6}$$
$$= \frac{56 - 36}{28 - 18} \quad \text{Multiply.}$$
$$= \frac{20}{10} \quad \text{Subtract.}$$
$$= 2 \quad \text{Divide.}$$

Find the value of each expression for $x = 8$ and $y = 4$.

$$5x + 6y$$

_____

$$\frac{9x + 2y}{3x - 5y}$$

_____

Practice Exercises

Find the value of each expression if $x = 2$ and $y = 4$.

1. $9x - 3y + 2$

    1. _____

2. $\dfrac{2x + 3y}{3x - y + 2}$

    2. _____

3. $\dfrac{3y^2 + 2x^2}{5x + y^2}$

    3. _____

Objective 5.R.2 Solve applied problems that involve real number operations.

Practice Exercises

Perform the indicated operations.

4.　Find the sum of the numbers 13, 7, –1, –6, and –8.　　4. _____

Find the average of each group of numbers.

5.　13, 7, –1, –6, and –8.　　　　　　　　　　　　5. _____

6.　All integers between –7 and 3, including both
　　–7 and 3.　　　　　　　　　　　　　　　　6. _____

Objective 5.R.3 Simplify expressions.

Video Examples

Review these examples:

Simplify each expression.

$$7 + 5(9k + 6)$$

$$7 + 5(9k + 6) = 7 + 5(9k) + 5(6)$$
$$= 7 + 45k + 30$$
$$= 37 + 45k$$

$$9 - (4y - 6)$$

$$9 - (4y - 6) = 9 - 1(4y - 6)$$
$$= 9 - 4y + 6$$
$$= 15 - 4y$$

Now Try:

Simplify each expression.

$$8 + 9(2x + 7)$$

_____

$$8 - (7x - 3)$$

_____

Practice Exercises

Simplify each expression.

7. $4(2x + 5) + 7$

7. _____

8. $-4 + s - (12 - 21)$

8. _____

9. $-2(-5x + 2) + 7$

9. _____

Objective 5.R.4 Solve a formula for one variable, given the values of the other variables.

Video Examples

Review this example:

Find the value of the remaining variable in the formula.

$A = LW;\quad A = 54,\ L = 8$

Substitute the given values for A and L into the formula.

$$A = LW$$
$$54 = 8W$$
$$\frac{54}{8} = \frac{8W}{8}$$
$$6.75 = W$$

The width is 6.75. Since $8(6.75) = 54$, the answer checks.

Now Try:

Find the value of the remaining variable in the formula.

$A = LW;\quad A = 88,\ L = 16$

_____

Practice Exercises

In the following exercises, a formula is given, along with the values of all but one of the variables in the formula. Find the value of the variable that is not given.

10. $S = \dfrac{a}{1-r};\ S = 60,\ r = 0.4$

10. _____

11. $I = prt;\ I = 288,\ r = 0.04,\ t = 3$

11. _____

12. $A = \frac{1}{2}(b+B)h;\ b = 6,\ B = 16,\ A = 132$

12. _____

Chapter 6 FACTORING AND APPLICATIONS

Learning Objectives

6.R.1   Write numbers in prime factored form.
6.R.2   Solve multiplication and division problems involving 0.
6.R.3   Solve a formula for a specified variable.
6.R.4   Know the vocabulary for polynomials.
6.R.5   Multiply two polynomials.
6.R.6   Multiply binomials using the FOIL method.
6.R.7   Square binomials.
6.R.8   Find the product of the sum and difference of two terms.
6.R.9   Divide a polynomial by a polynomial.

Key Terms

Use the vocabulary terms listed below to complete each statement in exercises 1−22.

prime number	composite number	prime factorization	
quotient	dividend	divisor	reciprocals
formula	term	like terms	polynomial
descending powers	degree of a term	degree of a polynomial	
monomial	binomial	trinomial	FOIL
outer product	inner product	conjugate	binomial

1. A _____ has at least one factor other than itself and 1.

2. In a _____ every factor is a prime number.

3. The factors of a _____ are itself and 1.

4. The answer to a division problem is called the _____.

5. Pairs of numbers whose product is 1 are called _____.

6. In the division $x \div y$, x is called the _____.

7. In the division $x \div y$, y is called the _____.

8. An equation in which variables are used to describe a relationship is called a(n)
    _____.

9. The _____ is the sum of the exponents on the variables in
    that term.

10. A polynomial in x is written in _____ if the exponents on
    x in its terms are decreasing order.

11. A _____ is a number, a variable, or a product or quotient of a number and one or more variables raised to powers.

12. A polynomial with exactly three terms is called a _____.

13. A _____ is a term, or the sum of a finite number of terms with whole number exponents.

14. A polynomial with exactly one term is called a _____.

15. The _____ is the greatest degree of any term of the polynomial.

16. A _____ is a polynomial with exactly two terms.

17. Terms with exactly the same variables (including the same exponents) are called _____.

18. The _____ of $(2y-5)(y+8)$ is $-5y$.

19. _____ is a shortcut method for finding the product of two binomials.

20. The _____ of $(2y-5)(y+8)$ is $16y$.

21. A polynomial with two terms is called a _____.

22. The _____ of $a+b$ is $a-b$.

Objective 6.R.1 Write numbers in prime factored form.

Video Examples

Review these examples:

Write each number in prime factored form.

26

We factor using prime factors 2 and 13, as
$26 = 2 \cdot 13$.

54

We use a factor tree, as shown below. The prime factors are boxed.

Divide by the least prime factor of 54, which is 2.	$54 = 2 \cdot 27$	
Divide 27 by 3 to find two factors of 27.	$54 = 2 \cdot 3 \cdot 9$	
Now factor 9 as $3 \cdot 3$.	$54 = 2 \cdot 3 \cdot 3 \cdot 3$	

$$54$$
$$/ \ \backslash$$
$$\boxed{2} \cdot 27$$
$$/ \ \backslash$$
$$\boxed{3} \cdot 9$$
$$/ \ \backslash$$
$$\boxed{3} \cdot \boxed{3}$$

Now Try:

Write each number in prime factored form.

55

_____

210

_____

Practice Exercises

Write each number in prime factored form.

1. 98

2. 256

3. 546

1. _____

2. _____

3. _____

Objective 6.R.2 Solve multiplication and division problems involving 0.

Practice Exercises

Find the product.

4. $18,000(0)$

4. _____

Find each quotient.

5. $\dfrac{0}{-2}$

5. _____

6. $\dfrac{10}{0}$

6. _____

Objective 6.R.3 Solve a formula for a specified variable.

Video Examples

Review these examples:

Solve $A = \frac{1}{2}bh$ for h.

$$A = \frac{1}{2}bh$$

$$2A = bh$$

$$\frac{2A}{b} = \frac{bh}{b}$$

$$\frac{2A}{b} = h \quad \text{or} \quad h = \frac{2A}{b}$$

Solve $A = p + prt$ for r.

$$A = p + prt$$

$$A - p = p + prt - p$$

$$A - p = prt$$

$$\frac{A-p}{pt} = \frac{prt}{pt}$$

$$\frac{A-p}{pt} = r \quad \text{or} \quad r = \frac{A-p}{pt}$$

Solve the equation for y.

$$4x \mid 5y - 15$$

$$-4x + 5y = 15$$

$$-4x + 5y + 4x = 15 + 4x$$

$$5y = 4x + 15$$

$$\frac{5y}{5} = \frac{4x+15}{5}$$

$$y = \frac{4x}{5} + \frac{15}{5}$$

$$y = \frac{4}{5}x + 3$$

Now Try:

Solve $d = rt$ for t.

_____

Solve $P = a + b + c$ for a.

_____

Solve the equation for y.

$$-18x + 3y = 15$$

_____

Practice Exercises

Solve each formula for the specified variable.

7. $V = LWH$ for H 7. _____

8. $S = (n - 2)180$ for n 8. _____

9. $V = \frac{1}{3}\pi r^2 h$ for h 9. _____

Objective 6.R.4 Know the vocabulary for polynomials.

Video Examples

Review these examples:	Now Try:
Simplify each polynomial if possible. Then give the degree and tell whether the polynomial is a monomial, a binomial, a trinomial, or none of these.	Simplify each polynomial if possible. Then give the degree and tell whether the polynomial is a monomial, a binomial, a trinomial, or none of these.

$$5x^4 + 7x$$

We cannot simplify further. This is a binomial of degree 4.

$$8x^3 + 4x^2 + 6$$

_____

$$9x - 7x + 3x$$

$$9x - 7x + 3x = 5x$$
The degree is 1. The simplified polynomial is a monomial.

$$x^5 + 3x^5$$

_____

Practice Exercises

For each polynomial, first simplify, if possible, and write the resulting polynomial in descending powers of the variable. Then give the degree of this polynomial, and tell whether it is a monomial, *a* binomial, *a* trinomial, *or none of these.*

10. $3n^8 - n^2 - 2n^8$

10. _____

degree: _____

type: _____

11. $-d^2 + 3.2d^3 - 5.7d^8 - 1.1d^5$

11. _____

degree: _____

type: _____

12. $-6c^4 - 6c^2 + 9c^4 - 4c^2 + 5c^5$

12. _____

degree: _____

type: _____

Objective 6.R.5 Multiply two polynomials.

Video Examples

Review these examples:

Multiply $(x^2 + 6)(5x^3 - 4x^2 + 3x)$.

Multiply each term of the second polynomial by
each term of the first.

$(x^2 + 6)(5x^3 - 4x^2 + 3x)$

$= x^2(5x^3) + x^2(-4x^2) + x^2(3x)$

$\quad + 6(5x^3) + 6(-4x^2) + 6(3x)$

$= 5x^5 - 4x^4 + 3x^3 + 30x^3 - 24x^2 + 18x$

$= 5x^5 - 4x^4 + 33x^3 - 24x^2 + 18x$

Multiply $(2x^3 + 7x^2 + 5x - 1)(4x + 6)$
vertically.

Write the polynomials vertically.
$$\begin{array}{r} 2x^3 + 7x^2 + 5x - 1 \\ 4x \ + 6 \\ \hline \end{array}$$

Begin by multiplying each term in the top row
by 6.
$$\begin{array}{r} 2x^3 \ + 7x^2 \ + 5x \ - 1 \\ 4x \ + 6 \\ \hline 12x^3 + 42x^2 + 30x - 6 \end{array}$$

Now multiply each term in the top row by $4x$.
Then add like terms.
$$\begin{array}{r} 2x^3 \ \ + 7x^2 \ \ + 5x \ - 1 \\ 4x \ \ + 6 \\ \hline 12x^3 \ + 42x^2 + 30x - 6 \\ 8x^4 + 28x^3 + 20x^2 \ - 4x \\ \hline 8x^4 + 40x^3 + 62x^2 + 26x - 6 \end{array}$$

The product is $8x^4 + 40x^3 + 62x^2 + 26x - 6$.

Now Try:

Multiply
$(x^3 + 9)(4x^4 - 2x^2 + x)$

_____

Multiply
$(4x^3 - 3x^2 + 6x + 5)(7x - 3)$
vertically.

_____

Practice Exercises

Find each product.

13. $(x+3)(x^2-3x+9)$

13. _____

14. $(2m^2+1)(3m^3+2m^2-4m)$

14. _____

15. $(3x^2+x)(2x^2+3x-4)$

15. _____

Objective 6.R.6 Multiply binomials using the FOIL method.

Video Examples

Review these examples:

Use the FOIL method to find the product $(x+7)(x-5)$.

Step 1 F Multiply the first terms: $x(x) = x^2$.

Step 2 O Find the outer product: $x(-5) = -5x$.

Step 3 I Find the inner product: $7(x) = 7x$.
Add the outer and inner products mentally:
$$-5x + 7x = 2x$$

Step 4 L Multiply the last terms: $7(-5) = -35$.

The product $(x+7)(x-5)$ is $x^2 + 2x - 35$.

Multiply $(7x-3)(4y+5)$.

First　　$7x(4y) = 28xy$

Outer　　$7x(5) = 35x$

Inner　　$-3(4y) = -12y$

Last　　$-3(5) = -15$

The product $(7x-3)(4y+5)$ is
$28xy + 35x - 12y - 15$.

Find the product.
$(8p-5q)(8p+q)$

$(8p-5q)(8p+q)$
$= 64p^2 + 8pq - 40pq - 5q^2$
$= 64p^2 - 32pq - 5q^2$

Now Try:

Use the FOIL method to find the product $(x+9)(x-6)$.

_____

Multiply $(8y-7)(2x+9)$.

_____

Find the product.
$(9p+4q)(5p-q)$

_____

Practice Exercises

Find each product.

16. $(5a - b)(4a + 3b)$ 16. _____

17. $(3 + 4a)(1 + 2a)$ 17. _____

18. $(2m + 3n)(-3m + 4n)$ 18. _____

Objective 6.R.7 Square binomials.

Video Examples

Review these examples:	Now Try:
Find $(m+5)^2$.	Find $(x+6)^2$.

$$(m+5)^2 = (m+5)(m+5)$$
$$= m^2 + 5m + 5m + 25$$
$$= m^2 + 10m + 25$$

_____

Square each binomial.

$$(7z-4)^2$$

$$(7z-4)^2 = (7z)^2 - 2(7z)(4) + (-4)^2$$
$$= 7^2 z^2 - 56z + 16$$
$$= 49z^2 - 56z + 16$$

Square each binomial.

$$(8a-3b)^2$$

_____

$$(6x+3y)^2$$

$$(6x+3y)^2 = (6x)^2 + 2(6x)(3y) + (3y)^2$$
$$= 36x^2 + 36xy + 9y^2$$

$$(2a+9k)^2$$

_____

Practice Exercises

Find each square by using the pattern for the square of a binomial.

19. $(7+x)^2$

19. _____

20. $(2m-3p)^2$

20. _____

21. $(4y-0.7)^2$

21. _____

Objective 6.R.8 Find the product of the sum and difference of two terms.

Video Examples

Review these examples:	**Now Try:**
Find the product.	Find the product.
$(x+5)(x-5)$	$(x+9)(x-9)$
Use the rule for the product of the sum and difference of two terms.	
$(x+5)(x-5) = x^2 - 5^2$	_____
$\qquad\qquad = x^2 - 25$	
Find each product.	Find each product.
$\left(z - \dfrac{6}{7}\right)\left(z + \dfrac{6}{7}\right)$	$\left(x + \dfrac{3}{5}\right)\left(x - \dfrac{3}{5}\right)$
$\left(z - \dfrac{6}{7}\right)\left(z + \dfrac{6}{7}\right) = z^2 - \left(\dfrac{6}{7}\right)^2$	_____
$\qquad\qquad = z^2 - \dfrac{36}{49}$	
$(6x + w)(6x - w)$	$(11x - y)(11x + y)$
$(6x + w)(6x - w) = (6x)^2 - w^2$	
$\qquad\qquad = 36x^2 - w^2$	_____
$3q\left(q^2 + 4\right)\left(q^2 - 4\right)$	$4p\left(p^2 + 6\right)\left(p^2 - 6\right)$
First, multiply the conjugates.	
$3q\left(q^2 + 4\right)\left(q^2 - 4\right) = 3q\left(q^4 - 16\right)$	_____
$\qquad\qquad = 3q^5 - 48q$	

Practice Exercises

Find each product by using the pattern for the sum and difference of two terms.

22. $(12 + x)(12 - x)$

22. _____

23. $(8k + 5p)(8k - 5p)$

23. _____

24. $\left(\dfrac{4}{7}t + 2u\right)\left(\dfrac{4}{7}t - 2u\right)$

24. _____

Objective 6.R.9 Divide a polynomial by a polynomial.

Video Examples

Review these examples:

Divide $8x + 9x^3 - 7 - 9x^2$ by $3x - 1$.

Write the dividend in descending powers as $9x^3 - 9x^2 + 8x - 7$.

Step 1 $9x^3$ divided by $3x$ is $3x^2$.
$3x^2(3x - 1) = 9x^3 - 3x^2$

Step 2 Subtract. Bring down the next term.

Step 3 $-6x^2$ divided by $3x$ is $-2x$.
$-2x(3x - 1) = -6x^2 + 2x$

Step 4 Subtract. Bring down the next term.

Step 5 $6x$ divided by $3x$ is 2.
$2(3x - 1) = 6x - 2$

$$
\begin{array}{r}
3x^2 - 2x + 2 \\
3x - 1 \overline{)\,9x^3 - 9x^2 + 8x - 7} \\
\underline{9x^3 - 3x^2} \\
-6x^2 + 8x \\
\underline{-6x^2 + 2x} \\
6x - 7 \\
\underline{6x - 2} \\
-5
\end{array}
$$

$$\frac{9x^3 - 9x^2 + 8x - 7}{3x - 1} = 3x^2 - 2x + 2 + \frac{-5}{3x - 1}$$

Step 7 Multiply to check.

Check $(3x - 1)\left(3x^2 - 2x + 2 + \dfrac{-5}{3x - 1}\right)$

$\qquad = (3x - 1)(3x^2) + (3x - 1)(-2x)$

$\qquad\quad + (3x - 1)(2) + (3x - 1)\left(\dfrac{-5}{3x - 1}\right)$

$\qquad = 9x^3 - 3x^2 - 6x^2 + 2x + 6x - 2 - 5$

$\qquad = 9x^3 - 9x^2 + 8x - 7$

Now Try:

Divide $-12x^2 + 10x^3 - 3 - 8x$ by $5x - 1$.

_____

Divide $x^3 - 64$ by $x - 4$.

Here the dividend is missing the x^2-term and the x-term. We use 0 as the coefficient for each missing term.

$$\begin{array}{r} x^2 + 4x + 16 \\ x-4 \overline{\smash{)}\, x^3 + 0x^2 + 0x - 64} \\ \underline{x^3 - 4x^2} \\ 4x^2 + 0x \\ \underline{4x^2 - 16x} \\ 16x - 64 \\ \underline{16x - 64} \\ 0 \end{array}$$

The remainder is 0. The quotient is $x^2 + 4x + 16$.

Check $(x-4)(x^2 + 4x + 16)$

$$= x^3 + 4x^2 + 16x - 4x^2 - 16x - 64$$

$$= x^3 - 64$$

Divide $x^4 - 3x^3 + 7x^2 - 8x + 14$ by $x^2 + 2$.

Since $x^2 + 2$ is missing the x-term, we write it as $x^2 + 0x + 2$.

$$\begin{array}{r} x^2 - 3x + 5 \\ x^2 + 0x + 2 \overline{\smash{)}\, x^4 - 3x^3 + 7x^2 - 8x + 14} \\ \underline{x^4 + 0x^3 + 2x^2} \\ -3x^3 + 5x^2 - 8x \\ \underline{-3x^3 + 0x^2 - 6x} \\ 5x^2 - 2x + 14 \\ \underline{5x^2 + 0x + 10} \\ -2x + 4 \end{array}$$

The quotient is $x^2 - 3x + 5 + \dfrac{-2x+4}{x^2+2}$

The check shows that the quotient multiplied by the divisor gives the original dividend.

Divide $x^3 - 1000$ by $x - 10$.

_____

Divide $3x^4 + 5x^3 - 7x^2 - 12x + 9$ by $x^2 - 4$.

_____

Divide $5x^3 + 8x^2 + 12x - 1$ by $5x + 5$.

$$5x + 5 \overline{\smash{\big)} \begin{array}{l} x^2 + \dfrac{3}{5}x + \dfrac{9}{5} \\[4pt] 5x^3 + 8x^2 + 12x - 1 \end{array}}$$

$$\underline{5x^3 + 5x^2}$$
$$3x^2 + 12x$$
$$\underline{3x^2 + 3x}$$
$$9x - 1$$
$$\underline{9x + 9}$$
$$-10$$

The answer is $x^2 + \dfrac{3}{5}x + \dfrac{9}{5} - \dfrac{10}{5x + 5}$.

Divide $8x^3 - 7x^2 + 4x + 1$ by $8x - 8$.

_____

Practice Exercises

Perform each division.

25. $\dfrac{-6x^2 + 23x - 20}{2x - 5}$

25. _____

26. $\dfrac{6x^4 - 12x^3 + 13x^2 - 5x - 1}{2x^2 + 3}$

26. _____

27. $\dfrac{2a^4 + 5a^2 + 3}{2a^2 + 3}$

27. _____

Chapter 7 FACTORING EXPRESSIONS AND APPLICATIONS

Learning Objectives

7.R.1   Write fractions in lowest terms.
7.R.2   Multiply and divide fractions and mixed numbers.
7.R.3   Add and subtract fractions.
7.R.4   Use the distributive property.
7.R.5   Solve equations with fractions or decimals as coefficients.
7.R.6   Solve problems involving unknown numbers.
7.R.7   Solve a formula for a specified variable.
7.R.8   Use combinations of the rules for exponents that involve integer exponents.
7.R.9   Factor any polynomial.

Key Terms

Use the vocabulary terms listed below to complete each statement in exercises 1−15.

numerator	denominator	proper fraction
improper fraction	equivalent fractions	lowest terms
prime number	composite number	prime factorization
exponent	base	product rule for exponents
power rule for exponents	FOIL	factoring by grouping

1.   Two fractions are _____ when they represent the same portion of a whole.

2.   A fraction whose numerator is larger than its denominator is called an _____.

3.   In the fraction $\frac{2}{9}$, the 2 is the _____.

4.   A fraction whose denominator is larger than its numerator is called a _____.

5.   The _____ of a fraction shows the number of equal parts in a whole.

6.   A _____ has at least one factor other than itself and 1.

7.   In a _____ every factor is a prime number.

8.   The factors of a _____ are itself and 1.

9.   A fraction is written in _____ when its numerator and denominator have no common factor other than 1.

10. An equation in which variables are used to describe a relationship is called a(n)
_____.

11. The statement "If m and n are any integers, then $\left(a^m\right)^n = a^{mn}$" is an example of the
_____.

12. In the expression a^m, a is the _____ and m is the _____.

13. The statement "If m and n are any integers, then $a^m \cdot a^n = a^{m+n}$ is an example of the
_____.

14. When there are more than three terms in a polynomial, use a process called
_____ to factor the polynomial.

15. _____ is a shortcut method for finding the product of two binomials.

Objective 7.R.1 Write fractions in lowest terms.

Video Examples

Review this example:
Write the fraction in lowest terms.

$$\frac{75}{100}$$

$$\frac{75}{100} = \frac{3 \cdot 25}{4 \cdot 25} = \frac{3}{4} \cdot 1 = \frac{3}{4}$$

Now Try:
Write the fraction in lowest terms.
$$\frac{81}{108}$$

_____

Practice Exercises

Write each fraction in lowest terms.

1. $\dfrac{42}{150}$

1. _____

2. $\dfrac{180}{216}$

2. _____

3. $\dfrac{132}{292}$

3. _____

Objective 7.R.2 Multiply and divide fractions and mixed numbers.

Video Examples

Review these examples:
Find each quotient, and write it in lowest terms.

$$\frac{2}{5} \div \frac{8}{7}$$

$$\frac{2}{5} \div \frac{8}{7} = \frac{2}{5} \cdot \frac{7}{8} \quad \text{Multiply by the reciprocal.}$$

$$= \frac{2 \cdot 7}{5 \cdot 4 \cdot 2} \quad \text{Multiply and factor.}$$

$$= \frac{7}{20}$$

$$\frac{7}{9} \div 14$$

$$\frac{7}{9} \div 14 = \frac{7}{9} \cdot \frac{1}{14} \quad \text{Multiply by the reciprocal.}$$

$$= \frac{7 \cdot 1}{9 \cdot 2 \cdot 7} \quad \text{Multiply and factor.}$$

$$= \frac{1}{18}$$

Now Try:
Find each quotient, and write it in lowest terms.

$$\frac{6}{7} \div \frac{9}{8}$$

_____

$$\frac{4}{5} \div 8$$

_____

Practice Exercises

Find each product or quotient, and write it in lowest terms.

4. $\dfrac{25}{11} \cdot \dfrac{33}{10}$

4. _____

5. $\dfrac{5}{4} \div \dfrac{25}{28}$

5. _____

6. $4\dfrac{3}{8} \cdot 2\dfrac{4}{7}$

6. _____

Objective 7.R.3 Add and subtract fractions and mixed numbers.

Video Examples

Review these examples:

Add. Write sum in lowest terms.

$$\frac{5}{21}+\frac{3}{14}$$

Step 1 To find the LCD, factor the denominators to prime factored form.

$21 = 3\cdot 7$ and $14 = 2\cdot 7$

7 is a factor of both denominators.

$$\begin{matrix}21 & 14\\ \wedge & \wedge\end{matrix}$$

Step 2 LCD $= 3\cdot 7\cdot 2 = 42$

In this example, the LCD needs one factor of 3, one factor of 7 and one factor of 2.

Step 3 Now we can use the second property of 1 to write each fraction with 42 as the denominator.

$$\frac{5}{21}=\frac{5}{21}\cdot\frac{2}{2}=\frac{10}{42}\quad\text{and}\quad\frac{3}{14}=\frac{3}{14}\cdot\frac{3}{3}=\frac{9}{42}$$

Now add the two equivalent fractions to get the sum.

$$\frac{5}{21}+\frac{3}{14}=\frac{10}{42}+\frac{9}{42}$$
$$=\frac{19}{42}$$

Subtract. Write difference in lowest terms.

$$\frac{14}{15}-\frac{5}{8}$$

Since 15 and 8 have no common factors greater than 1, the LCD is $15\cdot 8 = 120$.

$$\frac{14}{15}-\frac{5}{8}=\frac{14}{15}\cdot\frac{8}{8}-\frac{5}{8}\cdot\frac{15}{15}$$
$$=\frac{112}{120}-\frac{75}{120}$$
$$=\frac{37}{120}$$

Now Try:

Add. Write sum in lowest terms.

$$\frac{7}{12}+\frac{3}{8}$$

_____

Subtract. Write difference in lowest terms.

$$\frac{17}{6}-\frac{13}{7}$$

_____

Practice Exercises

Find each sum or difference, and write it in lowest terms.

7. $\dfrac{23}{45} + \dfrac{47}{75}$

7. _____

8. $2\dfrac{3}{4} + 7\dfrac{2}{3}$

8. _____

9. $12\dfrac{5}{6} - 7\dfrac{7}{8}$

9. _____

Objective 7.R.4 Use the distributive property.

Video Examples

Review these examples:

Use the distributive property to rewrite the expression.

$4 \cdot 9 + 4 \cdot 2$

Use the distributive property in reverse.

$4 \cdot 9 + 4 \cdot 2 = 4(9+2)$

$\qquad = 4(11)$

$\qquad = 44$

Write the expression without parentheses.

$-(-p - 5r + 9x)$

$-(-p - 5r + 9x)$
$= -1 \cdot (-1p - 5r + 9x)$
$= -1 \cdot (-1p) - 1 \cdot (-5r) - 1 \cdot (9x)$
$= p + 5r - 9x$

Now Try:

Use the distributive property to rewrite the expression.

$25 \cdot 9 + 25 \cdot 6$

_____

Write the expression without parentheses.

$-(-4x - 5y + z)$

_____

Practice Exercises

Use the distributive property to rewrite each expression. Simplify if possible.

10. $n(2a - 4b + 6c)$

10. _____

11. $-2(5y - 9z)$

11. _____

12. $-(-2k + 7)$

12. _____

Objective 7.R.5 Solve equations with fractions or decimals as coefficients.

Video Examples

Review these examples:

Solve $\frac{3}{4}x - \frac{1}{2}x = -\frac{1}{8}x - 6$.

Multiply each side by 8, the LCD.

$$\frac{3}{4}x - \frac{1}{2}x = -\frac{1}{8}x - 6$$

Step 1 $\quad 8\left(\frac{3}{4}x - \frac{1}{2}x\right) = 8\left(-\frac{1}{8}x - 6\right)$

$$8\left(\frac{3}{4}x\right) + 8\left(-\frac{1}{2}x\right) = 8\left(-\frac{1}{8}x\right) + 8(-6)$$

$$6x - 4x = -x - 48$$

$$2x = -x - 48$$

Step 2 $\quad 2x + x = -x - 48 + x$

$$3x = -48$$

Step 3 $\quad \frac{3x}{3} = -\frac{48}{3}$

$$x = -16$$

Step 4 $\quad \frac{3}{4}x - \frac{1}{2}x = -\frac{1}{8}x - 6$

$$\frac{3}{4}(-16) - \frac{1}{2}(-16) \overset{?}{=} -\frac{1}{8}(-16) - 6$$

$$-12 + 8 \overset{?}{=} 2 - 6$$

$$-4 = -4 \quad \text{True}$$

The solution set is $\{-16\}$.

Solve $0.2x + 0.04(10 - x) = 0.06(4)$.

To clear decimals, multiply by 100.

$$0.2x + 0.04(10 - x) = 0.06(4)$$

Step 1 $\quad 100[0.2x + 0.04(10 - x)] = 100[0.06(4)]$

$$100(0.2x) + 100[0.04(10 - x)] = 100[0.06(4)]$$

$$20x + 4(10) + 4(-x) = 24$$

$$20x + 40 - 4x = 24$$

$$16x + 40 = 24$$

Step 2 $\quad 16x + 40 - 40 = 24 - 40$

$$16x = -16$$

Step 3 $\quad \frac{16x}{16} = \frac{-16}{16}$

$$x = -1$$

Step 4 Check to confirm that $\{-1\}$ is the solution set.

Now Try:

Solve $\frac{2}{9}x - \frac{1}{6}x = \frac{2}{3}x + 11$.

_____

Solve
$0.5x + 0.04(5 - 8x) = 0.07(8)$.

_____

Practice Exercises

Solve each equation and check your solution.

13. $\dfrac{3}{8}x - \dfrac{1}{3}x = \dfrac{1}{12}$

13. _____

14. $\dfrac{1}{3}(2m - 1) \quad \dfrac{3}{4}m - \dfrac{5}{6}$

14. _____

15. $0.45a - 0.35(20 - a) = 0.02(50)$

15. _____

Objective 7.R.6 Solve problems involving unknown numbers.

Video Examples

Review this example:

The product of 5, and a number decreased by 8, is 150. What is the number?

Step 1 Read the problem carefully. We are asked to find a number.

Step 2 Assign a variable to represent the unknown quantity.

Let x = the number.

Step 3 Write an equation.

The product of 5, and a number decreased by 8, is 150.

$$5 \cdot (x - 8) = 150$$

Step 4 Solve the equation.

$$5(x - 8) = 150$$
$$5x - 40 = 150$$
$$5x - 40 + 40 = 150 + 40$$
$$5x = 190$$
$$\frac{5x}{5} = \frac{190}{5}$$
$$x = 38$$

Step 5 State the answer. The number is 38.

Step 6 Check. The number 38 decreased by 8 is 30. The product of 5 and 30 is 150. The answer, 38, is correct.

Now Try:

The product of 8, and a number decreased by 11, is 40. What is the number?

_____

Practice Exercises

Write an equation for each of the following and then solve the problem. Use x as the variable.

16. If 4 is added to 3 times a number, the result is 7. Find the number.

16. _____

_____

17. If –2 is multiplied by the difference between 4 and a number, the result is 24. Find the number.

17. _____

_____

18. If four times a number is added to 7, the result is five less than six times the number. Find the number.

18. _____

_____

Objective 7.R.7 Solve a formula for a specified variable.

Video Examples

Review these examples:

Solve $A = \frac{1}{2} bh$ for h.

$$A = \frac{1}{2} bh$$

$$2A = bh$$

$$\frac{2A}{b} = \frac{bh}{b}$$

$$\frac{2A}{b} = h \quad \text{or} \quad h = \frac{2A}{b}$$

Solve $A = p + prt$ for r.

$$A = p + prt$$

$$A - p = p + prt - p$$

$$A - p = prt$$

$$\frac{A - p}{pt} = \frac{prt}{pt}$$

$$\frac{A - p}{pt} = r \quad \text{or} \quad r = \frac{A - p}{pt}$$

Solve the equation for y.

$$-4x + 5y = 15$$

$$-4x + 5y = 15$$

$$-4x + 5y + 4x = 15 + 4x$$

$$5y = 4x + 15$$

$$\frac{5y}{5} = \frac{4x + 15}{5}$$

$$y = \frac{4x}{5} + \frac{15}{5}$$

$$y = \frac{4}{5} x + 3$$

Now Try:

Solve $d = rt$ for t.

_____

Solve $P = a + b + c$ for a.

_____

Solve the equation for y.

$$-18x + 3y = 15$$

_____

Practice Exercises

Solve each formula for the specified variable.

19. $V = LWH$ for H

19. _____

20. $S = (n-2)180$ for n

20. _____

21. $V = \frac{1}{3}\pi r^2 h$ for h

21. _____

Objective 7.R.8 Use combinations of the rules for exponents that involve integer exponents.

Video Examples

Review these examples:
Simplify each expression. Assume that all variables represent nonzero real numbers.

$$\left(\frac{3x^4}{4}\right)^{-5}$$

$$\left(\frac{3x^4}{4}\right)^{-5} = \left(\frac{4}{3x^4}\right)^{5}$$

$$= \frac{4^5}{3^5 x^{20}}$$

$$= \frac{1024}{243x^{20}}$$

$$\left(\frac{5x^{-3}}{2^{-1}y^4}\right)^{-2}$$

$$\left(\frac{5x^{-3}}{2^{-1}y^4}\right)^{-2} = \frac{5^{-2}x^6}{2^2 y^{-8}}$$

$$= \frac{x^6 y^8}{5^2 \cdot 2^2}$$

$$= \frac{x^6 y^8}{100}$$

Now Try:
Simplify each expression. Assume that all variables represent nonzero real numbers.

$$\left(\frac{2p^4}{3}\right)^{-5}$$

_____

$$\left(\frac{4x^{-4}}{5^{-1}y^5}\right)^{-3}$$

_____

Practice Exercises

Simplify each expression, and write it using only positive exponents. Assume that all variables represent nonzero real numbers.

22. $(9xy)^7 (9xy)^{-8}$

22. _____

23. $\dfrac{\left(a^{-1}b^{-2}\right)^{-4}\left(ab^2\right)^6}{\left(a^3b\right)^{-2}}$

23. _____

24. $\left(\dfrac{k^3t^4}{k^2t^{-1}}\right)^{-4}$

24. _____

Objective 7.R.9 Factor any polynomial.

Video Examples

Review these examples:	**Now Try:**
Factor the polynomial.	Factor the polynomial.
$9x(a+c)-z(a+c)$	$7x(y+z)-5(y+z)$
Factor out $(a+c)$	
$9x(a+c)-z(a+c)=(a+c)(9x-z)$	_____

Factor each binomial if possible.	Factor each binomial if possible.
$16y^2+49$	$36p^2+169$
$16y^2+49$ is prime. It is the sum of squares. There is no common factor.	_____
$216p^3-125z^3$	$27t^3-64w^3$
Difference of cubes	_____
$216p^3-125z^3$	
$\quad=(6p)^3-(5z)^3$	
$\quad=(6p-5z)\left[(6p)^2+(6p)(5z)+(5z)^2\right]$	
$\quad=(6p-5z)(36p^2+30pz+25z^2)$	

Factor the trinomial.	Factor the trinomial.
y^2-3y-4	$4k^2-7k-2$
$y^2-3y-4=(y-4)(y+1)$	_____
Factor the polynomial.	Factor the polynomial.
$ad-8d+af-8f$	$cm-3m+cx-3x$
Group the terms and factor each group.	_____
$ad-8d+af-8f$	
$\quad=(ad-8d)+(af-8f)$	
$\quad=d(a-8)+f(a-8)$	
$\quad=(a-8)(d+f)$	

Practice Exercises

Factor completely.

25. $-12x^2 - 6x$

25. _____

26. $12a^2b^2 + 3a^2b - 9ab^2$

26. _____

27. $(x+1)(2x+3) - (x+1)$

27. _____

28. $2x^3y^4 - 72xy^2$

28. _____

29. $128x^3 - 2y^3$

29. _____

30. $y^6 + 1$

30. _____

31. $4x^2 - 12xy + 9y^2$

31. _____

32. $2a^2 - 17a + 30$

32. _____

33. $14w^2 + 6wx - 35wx - 15x^2$

33. _____

34. $x^3 - 3x^2 + 7x - 21$

34. _____

35. $a^2 - 6ab + 9b^2 - 25$

35. _____

Chapter 8 EQUATIONS, INEQUALITIES, GRAPHS, AND SYSTEMS REVISITED

Learning Objectives

8.R.1 Solve linear equations using more than one property of equality.
8.R.2 Solve linear inequalities by using both properties of incquality.
8.R.3 Use inequalities to solve applied problems.
8.R.4 Solve linear inequalities with three parts.
8.R.5 Graph linear equations by plotting ordered pairs.
8.R.6 Solve problems about quantities and their costs.
8.R.7 Solve problems about mixtures.
8.R.8 Solve problems about distance, rate (or speed), and time.

Key Terms

Use the vocabulary terms listed below to complete each statement in exercises 1–14.

conditional equation	identity	contradiction
inequalities	interval	interval notation
linear inequality	three-part inequality	graph
graphing	y-intercept	x-intercept
system of linear equations		$d = rt$

1.   An equation with no solution is called a(n) _____.

2.   A(n) _____ is an equation that is true for some values of the variable and false for other values.

3.   An equation that is true for all values of the variable is called a(n) _____.

4.   An inequality that says that one number is between two other numbers is a(n)_____.

5.   A portion of a number line is called a(n) _____.

6.   A(n) _____ can be written in the form $Ax + B < C$, $Ax + B \leq C$, $Ax + B > C$, or $Ax + B \geq C$, where A, B, and C are real numbers with $A \neq 0$.

7.   Algebraic expressions related by $<$, $\leq$, $>$, or $\geq$ are called _____.

8.   The _____ for $a \leq x < b$ is $[a, b)$.

9.   If a graph intersects the y-axis at k, then the _____ is $(0, k)$.

10.   If a graph intersects the x-axis at k, then the _____ is $(k, 0)$.

11.   The process of plotting the ordered pairs that satisfy a linear equation and drawing a line through them is called _____.

12.   The set of all points that correspond to the ordered pairs that satisfy the equation is called the _____ of the equation.

13.   The formula that relates distance, rate, and time is _____.

14.   A _____ consists of at least two linear equations with different variables.

Objective 8.R.1 Solve linear equations using more than one property of equality.

Video Examples

Review this example:

Solve $5(k-4)-k=k-2$.

Step 1 Clear parentheses using the distributive property.

$$5(k-4)-k=k-2$$
$$5(k)+5(-4)-k=k-2$$
$$5k-20-k=k-2$$
$$4k-20=k-2$$

Step 2 $\quad 4k-20-k=k-2-k$
$$3k-20=-2$$
$$3k-20+20=-2+20$$
$$3k=18$$

Step 3 $\qquad \dfrac{3k}{3}=\dfrac{18}{3}$
$$k=6$$

Step 4 Check by substituting 6 for k in the original equation.
$$5(k-4)-k=k-2$$
$$5(6-4)-6\overset{?}{=}6-2$$
$$5(2)-6\overset{?}{=}4$$
$$10-6\overset{?}{=}4$$
$$4=4 \quad \text{True}$$
The solution, 6, checks, so the solution set is $\{6\}$.

Solve $6x-18=6(x-3)$.

$$6x-18=6(x-3)$$
$$6x-18=6x-18$$
$$6x-18-6x=6x-18-6x$$
$$-18=-18$$
$$-18+18=-18+18$$
$$0=0$$
The solution set is {all real numbers}.

Now Try:

Solve $9(k-2)-k=k+10$.

_____

Solve $3x+4(x-5)=7x-20$.

_____

Solve $3x + 4(x - 5) = 7x + 5$.

$3x + 4(x - 5) = 7x + 5$

$3x + 4x - 20 = 7x + 5$

$\qquad 7x - 20 = 7x + 5$

$7x - 20 - 7x = 7x + 5 - 7x$

$\qquad\qquad -20 = 5 \quad$ False

There is no solution. The solution set is $\varnothing$.

Solve $-5x + 17 = x - 6(x + 3)$.

_____

Practice Exercises

Solve each equation and check your solution.

1. $\quad 7t + 6 = 11t - 4$

1. _____

2. $\quad 3a - 6a + 4(a - 4) = -2(a + 2)$

2. _____

3. $\quad 3(t + 5) = 6 - 2(t - 4)$

3. _____

Solve each equation and check your solution.

4. $3(6x-7)=2(9x-6)$

4. _____

5. $6y-3(y+2)-3(y-2)$

5. _____

6. $3(r-2)-r+4-2r+6$

6. _____

Objective 8.R.2 Solve linear inequalities by using both properties of inequality.

Video Examples

Review this example:

Solve $4x+3-7>-2x+8+3x$. Graph the solution set.

Step 1 Combine like terms and simplify.
$$4x+3-7>-2x+8+3x$$
$$4x-4>x+8$$

Step 2 Use the addition property of inequality.
$$4x-4-x>x+8-x$$
$$3x-4>8$$
$$3x-4+4>8+4$$
$$3x>12$$

Step 3 Use the multiplication property of inequality.
$$\frac{3x}{3}>\frac{12}{3}$$
$$x>4$$

The solution set is $(4, \infty)$. The graph is shown below.

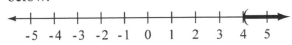

Now Try:

Solve $8x-5+4\geq 6x-3x+9$. Graph the solution set.

Practice Exercises

Solve each inequality. Write the solution set in interval notation and then graph it.

7. $4(y-3)+2>3(y-2)$

7. _____

8. $-3(m+2)+3\leq -4(m-2)-6$

8. _____

9. $7(2-x)\leq -2(x-3)-x$

9. _____

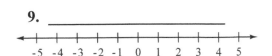

Objective 8.R.3 Use inequalities to solve applied problems.

Video Examples

Review this example:

Ruth tutors mathematics in the evenings in an office for which she pays $600 per month rent. If rent is her only expense and she charges each student $40 per month, how many students must she teach to make a profit of at least $1600 per month?

Step 1 Read the problem again.

Step 2 Assign a variable.
   Let x – the number of students.

Step 3 Write an inequality.
   $40x - 600 \geq 1600$

Step 4 Solve.
   $40x - 600 + 600 \geq 1600 + 600$

$$40x \geq 2200$$

$$\frac{40x}{40} \geq \frac{2200}{40}$$

$$x \geq 55$$

Step 5 State the answer. Ruth must have 55 or more students to have at least $1600 profit.

Step 6 Check. $40(55) - 600 = 1600$ Also, any number greater than 55 makes the profit greater than $1600.

Now Try:

Two sides of a triangle are equal in length, with the third side 8 feet longer than one of the equal sides. The perimeter of the triangle cannot be more than 38 feet. Find the largest possible value for the length of the equal sides.

_____

Practice Exercises

Solve each problem.

10.   Lauren has grades of 98 and 86 on her first two chemistry quizzes. What must she score on her third quiz to have an average of at least 91 on the three quizzes?

10. _____

11. Nina has a budget of $230 for gifts for this year. So
     far she has bought gifts costing $47.52, $38.98, and
     $26.98. If she has three more gifts to buy, find the
     average amount she can spend on each gift and still
     stay within her budget.

11. _____

12. If twice the sum of a number and 7 is subtracted
     from three times the number, the result is more
     than –9. Find all such numbers.

12. _____

Objective 8.R.4 Solve linear inequalities with three parts.

Video Examples

Review these examples:

Write the inequality $-4 \leq x < 3$ in interval notation, and graph the interval.

$-4 \leq x < 3$

x is between -4 and 3 (excluding 3).

In interval notation, we write $[-4, 3)$.

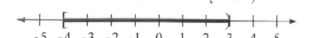

Solve the inequality, and graph the solution set.

$3 \leq 4x - 5 < 7$

$$3 \leq 4x - 5 < 7$$
$$3 + 5 \leq 4x - 5 + 5 < 7 + 5$$
$$8 \leq 4x < 12$$
$$\frac{8}{4} < \frac{4x}{4} < \frac{12}{4}$$
$$2 \leq x < 3$$

The solution set is $[2, 3)$. The graph is shown below.

Now Try:

Write the inequality $-5 < x \leq -1$ in interval notation, and graph the interval.

_____

Solve the inequality, and graph the solution set.

$8 \leq 6x - 4 < 20$

_____

Practice Exercises

Solve each inequality. Write the solution set in interval notation and then graph it.

13. $7 < 2x + 3 \leq 13$

13. _____

14. $-17 \leq 3x - 2 < -11$

14. _____

15. $1 < 3z + 4 < 19$

15. _____

Objective 8.R.5 Graph linear equations by plotting ordered pairs.

Video Examples

Review this example:

Graph $2x + 3y = 6$.

First let $x = 0$ and then let $y = 0$ to determine two ordered pairs.

$$2(0) + 3y = 6 \quad \mid \quad 2x + 3(0) = 6$$
$$0 + 3y = 6 \quad \mid \quad 2x + 0 = 6$$
$$3y = 6 \quad \mid \quad 2x = 6$$
$$y = 2 \quad \mid \quad x = 3$$

The ordered pairs are (0, 2) and (3, 0). Find a third ordered pair by choosing a number other than 0 for x or y. We choose $y = 4$.

$$2x + 3(4) = 6$$
$$2x + 12 = 6$$
$$2x = -6$$
$$x = -3$$

This gives the ordered pair (−3, 4). We plot the three ordered pairs (0, 2), (3, 0), and (−3, 4) and draw a line through them.

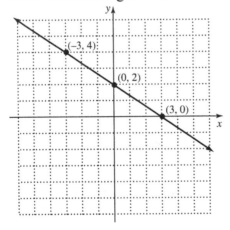

Now Try:

Graph $x + y = 3$.

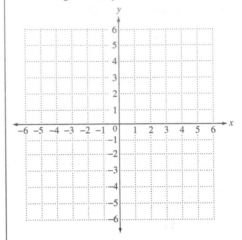

Practice Exercises

Complete the ordered pairs for each equation. Then graph the equation by plotting the points and drawing a line through them.

16. $y = 3x - 2$

   (0,)

   (, 0)

   (2,)

16.

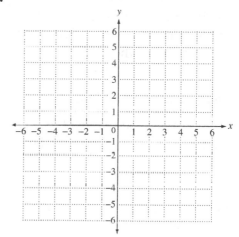

17. $x - y = 4$

   (0,)

   (, 0)

   (−2,)

17.

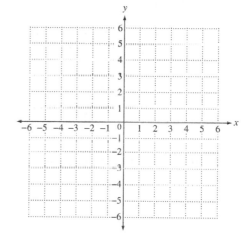

18. $x = 2y + 1$

   (0,)

   (, 0)

   (, −2)

18.

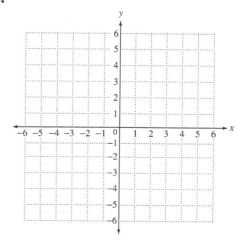

Objective 8.R.6 Solve problems about quantities and their costs.

Video Examples

Review this example:

The total receipts for a basketball game were $4690.50. There were 723 tickets sold, some for children and some for adults. If the adult tickets cost $9.50 and the children's tickets cost $4, how many of each type were there?

Step 1 Read the problem.

Step 2 Assign variables. Let x = the number of adult tickets and y = the number of children's tickets.

Step 3 Write two equations. The total number of tickets is 723. The total value of the tickets is $4690.50.

$$x + y = 723 \qquad (1)$$
$$9.50x + 4y = 4690.5 \quad (2)$$

Step 4 Solve the system. Solve using elimination. Multiply equation (1) by –4.

$$-4x - 4y = -2892$$
$$\underline{9.50x + 4y = 4690.5}$$
$$5.50x \qquad = 1798.50$$
$$x = 327$$

Substitute 327 for x in equation (1) to find y.

$$x + y = 723$$
$$327 + y = 723$$
$$y = 396$$

Step 5 State the answer. The number of adult tickets is 327 and the number of children's tickets is 396.

Step 6 Check. The sum of the tickets is 723. The value of the tickets is

$$9.50(327) + 4(396) = 4690.50$$

This checks.

Now Try:

Twice as many general admission tickets to a basketball game were sold as reserved seat tickets. General admission tickets cost $10 and reserved seat tickets cost $15. If the total value of both kinds of tickets was $26,250, how many tickets of each kind were sold?

_____

Practice Exercises

Write a system of equations for each problem, then solve the problem.

19. There were 411 tickets sold for a soccer game, some
for students and some for nonstudents. Student
tickets cost $4.25 and nonstudent tickets cost $8.50
each. The total receipts were $3021.75. How many
of each type were sold?

19.

student tix_____

nonstudent tix_____

20. A cashier has some $5 bills and some $10 bills. The
total value of the money is $750. If the number of
tens is equal to twice the number of fives, how many
of each type are there?

20.

$5 bills _____

$10 bills _____

21. Luke plans to buy 10 ties with exactly $162. If some
ties cost $14, and the others cost $25, how many ties
of each price should he buy?

21.

$14 ties_____

$25 ties_____

Objective 8.R.7 Solve problems about mixtures.

Video Examples

Review this example:

A mixture of 75% solution should be mixed with a 55% solution to get 70 liters of 63% solution. Determine the number of liters required of the 55% and 75% solutions.

Step1 Read the problem carefully.

Step 2 Assign variables. Let x = the number of liters of the 75% liquid and y = the number of liters of the 55% liquid.

Step 3 Write two equations. The total amount of liquid of the final mixture is 70 liters. The amount of 75% solution mixed with 55% solution will equal the 70 liters of 63% solution.

$$x + y = 70 \qquad (1)$$
$$0.75x + 0.55y = 0.63(70) \quad (2)$$

Step 4 Solve the system. Solve by substitution. Solving equation (1) for x results in $70 - y$.

$$0.75x + 0.55y = 0.63(70)$$
$$0.75(70 - y) + 0.55y = 44.1$$
$$52.5 - 0.75y + 0.55y = 44.1$$
$$-0.20y + 52.5 = 44.1$$
$$-0.20y = -8.4$$
$$y = 42$$

Substitute 42 for y in equation (1).
$$x = 70 - 42 = 28$$

Step 5 State the answer. There should be 28 liters of 75% solution and 42 liters of 55% solution.

Step 6 Check the answer in the original problems. $28 + 42 = 70$ and
$0.75(28) + 0.55(42) = 44.1$ The answer checks.

Now Try:

A mixture of 85% solution should be mixed with a 65% solution to get 80 ounces of 77% solution. Determine the number of ounces required of the 85% and 65% solutions.

_____

Practice Exercises

Write a system of equations for each problem, then solve the problem.

22.   Jorge wishes to make 150 pounds of coffee blend
        that can be sold for $8 per pound. The blend will be
        a mixture of coffee worth $6 per pound and coffee
        worth $12 per pound. How many pounds of each
        kind of coffee should be used in the mixture?

22.

$6 coffee_____

$12 coffee_____

23.   A solution of 50% acid is mixed with a 10% acid
        solution to get a 500 mL solution that is 40% acid.
        Determine the number of milliliters required of the
        50% and 10% solutions.

23.

50% solution _____

10% solution _____

24.   Ben wishes to blend candy selling for $1.60 a pound
        with candy selling for $2.50 a pound to get a mixture
        that will be sold for $1.90 a pound. How many
        pounds of the $1.60 and the $2.50 candy should be
        used to get 30 pounds of the mixture?

24.

$1.60 candy _____

$2.50 candy _____

Objective 8.R.8 Solve problems about distance, rate (or speed), and time.

Video Examples

Review this example:

Bill and Hillary start in Washington and fly in opposite directions. At the end of 4 hours, they are 4896 kilometers apart. If Bill flies 60 kilometers per hour faster than Hillary, what are their speeds?

Step 1 Read the problem carefully.

Step 2 Assign variables. Let x = Bill's rate of speed and y = Hillary's rate of speed.

Step 3 Write two equations.

$$x = 60 + y \qquad (1)$$

$$4x + 4y = 4896 \quad (2)$$

Step 4 Solve the system. . Solve by substitution.

$$4(60 + y) + 4y = 4896$$

$$240 + 4y + 4y = 4896$$

$$240 + 8y = 4896$$

$$8y = 4656$$

$$y = 582$$

Substitute 582 for y in equation (1).

$$x = 60 + 582 = 642$$

Step 5 State the answer. Bill's rate is 642 kmh and Hillary's rate is 582 kmh.

Step 6 Check. Since 4(642) + 4(582) = 4896 and $642 = 582 + 60$ the answers check.

Now Try:

Enid leaves Cherry Hill, driving by car toward New York, which is 90 miles away. At the same time, Jerry, riding his bicycle, leaves New York cycling toward Cherry Hill. Enid is traveling 28 miles per hour faster than Jerry. They pass each other $1\frac{1}{2}$ hours later. What are their speeds?

_____

Practice Exercises

Write a system of equations for each problem, and then solve the problem.

25. It takes Carla's boat $\frac{1}{2}$ hour to go 8 miles downstream and 1 hour to make the return trip upstream. Find the speed of the current and the speed of Carla's boat in still water.

    25.

    boat speed_____

    current speed _____

26. Two planes left Philadelphia traveling in opposite directions. Plane A left 15 minutes before plane B After plane B had been flying for 1 hour, the planes were 860 miles apart. What were the speeds of the two planes if plane A was flying 40 miles per hour faster than plane B?

    26.

    plane A_____

    plane B _____

27. At the beginning of a fund-raising walk, Steve and Vic are 30 miles apart. If they leave at the same time and walk in the same direction, Steve would overtake Vic in 15 hours. If they walked toward each other, they would meet in 3 hours. What are their speeds?

    27.

    Steve   _____

    Vic   _____

Chapter 9 RELATIONS AND FUNCTIONS

Learning Objectives
9.R.1 Solve a formula for a specified variable.
9.R.2 Decide whether a given ordered pair is a solution of a given equation.
9.R.3 Complete a table of values.
9.R.4 Graph linear equations by plotting ordered pairs.
9.R.5 Graph linear equations of the form $y = b$ or $x = a$.
9.R.6 Evaluate polynomials.
9.R.7 Perform operations with polynomials.

Key Terms

Use the vocabulary terms listed below to complete each statement in exercises 1–26.

formula	line graph	linear equation in two variables
ordered pair		table of values x-axis
y-axis	rectangular (Cartesian) coordinate system	
origin	quadrants	plane coordinates
plot	graph	graphing y-intercept
x-intercept term		like terms polynomial
descending powers		degree of a term monomial
degree of a polynomial		binomial trinomial

1. An equation in which variables are used to describe a relationship is called a(n) _____.

2. A _____ uses dots connected by lines to show trends.

3. An equation that can be written in the form $Ax + By = C$, where A, B, and C are real numbers and A, $B \neq 0$, is called a _____.

4. _____ are the numbers in the ordered pair that specify the location of a point on a rectangular coordinate system.

5. In a coordinate system, the horizontal axis is called the _____.

6. In a coordinate system, the vertical axis is called the _____.

7. A pair of numbers written between parentheses in which order is important is called a(n) _____.

8. Together, the x-axis and the y-axis form a _____.

9. A coordinate system divides the plane into four regions called _____.

10.   The axis lines in a coordinate system intersect at the _____.

11.   To _____ an ordered pair is to find the corresponding point on a
coordinate system.

12.   A table showing selected ordered pairs of numbers that satisfy an equation is called a
_____.

13.   A flat surface determined by two intersecting lines is a _____.

14.   If a graph intersects the y-axis at k, then the _____ is (0, k).

15.   If a graph intersects the x-axis at k, then the _____ is (k, 0).

16.   The process of plotting the ordered pairs that satisfy a linear equation and drawing a
line through them is called _____.

17.   The set of all points that correspond to the ordered pairs that satisfy the equation is
called the _____ of the equation.

18.   The _____ is the sum of the exponents on the variables in
that term.

19.   A polynomial in x is written in _____ if the exponents on
x in its terms are decreasing order.

20.   A _____ is a number, a variable, or a product or quotient of a number
and one or more variables raised to powers.

21.   A polynomial with exactly three terms is called a _____.

22.   A _____ is a term, or the sum of a finite number of terms with
whole number exponents.

23.   A polynomial with exactly one term is called a _____.

24.   A _____ is a polynomial with exactly two terms.

25.   Terms with exactly the same variables (including the same exponents) are called
_____.

26.   The _____ is the greatest degree of any term of the
polynomial.

Objective 9.R.1 Solve a formula for a specified variable.

Video Examples

Review these examples:

Solve $A = \frac{1}{2}bh$ for h.

$$A = \frac{1}{2}bh$$

$$2A = bh$$

$$\frac{2A}{b} = \frac{bh}{b}$$

$$\frac{2A}{b} = h \quad \text{or} \quad h = \frac{2A}{b}$$

Solve $A = p + prt$ for r.

$$A = p + prt$$

$$A - p = p + prt - p$$

$$A - p = prt$$

$$\frac{A-p}{pt} = \frac{prt}{pt}$$

$$\frac{A-p}{pt} = r \quad \text{or} \quad r = \frac{A-p}{pt}$$

Solve the equation for y.

$$-4x + 5y = 15$$

$$-4x + 5y = 15$$

$$-4x + 5y + 4x = 15 + 4x$$

$$5y = 4x + 15$$

$$\frac{5y}{5} = \frac{4x+15}{5}$$

$$y = \frac{4x}{5} + \frac{15}{5}$$

$$y = \frac{4}{5}x + 3$$

Now Try:

Solve $d = rt$ for t.

_____

Solve $P = a + b + c$ for a.

_____

Solve the equation for y.

$$-18x + 3y = 15$$

_____

Practice Exercises

Solve each formula for the specified variable.

1. $V = LWH$ for H

1. _____

2. $S = (n-2)180$ for n

2. _____

3. $V = \frac{1}{3}\pi r^2 h$ for h

3. _____

Objective 9.R.2 Decide whether a given ordered pair is a solution of a given equation.

Video Examples

Review these examples:

Decide whether each ordered pair is a solution of the equation $4x + 5y = 40$.

(5, 4)

Substitute 5 for x and 4 for y in the given equation.

$$4x + 5y = 40$$
$$4(5) + 5(4) \overset{?}{=} 40$$
$$20 + 20 \overset{?}{=} 40$$
$$40 = 40 \quad \text{True}$$

This result is true, so (5, 4) is a solution of $4x + 5y = 40$.

(−3, 6)

Substitute −3 for x and 6 for y in the given equation.

$$4x + 5y = 40$$
$$4(-3) + 5(6) \overset{?}{=} 40$$
$$-12 + 30 \overset{?}{=} 40$$
$$18 = 40 \quad \text{False}$$

This result is false, so (−3, 6) is not a solution of $4x + 5y = 40$.

Now Try:

Decide whether each ordered pair is a solution of the equation $3x - 4y = 12$.

(8, 3)

_____

(5, −4)

_____

Practice Exercises

Decide whether the given ordered pair is a solution of the given equation.

4. $4x - 3y = 10$; $(1, \ 2)$

4. _____

5. $2x - 3y = 1$; $\left(0, \ \frac{1}{3}\right)$

5. _____

6. $x = -7$; $(-7, \ 9)$

6. _____

Objective 9.R.3 Complete a table of values.

Video Examples

Review this example:

Complete the table of values for the equation. Then write the results as ordered pairs.

$2x - 3y = 6$

x	y
9	
6	
	-2
	8

From the table, we can write the ordered pairs:
(9, ____), (6, ____), (____, -2), (____, 8).

From the first row of the table, let $x = 9$ in the equation. From the second row of the table, let $x = 6$.

If $x = 9$,	If $x = 6$,
$2x - 3y = 6$	$2x - 3y = 6$
$2(9) - 3y = 6$	$2(6) - 3y = 6$
$18 \quad 3y = 6$	$12 \quad 3y = 6$
$-3y = -12$	$-3y = -6$
$y = 4$	$y = 2$

The first two ordered pairs are (9, 4) and (6, 2).

From the third and fourth rows of the table, let $y = -2$ and $y = 8$, respectively.

If $y = -2$,	If $y = 8$,
$2x - 3y = 6$	$2x - 3y = 6$
$2x - 3(-2) = 6$	$2x \quad 3(8) = 6$
$2x + 6 = 6$	$2x - 24 = 6$
$2x = 0$	$2x = 30$
$x = 0$	$x = 15$

The last two ordered pairs are (0, -2) and (15, 8). The completed table and corresponding ordered pairs follow.

x	y	Ordered pairs
9	4	→ (9, 4)
6	2	→ (6, 2)
0	-2	→ (0, -2)
15	8	→ (15, 8)

Now Try:

Complete the table of values for the equation. Then write the results as ordered pairs.

$4x - y = 8$

x	y
1	
5	
	0
	4

_____

Practice Exercises

Complete each table of values. Write the results as ordered pairs.

7. $2x + 5 = 7$

x	y
	-3
	0
	5

7. _____

8. $y - 4 = 0$

x	y
-4	
0	
6	

8. _____

9. $4x + 3y = 12$

x	y
0	
	0
	-1

9. _____

Objective 9.R.4 Graph linear equations by plotting ordered pairs.

Video Examples

Review this example:

Graph $2x + 3y = 6$.

First let $x = 0$ and then let $y = 0$ to determine two ordered pairs.

$$
\begin{array}{c|c}
2(0) + 3y = 6 & 2x + 3(0) = 6 \\
0 + 3y = 6 & 2x + 0 = 6 \\
3y = 6 & 2x = 6 \\
y = 2 & x = 3
\end{array}
$$

The ordered pairs are (0, 2) and (3, 0). Find a third ordered pair by choosing a number other than 0 for x or y. We choose $y = 4$.

$$2x + 3(4) = 6$$
$$2x + 12 = 6$$
$$2x = -6$$
$$x = -3$$

This gives the ordered pair (–3, 4). We plot the three ordered pairs (0, 2), (3, 0), and (–3, 4) and draw a line through them.

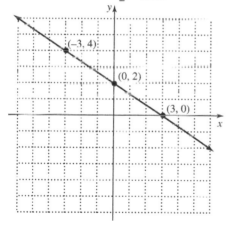

Now Try:

Graph $x + y = 3$.

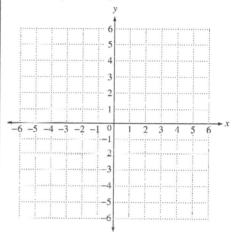

Practice Exercises

Complete the ordered pairs for each equation. Then graph the equation by plotting the points and drawing a line through them.

10. $y = 3x - 2$

(0,)

(, 0)

(2,)

10.

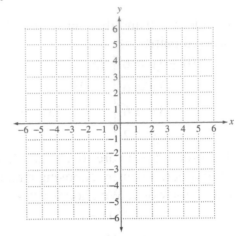

11. $x - y = 4$

(0,)

(, 0)

(−2,)

11.

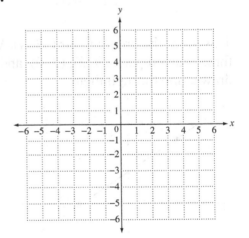

12. $x = 2y + 1$

(0,)

(, 0)

(, −2)

12.

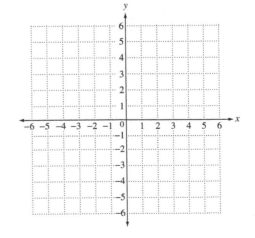

Objective 9.R.5 Graph linear equations of the form $y = b$ or $x = a$.

Video Examples

Review these examples:

Graph $y = -2$.

For any value of x, y is always -2. Three ordered pairs that satisfy the equation are $(-4, -2)$, $(0, -2)$ and $(2, -2)$. Drawing a line through these points gives the horizontal line. The y-intercept is $(0, -2)$. There is no x-intercept.

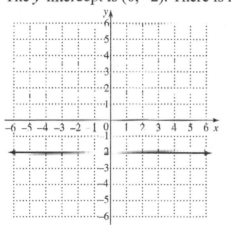

Graph $x + 4 = 0$.

First we subtract 4 from each side of the equation to get the equivalent equation $x = -4$. All ordered-pair solutions of this equation have x-coordinate -4.

Three ordered pairs that satisfy the equation are $(-4, -1)$, $(-4, 0)$, and $(-4, 3)$. The graph is a vertical line. The x-intercept is $(-4, 0)$. There is no y-intercept.

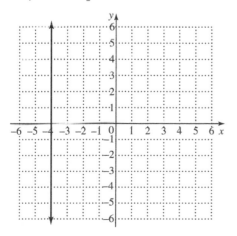

Now Try:

Graph $y = 4$.

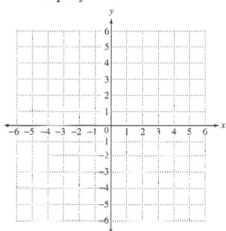

Graph $x = 0$.

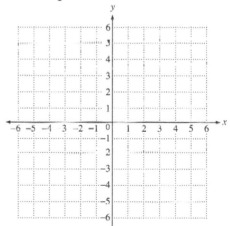

Practice Exercises

Graph each equation.

13. $x - 1 = 0$

13.

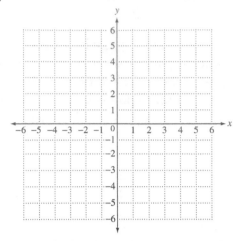

14. $y + 3 = 0$

14.

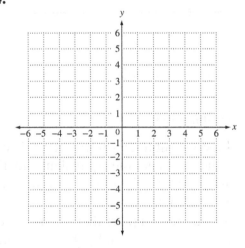

Copyright © 2018 Pearson Education, Inc.

Objective 9.R.6 Evaluate polynomials.

Video Examples

Review these examples:

Find the value of $4x^3 + 6x^2 - 5x - 5$ for

$x = -3$

$4x^3 + 6x^2 - 5x - 5$

$= 4(-3)^3 + 6(-3)^2 - 5(-3) - 5$

$= 4(-27) + 6(9) - 5(-3) - 5$

$= -108 + 54 + 15 - 5$

$= -44$

$x = 2$

$4x^3 + 6x^2 - 5x - 5$

$= 4(2)^3 + 6(2)^2 - 5(2) - 5$

$= 4(8) + 6(4) - 5(2) - 5$

$= 32 + 24 - 10 - 5$

$= 41$

Now Try:

Find the value of

$5x^4 + 3x^2 - 9x - 7$ for

$x = 4$

_____

$x = -4$

_____

Practice Exercises

Find the value of each polynomial (a) *when x = 2 and* (b) *when x = 3.*

15. $3x^3 + 4x - 19$

15. a._____

b._____

16. $-4x^3 + 10x^2 - 1$

16. a._____

b._____

17. $x^4 - 3x^2 - 8x + 9$

17. a._____

b._____

Objective 9.R.7 Perform operations with polynomials.

Video Examples

Review these examples:

Add vertically.

$5x^4 - 7x^3 + 9$ and $-3x^4 + 8x^3 - 7$

Write like terms in columns.

$5x^4 - 7x^3 + 9$
$\underline{-3x^4 + 8x^3 - 7}$

Now add, column by column.

$5x^4 \quad -7x^3 \quad 9$
$\underline{-3x^4} \quad \underline{8x^3} \quad \underline{-7}$
$2x^4 \quad \quad x^3 \quad \quad 2$

Add the three sums together to obtain the answer.

$2x^4 + x^3 + 2$

Find the sum.

$\left(5x^4 - 7x^2 + 6x\right) + \left(-3x^3 + 4x^2 - 7\right)$

$\left(5x^4 - 7x^2 + 6x\right) + \left(-3x^3 + 4x^2 - 7\right)$
$= 5x^4 - 3x^3 - 7x^2 + 4x^2 + 6x - 7$
$= 5x^4 - 3x^3 - 3x^2 + 6x - 7$

Perform the subtraction.

Subtract $8x^3 - 5x^2 + 8$ from $9x^3 + 6x^2 - 7$.

$\left(9x^3 + 6x^2 - 7\right) - \left(8x^3 - 5x^2 + 8\right)$
$= \left(9x^3 + 6x^2 - 7\right) + \left(-8x^3 + 5x^2 - 8\right)$
$= x^3 + 11x^2 - 15$

Add or subtract as indicated.
$\left(3x^2y + 5xy + y^2\right) - \left(4x^2y + xy - 3y^2\right)$

$\left(3x^2y + 5xy + y^2\right) - \left(4x^2y + xy - 3y^2\right)$
$= 3x^2y + 5xy + y^2 - 4x^2y - xy + 3y^2$
$= -x^2y + 4xy + 4y^2$

Now Try:

Add vertically.

$8x^3 - 9x^2 + x$ and

$-3x^3 + 4x^2 + 3x$

_____

Find the sum.

$\left(8x^2 - 6x + 4\right) + \left(7x^3 - 8x - 5\right)$

_____

Perform the subtraction.
$\left(7x^3 - 3x - 5\right) - \left(18x^3 + 4x - 6\right)$

_____

Add or subtract as indicated.
$\left(7x^2y + 3xy + 4y^2\right)$
$-\left(6x^2y - xy + 4y^2\right)$

_____

Practice Exercises

Add.

18. $9m^3 + 4m^2 - 2m + 3$

$\underline{-4m^3 - 6m^2 - 2m + 1}$

18. _____

19. $\left(x^2 + 6x - 8\right) + \left(3x^2 - 10\right)$

19. _____

20. $\left(3r^3 + 5r^2 - 6\right) + \left(2r^2 - 5r + 4\right)$

20. _____

Subtract.

21. $\left(-8w^3 + 11w^2 - 12\right) - \left(-10w^2 + 3\right)$

21. _____

22. $\left(8b^4 - 4b^3 + 7\right) - \left(2b^2 + b + 9\right)$

22. _____

23. $\left(9x^3 + 7x^2 - 6x + 3\right) - \left(6x^3 - 6x + 1\right)$

23. _____

Add or subtract as indicated.

24. $\left(-2a^6 + 8a^4b - b^2\right) - \left(a^6 + 7a^4b + 2b^2\right)$ **24.** _____

25. $\left(4ab + 2bc - 9ac\right) + \left(3ca - 2cb - 9ba\right)$ **25.** _____

26. $\left(2x^2y + 2xy - 4xy^2\right) + \left(6xy + 9xy^2\right) - \left(9x^2y + 5xy\right)$ **26.** _____

Chapter 10 ROOTS, RADICALS, AND ROOT FUNCTIONS

Learning Objectives
10.R.1 Use a linear equation to model data.
10.R.2 Use combinations of the rules for exponents that involve integer exponents.
10.R.3 Solve problems by applying the Pythagorean theorem.
10.R.4 Graph basic polynomial functions.

Key Terms

Use the vocabulary terms listed below to complete each statement in exercises 1–13.

graph	graphing	y-intercept	x-intercept
exponent	base	product rule for exponents	
power rule for exponents		hypotenuse	legs
polynomial function of degree n			identity function
squaring function			cubing function

1. If a graph intersects the y-axis at k, then the _____ is $(0, k)$.

2. If a graph intersects the x-axis at k, then the _____ is $(k, 0)$.

3. The process of plotting the ordered pairs that satisfy a linear equation and drawing a line through them is called _____.

4. The set of all points that correspond to the ordered pairs that satisfy the equation is called the _____ of the equation.

5. The statement "If m and n are any integers, then $\left(a^m\right)^n = a^{mn}$" is an example of the _____.

6. In the expression a^m, a is the _____ and m is the _____.

7. The statement "If m and n are any integers, then $a^m \cdot a^n = a^{m+n}$" is an example of the _____.

8. In a right triangle, the sides that form the right angle are the _____.

9. The longest side of a right triangle is the _____.

10. The polynomial function $f(x) = x^2$ is the _____.

11. A function defined by $f(x) = a_n x^n + a_{n-1} x^{n-1} + \cdots + a_1 x + a_0$, where $a_n \neq 0$ and n is a whole number is a _____.

12. The polynomial function $f(x) = x^3$ is the _____.

13. The polynomial function $f(x) = x$ is the _____.

Objective 10.R.1 Use a linear equation to model data.

Video Examples

Review these examples:

Every year sea turtles return to a certain group of islands to lay eggs. The number of turtle eggs that hatch can be approximated by the equation $y = -70x + 3260$, where y is the number of eggs that hatch and $x = 0$ representing 1990.

a. Use this equation to find the number of eggs that hatched in 1995, 2000, and 2005, and 2015.

Substitute the appropriate value for each year x to find the number of eggs hatched in that year.

For 1995:

$y = -70(5) + 3260$ $1995 - 1990 = 5$

$y = 2910$ eggs   Replace x with 5.

For 2000:

$y = -70(10) + 3260$ $2000 - 1990 = 10$

$y = 2560$ eggs   Replace x with 10.

For 2005:

$y = -70(15) + 3260$ $2005 - 1990 = 15$

$y = 2210$ eggs   Replace x with 15.

For 2015:

$y = -70(25) + 3260$ $2015 - 1990 = 25$

$y = 1510$ eggs   Replace x with 25.

b. Write the information from part (a) as four ordered pairs, and use them to graph the given linear equation.

Since x represents the year and y represents the number of eggs, the ordered pairs are (5, 2910), (10, 2560), (15, 2210), and (25, 1510).

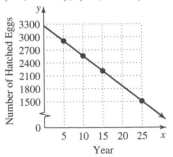

Now Try:

Suppose that the demand and price for a certain model of calculator are related by the equation $y = 45 - \frac{3}{5}x$, where y is the price (in dollars) and x is the demand (in thousands of calculators).

a. Assuming that this model is valid for a demand up to 50,000 calculators, use this equation to find the price of calculators at each level of demand.

0 calculators _____

5000 calculators _____

20,000 calculators _____

45,000 calculators _____

b. Write the information from part (a) as four ordered pairs, and use them to graph the given linear equation.

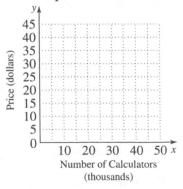

c. Use this graph and the equation to estimate the number of eggs that will hatch in 2010.

For 2010, $x = 20$. On the graph, find 20 on the horizontal axis, move up to the graphed line and then across to the vertical axis. It appears that in 2010, there were about 1900 eggs.

To use the equation, substitute 20 for x.

$$y = -70(20) + 3260$$

$$y = 1860 \text{ eggs}$$

This result for 2020 is close to our estimate of 1900 eggs from the graph.

c. Use this graph and the equation to estimate the price of 30,000 calculators.

_____

Practice Exercises

Solve each problem. Then graph the equation.

1. The profit y in millions of dollars earned by a small computer company can be approximated by the linear equation $y = 0.63x + 4.9$, where $x = 0$ corresponds to 2014, $x - 1$ corresponds to 2015, and so on. Use this equation to approximate the profit in each year from 2014 through 2017.

1. 2014 _____

   2015 _____

   2016 _____

   2017 _____

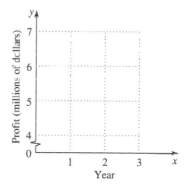

2. The number of band instruments sold by Elmer's Music Shop can be approximated by the equation $y = 325 + 42x$, where y is the number of instruments sold and x is the time in years, with $x = 0$ representing 2013. Use this equation to approximate the number of instruments sold in each year from 2013 through 2016.

2. 2013 _____

2014 _____

2015 _____

2016 _____

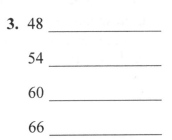

3. According to *The Old Farmer's Almanac*, the temperature in degrees Celsius can be determined by the equation $y = \frac{1}{3}x + 4$, where x is the number of cricket chirps in 25 seconds and y is the temperature in degrees Celsius. Use this equation to find the temperature when there are 48 chirps, 54 chirps, 60 chirps, and 66 chirps.

3. 48 _____

54 _____

60 _____

66 _____

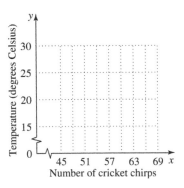

Objective 10.R.2 Use combinations of the rules for exponents.

Video Examples

Review these examples:

Simplify each expression. Assume that all variables represent nonzero real numbers.

$$\left(\frac{3x^4}{4}\right)^{-5}$$

$$\left(\frac{3x^4}{4}\right)^{-5} = \left(\frac{4}{3x^4}\right)^{5}$$

$$= \frac{4^5}{3^5 x^{20}}$$

$$= \frac{1024}{243x^{20}}$$

$$\left(\frac{5x^{-3}}{2^{-1}y^4}\right)^{2}$$

$$\left(\frac{5x^{-3}}{2^{-1}y^4}\right)^{-2} = \frac{5^{-2}x^6}{2^2 y^{-8}}$$

$$= \frac{x^6 y^8}{5^2 \cdot 2^2}$$

$$= \frac{x^6 y^8}{100}$$

Now Try:

Simplify each expression. Assume that all variables represent nonzero real numbers.

$$\left(\frac{2p^4}{3}\right)^{-5}$$

_____

$$\left(\frac{4x^{-4}}{5^{-1}y^5}\right)^{-3}$$

_____

Practice Exercises

Simplify each expression, and write it using only positive exponents. Assume that all variables represent nonzero real numbers.

4. $(9xy)^7 (9xy)^{-8}$

4. _____

5. $\dfrac{\left(a^{-1}b^{-2}\right)^{-4}\left(ab^2\right)^6}{\left(a^3b\right)^{-2}}$

5. _____

6. $\left(\dfrac{k^3t^4}{k^2t^{-1}}\right)^{-4}$

6. _____

Objective 10.R.3 Solve problems by applying the Pythagorean theorem.

Video Examples

Review this example:

Penny and Carla started biking from the same corner. Penny biked east and Carla biked south. When they were 26 miles apart, Carla had biked 14 miles further than Penny. Find the distance each biked.

Step 1 Read carefully. Find the two distances.

Step 2 Assign a variable.

    Let x = Penny's distance.

    Then $x + 14$ = Carla's distance.

Step 3 Write an equation. Substitute into the Pythagorean theorem.

$$a^2 + b^2 = c^2$$

$$x^2 + (x+14)^2 = 26^2$$

Step 4 Solve.

$$x^2 + x^2 + 28x + 196 = 676$$

$$2x^2 + 28x - 480 = 0$$

$$2(x^2 + 14x - 240) = 0$$

$$x^2 + 14x - 240 = 0$$

$$(x+24)(x-10) = 0$$

$$x + 24 = 0 \quad \text{or} \quad x - 10 = 0$$

$$x = -24 \quad \text{or} \quad x = 10$$

Step 5 State the answer. Since –24 cannot be a distance, 10 is the distance for Penny, and 10 + 14 = 24 is the distance for Carla.

Step 6 Check. Since $10^2 + 24^2 = 26^2$ is true, the answer is correct.

Now Try:

A ladder is leaning against a building. The distance from the bottom of the ladder to the building is 8 feet less than the length of the ladder. How high up the side of the building is the top of the ladder if that distance is 4 feet less than the length of the ladder?

_____

Practice Exercises

Solve each problem.

7. A field is in the shape of a right triangle. The shorter
   leg measures 45 meters. The hypotenuse measures
   45 meters less than twice the longer the leg. Find the
   dimensions of the lot.

7. _____

8. A train and a car leave a station at the same time, the
   train traveling due north and the car traveling west.
   When they are 100 miles apart, the train has traveled
   20 miles farther than the car. Find the distance each
   has traveled.

8. car_____

   train _____

9. Two ships left a dock at the same time. When they
   were 25 miles apart, the ship that sailed due south
   had gone 10 miles less than twice the distance
   traveled by the ship that sailed due west. Find the
   distance traveled by the ship that sailed due south.

9. _____

Objective 10.R.4 Graph basic polynomial functions.

Review these examples:

Graph each function. Give the domain and range.

$$f(x) = -3x + 2$$

Plot the points and join them with a straight line. The domain and range are both $(-\infty, \infty)$.

x	$f(x) = -3x + 2$
-1	5
0	2
1	-1
2	-4
3	-7

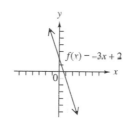

$$g(x) = -2x^2$$

The graph of $g(x)$ has the same shape as that of $f(x) = x^2$ but is narrower and opens downward. The domain is $(-\infty, \infty)$. The range is $(-\infty, 0]$.

x	$g(x) = -2x^2$
-2	-8
-1	-2
0	0
1	-2
2	-8

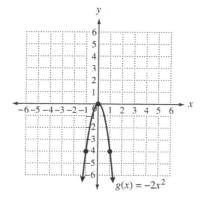

Now Try:

Graph each function. Give the domain and range.

$$f(x) = -2x - 3$$

_____

_____

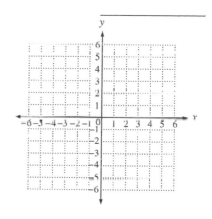

$$f(x) = x^2 - 1$$

_____

_____

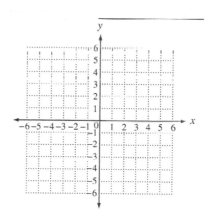

$f(x) = x^3 + 3.$

For this function, cube the input and add 3 to the result. The graph is the cubing function shifted 3 units up.

x	$f(x) = x^3 + 3$
-2	-5
-1	2
0	3
1	4
2	11

The domain and range is $(-\infty, \infty)$.

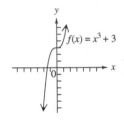

$f(x) = -x^3 + 1.$

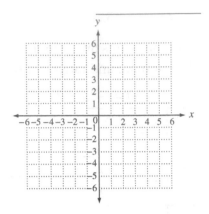

Practice Exercises

Graph each function. Give the domain and range.

10. $f(x) = \dfrac{1}{2}x + \dfrac{1}{2}$

10. domain _____

range _____

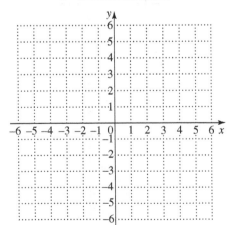

11. $f(x) = -2x^2$

11. domain _____

range _____

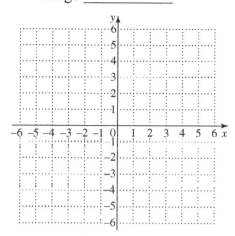

12. $f(x) = x^3 + 2$

12. domain _____

range _____

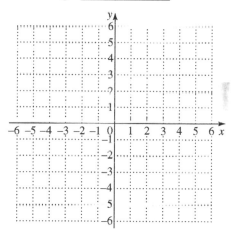

Chapter 11 QUADRATIC EQUATIONS, INEQUALITIES, AND FUNCTIONS

Learning Objectives

11.R.1   Graph linear inequalities in two variables.
11.R.2   Factor any polynomial.
11.R.3   Solve quadratic equations using the zero-factor property.
11.R.4   Solve application problems involving quadratic equations.
11.R.5   Solve problems using given quadratic models.
11.R.6   Solve application problems involving polynomial functions.
11.R.7   Graph basic polynomial functions.
11.R.8   Solve applications involving radical expressions and graphs.
11.R.9   Solve radical equations that require additional steps.
11.R.10 Solve applications and extensions involving complex numbers.

Key Terms

Use the vocabulary terms listed below to complete each statement in exercises 1–20.

linear inequality in two variables	boundary line
FOIL factoring by grouping	
quadratic equation standard form	
polynomial function of degree *n*	identity function
squaring function	cubing function
radical equation proposed solution	extraneous solution
complex number real part imaginary part	
pure imaginary number standard form (of a complex number)	
nonreal complex number complex conjugate	

1.   In the graph of a linear inequality, the _____ separates the region that satisfies the inequality from the region that does not satisfy the inequality.

2.   An inequality that can be written in the form $Ax + By < C$, $Ax + By > C$, $Ax + By \leq C$, or $Ax + By \geq C$ is called a _____.

3.   When there are more than three terms in a polynomial, use a process called _____ to factor the polynomial.

4.   _____ is a shortcut method for finding the product of two binomials.

5.   An equation written in the form $ax^2 + bx + c = 0$ is written in the _____ of a quadratic equation.

6. An equation that can written in the form $ax^2 + bx + c = 0$, with $a \neq 0$, is a
_____.

7. The polynomial function $f(x) = x^2$ is the _____.

8. A function defined by $f(x) = a_n x^n + a_{n-1} x^{n-1} + \cdots + a_1 x + a_0$, where $a_n \neq 0$
and n is a whole number is a _____.

9. The polynomial function $f(x) = x^3$ is the _____.

10. The polynomial function $f(x) = x$ is the _____.

11. A(n) _____ is a potential solution to an equation that does
not satisfy the equation.

12. An equation with a variable in the radicand is a(n) _____.

13. A value of a variable that appears to be a solution of an equation is a(n)
_____.

14. A _____ is a number that can be written in the
form $a + bi$, where a and b are real numbers.

15. The _____ of $a + bi$ is $a - bi$.

16. The _____ of $a + bi$ is bi.

17. The _____ of $a + bi$ is a.

18. A complex number is in _____ if it is written in the
form $a + bi$.

19. A complex number $a + bi$ with $a = 0$ and $b \neq 0$ is called a
_____.

20. A complex number $a + bi$ $b \neq 0$ is called a _____.

Objective 11.R.1 Graph linear inequalities in two variables.

Video Examples

Review these examples:

Graph $2x + 5y > -10$.

This equation does not include equality. Therefore, the points on the line $2x + 5y = -10$ do not belong to the graph. However, the line still serves as a boundary for two regions.

To graph the inequality, first graph the equation $2x + 5y > -10$. Use a dashed line to show that the points on the line are not solutions of the inequality $2x + 5y > -10$. Then choose a test point to see which region satisfies the inequality.

$$2x + 5y > -10$$
$$2(0) + 5(0) \overset{?}{>} -10$$
$$0 + 0 \overset{?}{>} -10$$
$$0 > -10 \quad \text{True}$$

Since $0 > -10$ is true, the graph of the inequality is the region that contains $(0, 0)$. Shade that region.

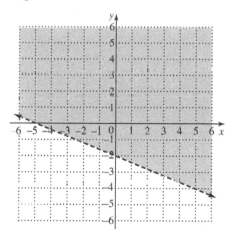

Now Try:

Graph $5x + 4y > 20$.

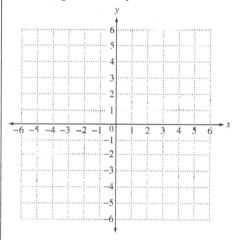

Graph $x - 4 \leq -1$.

First, solve the inequality for x.

   $x \leq 3$

Now graph the line $x = 3$, a vertical line through the point $(3, 0)$. Use a solid line, and choose $(0, 0)$ as a test point.

   $0 \leq 3$ True

Because $0 \leq 3$ is true, we shade the region containing $(0, 0)$.

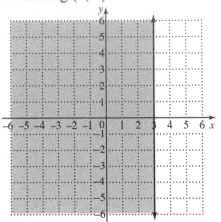

Graph $y \geq -1$.

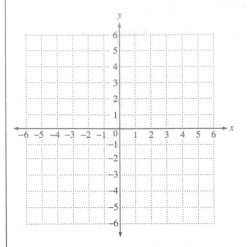

Practice Exercises

Graph each linear inequality.

1. $y \geq x - 1$

1.

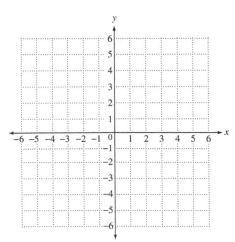

2. $y > -x + 2$

2.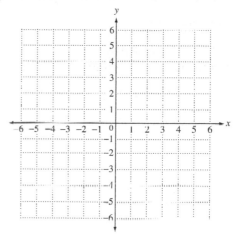

3. $3x - 4y - 12 > 0$

3.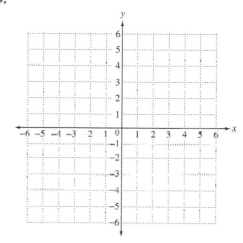

Objective 11.R.2 Factor any polynomial.

Video Examples

Review these examples:	**Now Try:**
Factor the polynomial.	Factor the polynomial.
$9x(a+c)-z(a+c)$	$7x(y+z)-5(y+z)$
Factor out $(a+c)$ $9x(a+c)-z(a+c)=(a+c)(9x-z)$	_____
Factor each binomial if possible.	Factor each binomial if possible.
$16y^2+49$	$36p^2+169$
$16y^2+49$ is prime. It is the sum of squares. There is no common factor.	_____
$216p^3-125z^3$	$27t^3-64w^3$
Difference of cubes $216p^3-125z^3$ $\quad=(6p)^3-(5z)^3$ $\quad=(6p-5z)\big[(6p)^2+(6p)(5z)+(5z)^2\big]$ $\quad=(6p-5z)(36p^2+30pz+25z^2)$	_____
Factor the trinomial.	Factor the trinomial.
y^2-3y-4	$4k^2-7k-2$
$y^2-3y-4=(y-4)(y+1)$	_____
Factor the polynomial.	Factor the polynomial.
$ad-8d+af-8f$	$cm-3m+cx-3x$
Group the terms and factor each group. $ad-8d+af-8f$ $\quad=(ad-8d)+(af-8f)$ $\quad=d(a-8)+f(a-8)$ $\quad=(a-8)(d+f)$	_____

Practice Exercises

Factor completely.

4. $\quad -12x^2 - 6x$

4. _____

5. $\quad 12a^2b^2 + 3a^2b - 9ab^2$

5. _____

6. $\quad (x+1)(2x+3) - (x+1)$

6. _____

7. $\quad 2x^3y^4 - 72xy^2$

7. _____

8. $\quad 128x^3 - 2y^3$

8. _____

9. $\quad y^6 + 1$

9. _____

10. $4x^2 - 12xy + 9y^2$ **10.** _____

11. $2a^2 - 17a + 30$ **11.** _____

12. $14w^2 + 6wx - 35wx - 15x^2$ **12.** _____

13. $x^3 - 3x^2 + 7x - 21$ **13.** _____

14. $a^2 - 6ab + 9b^2 - 25$ **14.** _____

Objective 11.R.3 Solve quadratic equations using the zero-factor property.

Video Examples

Review these examples:

Solve the equation.

$$(x+9)(5x-6)=0$$

By the zero-factor property, either $x+9=0$ or $5x-6=0$, or both.

$$x+9=0 \quad \text{or} \quad 5x-6=0$$
$$x=-9 \quad \text{or} \quad 5x=6$$
$$x=\frac{6}{5}$$

Check:

Let $x=-9$.

$$(x+9)(5x-6)=0$$
$$(-9+9)[5(-9)-6]\stackrel{?}{=}0$$
$$0(-51)\stackrel{?}{=}0$$
$$0=0 \quad \text{True}$$

Let $x=\frac{6}{5}$.

$$(x+9)(5x-6)=0$$
$$\left(\frac{6}{5}+9\right)\left[5\left(\frac{6}{5}\right)-6\right]\stackrel{?}{=}0$$
$$\left(\frac{51}{5}\right)(6-6)\stackrel{?}{=}0$$
$$0=0 \quad \text{True}$$

Both values check, so the solution set is $\left\{-9,\ \frac{6}{5}\right\}$.

Solve the equation.
$$x^2-6x=-5$$

First, write the equation in standard form.

$$x^2-6x=-5$$
$$x^2-6x+5=0$$

Factor and use the zero-factor property.

$$(x-1)(x-5)=0$$
$$x-1=0 \quad \text{or} \quad x-5=0$$
$$x=1 \quad \text{or} \quad x=5$$

Check these solutions by substituting each in the original equation. The solution set is $\{1, 5\}$.

Now Try:

Solve the equation.

$$(x+12)(4x-7)=0$$

_____

Solve the equation.
$$x^2-5x=24$$

_____

Solve $8p^2 + 30 = 46p$.

$$8p^2 + 30 = 46p$$

$8p^2 - 46p + 30 = 0$ Standard form

$2(4p^2 - 23p + 15) = 0$ Factor out 2.

$4p^2 - 23p + 15 = 0$ Divide each side by 2

$(4p - 3)(p - 5) = 0$ Factor.

$4p - 3 = 0$ or $p - 5 = 0$ Zero-factor

$p = \dfrac{3}{4}$ or $p = 5$ property

The solution set is $\left\{\dfrac{3}{4},\ 5\right\}$.

Solve $15p^2 + 36 = 57p$.

_____

Solve each equation.

$k(3k + 5) = 2$

$$k(3k + 5) = 2$$

$$3k^2 + 5k = 2$$

$$3k^2 + 5k - 2 = 0$$

$$(3k - 1)(k + 2) = 0$$

$3k - 1 = 0$ or $k + 2 = 0$

$3k = 1$ or $k = -2$

$$k = \dfrac{1}{3}$$

The solution set is $\left\{-2,\ \dfrac{1}{3}\right\}$.

$y^2 = 9y$

$$y^2 = 9y$$

$$y^2 - 9y = 0$$

$$y(y - 9) = 0$$

$y = 0$ or $y - 9 = 0$

$$y = 9$$

The solution set is $\{0, 9\}$.

Solve each equation.

$k(4k - 23) = 6$

_____

$y^2 = 11y$

_____

Practice Exercises

Solve each equation and check your solutions.

15. $2x^2 - 3x - 20 = 0$

15. _____

16. $25x^2 = 20x$

16. _____

17. $c(5c + 17) = 12$

17. _____

Objective 11.R.4 Solve application problems involving quadratic equations.

Practice Exercises

The surface area of a cube can be described using the formula

$$S = 6d^2,$$

where S is the surface area and d is the length of one side of the cube.

18. Use the formula and complete the following table.   18. _____

d, in feet	0	1	2	3	__	__
S, in square feet	0	6	__	__	150	294

19. When $d = 0$, $S = 0$. Explain this in the context
     of the problem.

19. _____

_____

_____

_____

Objective 11.R.5 Solve problems using given quadratic models.

Video Examples

Review this example:

Jeff threw a stone straight upward at 46 feet per second from a dock 6 feet above a lake. The height of the stone above the lake t seconds after it is thrown is given by $h = -16t^2 + 46t + 6$. How long will it take for the stone to reach a height of 39 feet?

Substitute 39 for h.

$$39 = -16t^2 + 46t + 6$$

Solve for t.

$$16t^2 - 46t + 33 = 0$$

$$(8t - 11)(2t - 3) = 0$$

$$8t - 11 = 0 \quad \text{or} \quad 2t - 3 = 0$$

$$t = \frac{11}{8} \quad \text{or} \quad t = \frac{3}{2}$$

Since we have found two acceptable answers, the stone will be at height of 39 feet twice (once on its way up and once on its way down) —at $\frac{11}{8}$ sec or $\frac{3}{2}$ sec.

Now Try:

A ball is dropped from the roof of a 19.6 meter high building. Its height h (in meters) t seconds later is given by the equation $h = -4.9t^2 + 19.6$. After how many seconds is the height 14.7 meters?

_____

Practice Exercises

Solve each problem.

20.   If an object is propelled upward from a height of 16 feet with an initial velocity of 48 feet per second, its height h (in feet) t seconds later is given by the equation $h = -16t^2 + 48t + 16$.

(a)   After how many seconds is the height 52 feet?

(b)   After how many seconds is the height 48 feet?

20. a._____

    b._____

21.   A company determines that its daily revenue R (in
      dollars) for selling x items is modeled by the
      equation $R = x(150 - x)$. How many items must be
      sold for its revenue to be $4400?

21. _____

22.   If a ball is batted at an angle of 35°, the distance that
      the ball travels is given approximately by
      $D = 0.029v^2 + 0.021v - 1$, where v is the bat speed
      in miles per hour and D is the distance traveled in
      feet. Find the distance a batted ball will travel if the
      ball is batted with a velocity of 90 miles per hour.
      Round your answer to the nearest whole number.

22. _____

Objective 11.R.6 Solve application problems involving quadratic equations.

Practice Exercises

Solve each problem.

23. The cost in dollars to produce x t-shirts is
     $C(x) = 3.5x + 40$. The revenue in dollars from
     sales of x t-shirts is $R(x) = 12.99x$.
     (a) Write and simplify a function P that gives
         profit in terms of x.
     (b) Find the profit if 100 t-shirts are produced
         and sold.

23. a._____

    b. _____

24. The cost in dollars to produce x hats is
     $C(x) = 5.2x + 85$. The revenue in dollars from
     sales of x hats is $R(x) = 22x$.
     (a) Write and simplify a function P that gives
         profit in terms of x.
     (b) Find the profit if 50 hats are produced
         and sold.

24. a._____

    b._____

Objective 11.R.7 Graph basic polynomial functions.

Video Examples

Review these examples:

Graph each function. Give the domain and range.

$$f(x) = -3x + 2$$

Plot the points and join them with a straight line. The domain and range are both $(-\infty, \infty)$.

x	$f(x) = -3x + 2$
-1	5
0	2
1	-1
2	-4
3	-7

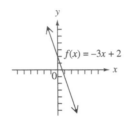

$$f(x) = -\frac{1}{4}x^2.$$

The graph has the same shape as that of $f(x) = x^2$ but is wider and opens downward. The domain is $(-\infty, \infty)$. The range is $(-\infty, 0]$.

x	$f(x) = -\frac{1}{4}x^2$
-4	-4
-2	-1
0	0
2	-1
4	-4

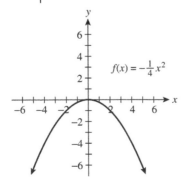

Now Try:

Graph each function. Give the domain and range.

$$f(x) = -2x - 3$$

_____

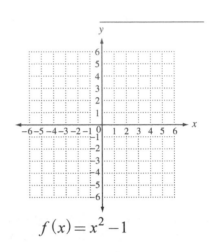

$$f(x) = x^2 - 1$$

_____

_____

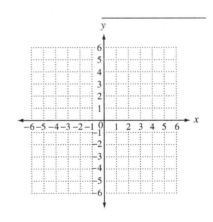

$f(x) = x^3 + 3.$

For this function, cube the input and add 3 to the result. The graph is the cubing function shifted 3 units up.

x	$f(x) = x^3 + 3$
-2	-5
-1	2
0	3
1	4
2	11

The domain and range is $(-\infty, \infty)$.

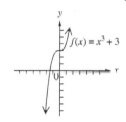

$f(x) = -x^3 + 1.$

_____

_____

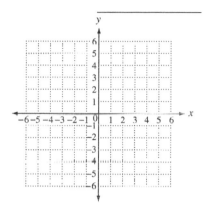

Practice Exercises

Graph each function. Give the domain and range.

25. $f(x) = \dfrac{1}{2}x + \dfrac{1}{2}$

25. domain _____

　　　range _____

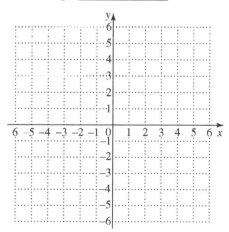

26. $f(x) = -2x^2$

26. domain _____

range _____

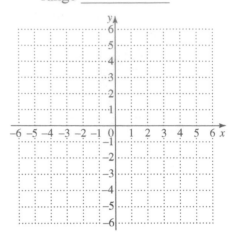

27. $f(x) = x^3 + 2$

27. domain _____

range _____

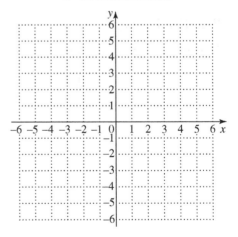

Objective 11.R.8 Solve applications involving radical expressions and graphs.

Practice Exercises

Solve each problem.

28. Heron's formula gives a method of finding the area
of a triangle if the lengths of its sides are known.
Suppose that a, b, and c are the lengths of the sides.
Let s denote one-half of the perimeter of the triangle
(called the semiperimeter)—that is,

$$s = \frac{1}{2}(a + b + c).$$

Then the area A of the triangle is

$$A = \sqrt{s(s-a)(s-b)(s-c)}.$$

Use Heron's formula to find the area of a triangle
with sides $a = 8$ m, $b = 16$ m, $c = 18$ m. Round
to the nearest square meter.

28. _____

29. The time t in seconds for one complete swing of a
simple pendulum, where L is the length of the

pendulum in feet is $t = 2\pi\sqrt{\dfrac{L}{32}}$. Find the time of a

complete swing of a 4-ft pendulum to the nearest
tenth of a second.

29. _____

Objective 11.R.9 Solve radical equations that require additional steps.

Video Examples

Review these examples:

Solve $\sqrt{x+3} = x-3$.

Step 1 The radical is isolated on the left side of the equation.

Step 2 Square each side.
$$\left(\sqrt{x+3}\right)^2 = (x-3)^2$$
$$x+3 = x^2-6x+9$$

Step 3 Write the equation in standard form and solve.
$$0 = x^2-7x+6$$
$$0 = (x-1)(x-6)$$
$$x-1=0 \quad \text{or} \quad x-6=0$$
$$x=1 \quad \text{or} \quad x=6$$

Step 4 Check each proposed solution in the original equation.

$$\sqrt{x+3} = x-3 \qquad\qquad \sqrt{x+3} = x-3$$
$$\sqrt{1+3} \overset{?}{=} 1-3 \qquad\qquad \sqrt{6+3} \overset{?}{=} 6-3$$
$$\sqrt{4} \overset{?}{=} -2 \qquad\qquad \sqrt{9} \overset{?}{=} 3$$
$$2 = -2 \quad \text{False} \qquad\qquad 3 = 3 \quad \text{True}$$

The solution set is $\{6\}$. The other proposed solution, 1, is extraneous.

Solve $\sqrt{3x} - 4 = \sqrt{x-2}$.

$$\left(\sqrt{3x}-4\right)^2 = \left(\sqrt{x-2}\right)^2$$
$$3x - 8\sqrt{3x} + 16 = x-2$$
$$-8\sqrt{3x} = -2x-18$$
$$\left(-8\sqrt{3x}\right)^2 = (-2x-18)^2$$
$$192x = 4x^2 + 72x + 324$$
$$0 = 4x^2 - 120x + 324$$
$$0 = 4\left(x^2 - 30x + 81\right)$$
$$0 = 4(x-3)(x-27)$$
$$x-3=0 \quad \text{or} \quad x-27=0$$
$$x=3 \quad \text{or} \quad x=27$$

Now Try:

Solve $\sqrt{x+11} = x-1$.

_____

Solve $\sqrt{3x+4} = \sqrt{9x} - 2$.

_____

Check

$$\sqrt{3x} - 4 = \sqrt{x-2}$$

$$\sqrt{3(3)} - 4 \stackrel{?}{=} \sqrt{3-2}$$

$$3 - 4 \stackrel{?}{=} \sqrt{1}$$

$$-1 = 1 \quad \text{False}$$

$$\sqrt{3x} - 4 = \sqrt{x-2}$$

$$\sqrt{3(27)} - 4 \stackrel{?}{=} \sqrt{27-2}$$

$$9 - 4 \stackrel{?}{=} \sqrt{25}$$

$$5 = 5 \quad \text{True}$$

The proposed solution, 27, is valid, but 3 is extraneous and must be rejected. The solution set is $\{27\}$.

Practice Exercises

Solve each equation.

30. $\sqrt{27 - 18v} = 2v - 3$

30. _____

31. $\sqrt{7r + 8} - \sqrt{r+1} = 5$

31. _____

32. $\sqrt{k + 10} + \sqrt{2k + 19} = 2$

32. _____

Objective 11.R.10 Solve applications and extensions involving complex numbers.

Practice Exercises

Ohm's law for the current I in a circuit with voltage E, resistance R, capacitance reactance X_C, and inductive reactance X_L is

$$I = \frac{E}{R + (X_L - X_C)i},$$

Use this law to work each problem.

33.   Find I if $E = 1 - 3i$, $R = 3$, $X_L = 5$, and $X_C = 4$.   **33.** _____

34.   Find E if $I = 2 - i$, $R = 2$, $X_L = 4$, and $X_C = 2$.   **34.** _____

Chapter 12 INVERSE, EXPONENTIAL, AND LOGARITHMIC FUNCTIONS

Learning Objectives
12.R.1 Use combinations of the rules for exponents that involve integer exponents.
12.R.2 Identify functions defined by graphs and equations.
12.R.3 Use function notation.
12.R.4 Solve applications involving linear functions.
12.R.5 Graph basic polynomial functions.
12.R.6 Solve applied problems using quadratic functions as models.

Key Terms

Use the vocabulary terms listed below to complete each statement in exercises 1–18.

exponent	base	product rule for exponents
power rule for exponents		dependent variable
independent variable	relation	function
domain	range	function notation
linear function	constant function	polynomial function of degree n
identity function	squaring function	cubing function
quadratic function		Pythagorean theorem

1. The statement "If m and n are any integers, then $\left(a^m\right)^n = a^{mn}$" is an example of the _____.

2. In the expression a^m, a is the _____ and m is the _____.

3. The statement "If m and n are any integers, then $a^m \cdot a^n = a^{m+n}$" is an example of the _____.

4. The _____ of a relation is the set of second components (y-values) of the ordered pairs of the relation.

5. A _____ is a set of ordered pairs of real numbers.

6. If the quantity y depends on x, then y is called the _____ in a relation between x and y.

7. The _____ of a relation is the set of first components (x-values) of the ordered pairs of the relation.

8. A _____ is a set of ordered pairs in which each value of the first component, x, corresponds to exactly one value of the second component, y.

9. If the quantity y depends on x, then x is called the _____
in a relation between x and y.

10. A function defined by an equation of the form $f(x) = ax + b$, for real numbers a
and b, is a _____.

11. _____ $f(x)$ represents the value of the function
at x, that is, the y-value that corresponds to x.

12. A _____ is a linear function of the form $f(x) = b$,
for a real number b.

13. The polynomial function $f(x) = x^2$ is the _____.

14. A function defined by $f(x) = a_n x^n + a_{n-1} x^{n-1} + \cdots + a_1 x + a_0$, where $a_n \neq 0$
and n is a whole number is a _____.

15. The polynomial function $f(x) = x^3$ is the _____.

16. The polynomial function $f(x) = x$ is the _____.

17. A function defined by $f(x) = ax^2 + bx + c$, for real numbers a, b, and c, with
$a \neq 0$, is a _____.

18. The _____ states that the sum of the squares of
the lengths of the legs of a right triangle equals the square of the length of the
hypotenuse.

Objective 12.R.1 Use combinations of the rules for exponents.

Video Examples

Review these examples:

Simplify each expression. Assume that all variables represent nonzero real numbers.

$$\left(\frac{3x^4}{4}\right)^{-5}$$

$$\left(\frac{3x^4}{4}\right)^{-5} = \left(\frac{4}{3x^4}\right)^{5}$$

$$= \frac{4^5}{3^5 x^{20}}$$

$$= \frac{1024}{243x^{20}}$$

$$\left(\frac{5x^{-3}}{2^{-1}y^4}\right)^{-2}$$

$$\left(\frac{5x^{-3}}{2^{-1}y^4}\right)^{-2} = \frac{5^{-2}x^6}{2^2 y^{-8}}$$

$$= \frac{x^6 y^8}{5^2 \cdot 2^2}$$

$$= \frac{x^6 y^8}{100}$$

Now Try:

Simplify each expression. Assume that all variables represent nonzero real numbers.

$$\left(\frac{2p^4}{3}\right)^{-5}$$

_____

$$\left(\frac{4x^{-4}}{5^{-1}y^5}\right)^{-3}$$

_____

Practice Exercises

Simplify each expression, and write it using only positive exponents. Assume that all variables represent nonzero real numbers.

1. $(9xy)^7 (9xy)^{-8}$

1. _____

2. $\dfrac{\left(a^{-1}b^{-2}\right)^{-4}\left(ab^2\right)^6}{\left(a^3b\right)^{-2}}$

2. _____

3. $\left(\dfrac{k^3t^4}{k^2t^{-1}}\right)^{-4}$

3. _____

Objective 12.R.2 Identify functions defined by graphs and equations.

Video Examples

Review these examples:

Use the vertical line test to determine whether the relation graphed is a function.

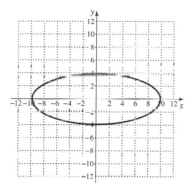

The graph is not a function.

Decide whether the relation defines y as a function of x. Give the domain.

$$y = 2x - 4$$

Each x value corresponds to just one y-value and the relation defines a function. Since x can be any real number, the domain is $(-\infty, \infty)$.

Now Try:

Use the vertical line test to determine whether the relation graphed is a function.

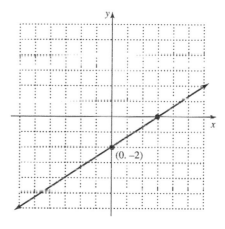

_____

Decide whether the relation defines y as a function of x. Give the domain.

$$y = 2x - 4$$

_____

_____

Practice Exercises

Use the vertical line test to determine whether the relation graphed is a function.

4.

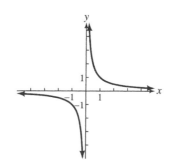

4. _____

Decide whether each equation defines y as a function of x. Give the domain.

5. $y^2 = x + 1$

5. _____

6. $y = \dfrac{3}{x+6}$

6. _____

Objective 12.R.3 Use function notation.

Video Examples

Review these examples:

Let $f(x) = 7x - 3$. Evaluate the function f for the following.

$x - 4$

Start with the given function. Replace x with 4.

$f(x) = 7x - 3$
$f(4) = 7(4) - 3$
$f(4) = 28 - 3$
$f(4) = 25$

Thus, $f(4) = 25$.

Let $g(x) = 5x + 6$. Find and simplify $g(n + 8)$.

Replace x with $n + 8$,

$g(x) = 5x + 6$

$g(n + 8) = 5(n + 8) + 6$

$g(n + 8) = 5n + 40 + 6$

$g(n + 8) = 5n + 46$

For the function, find $f(5)$.

$f = \{(7, -27), (5, -25), (3,\ -23), (1,\ -21)\}$

From the ordered pair (5, –25), we have
$f(5) = -25$.

Write the equation using function notation $f(x)$. Then find $f(-5)$.

$x - 5y = 8$

Step 1 $x - 5y = 8$
$-5y = -x + 8$
$y = \dfrac{1}{5}x - \dfrac{8}{5}$

Step 2 $f(x) = \dfrac{1}{5}x - \dfrac{8}{5}$

$f(-5) = \dfrac{1}{5}(-5) - \dfrac{8}{5}$

$f(-5) = -\dfrac{13}{5}$

Now Try:

Let $f(x) = 8x - 7$. Evaluate the function f for the following.

$x = 3$

_____

Let $g(x) = 4x - 7$. Find and simplify $g(a - 1)$.

_____

For the function, find $f(-6)$.

$f = \{(-2, 11), (-4, 17), (-6, 21), (-8, 24)\}$

_____

Write the equation using function notation $f(x)$. Then find $f(-3)$.

$2x + 3y = 7$

_____

Practice Exercises

For each function f, find (a) $f(-2)$*, (b)* $f(0)$*, and (c)* $f(-x)$*.*

7.　　$f(x) = 3x - 7$

7. a.＿＿＿＿＿＿＿＿

　　b.＿＿＿＿＿＿＿＿

　　c.＿＿＿＿＿＿＿＿

8.　　$f(x) = 2x^2 + x - 5$

8. a.＿＿＿＿＿＿＿＿

　　b.＿＿＿＿＿＿＿＿

　　c.＿＿＿＿＿＿＿＿

9.　　$f(x) = 9$

9. a.＿＿＿＿＿＿＿＿

　　b.＿＿＿＿＿＿＿＿

　　c.＿＿＿＿＿＿＿＿

Objective 12.R.4 Solve applications involving linear functions.

Practice Exercises

A rental truck costs $20, plus $0.20 per mile. Let x represent the number of miles the truck is driven and f(x) represent the total cost to rent the truck.

10.   Write a linear function that models this situation.   10. _____

11.   How much would it cost to drive 50 mi?   11. _____
       Interpret the answer in the context of the problem.

12.   Find the value of x if $f(x) = 70$. Express this   12. _____
       situation using function notation, and interpret it
       in the context of this problem.

Objective 12.R.5 Graph basic polynomial functions.

Video Examples

Review these examples:

Graph each function. Give the domain and range.

$$f(x) = -3x + 2$$

Plot the points and join them with a straight line.
The domain and range are both $(-\infty, \infty)$.

x	$f(x) = -3x + 2$
-1	5
0	2
1	-1
2	-4
3	-7

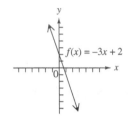

$$g(x) = -2x^2$$

The graph of $g(x)$ has the same shape as that of
$f(x) = x^2$ but is narrower and opens downward.
The domain is $(-\infty, \infty)$. The range is
$(-\infty, 0]$.

x	$g(x) = -2x^2$
-2	-8
-1	-2
0	0
1	-2
2	-8

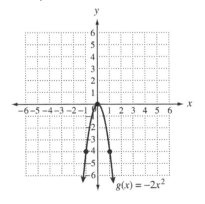

Now Try:

Graph each function. Give the
domain and range.

$$f(x) = -2x - 3$$

_____

_____

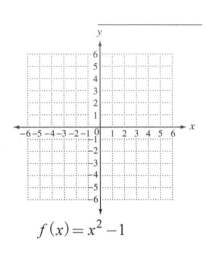

$$f(x) = x^2 - 1$$

_____

_____

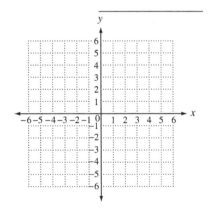

$f(x) = x^3 + 3.$

For this function, cube the input and add 3 to the result. The graph is the cubing function shifted 3 units up.

x	$f(x) = x^3 + 3$
-2	-5
-1	2
0	3
1	4
2	11

The domain and range is $(-\infty, \infty)$.

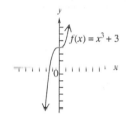

$f(x) = -x^3 + 1.$

_____

_____

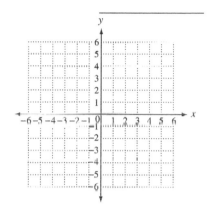

Practice Exercises

Graph each function. Give the domain and range.

13. $f(x) = \frac{1}{2}x + \frac{1}{2}$

13. domain _____

range _____

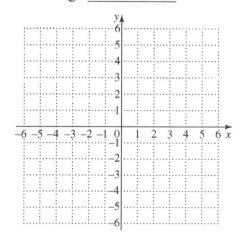

14. $f(x) = -2x^2$

14. domain _____

range _____

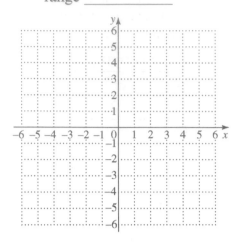

15. $f(x) = x^3 + 2$

15. domain _____

range _____

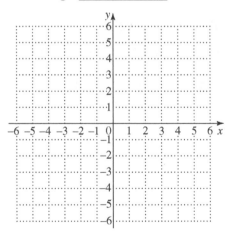

Copyright © 2018 Pearson Education, Inc.

Objective 12.R.6 Solve applied problems using quadratic functions as models.

Video Examples

Review this example:

A certain projectile is located at a distance of $d(t) = 3t^2 - 6t + 1$ feet from its starting point after t seconds. How many seconds will it take the projectile to travel 10 feet?

Let $d = 10$ in the formula and solve for t.

$$d = 3t^2 - 6t + 1$$
$$10 = 3t^2 - 6t + 1$$
$$9 = 3t^2 - 6t$$
$$3 = t^2 - 2t$$
$$3 + 1 = t^2 - 2t + 1$$
$$4 = (t-1)^2$$
$$\sqrt{4} = t - 1 \quad \text{or} \quad -\sqrt{4} = t - 1$$
$$2 = t - 1 \quad \text{or} \quad -2 = t - 1$$
$$3 = t \quad \text{or} \quad -1 = t$$

Since t represents time, we reject the negative solution. It will take 3 seconds for the projectile to travel 10 feet.

Now Try:

A baseball is thrown upward from a building 20 m high with a velocity of 15 m/sec. Its distance from the ground after t seconds is modeled by the function

$$f(t) = -4.9t^2 + 15t + 20.$$

When will the ball hit the ground? Round your answer to the nearest tenth.

_____

Practice Exercises

Solve each problem. Round answers to the nearest tenth.

16. A population of microorganisms grows according to the function $p(x) = 100 + 0.2x + 0.5x^2$, where x is given in hours. How many hours does it take to reach a population of 250 microorganisms?

16. _____

17. An object is thrown downward from a tower 280 feet   17. _____
high. The distance the object has fallen at time t in
seconds is given by $s(t) = 16t^2 + 68t$. How long will
it take the object to fall 100 feet?

18. A widget manufacturer estimates that her monthly   18. _____
revenue can be modeled by the function
$R(x) = -0.006x^2 + 32x - 10,000$. What is the
minimum number of items that must be sold for the
revenue to equal $30,000?

Chapter 13 NONLINEAR FUNCTIONS, CONIC SECTIONS, AND NONLINEAR SYSTEMS

Learning Objectives

13.R.1   Solve linear systems by graphing.
13.R.2   Solve linear systems by substitution.
13.R.3   Solve linear systems by elimination.
13.R.4   Solve systems of linear inequalities by graphing.
13.R.5   Graph basic polynomial functions.
13.R.6   Solve radical equations that require additional steps.
13.R.7   Graph parabolas with vertical axes.
13.R.8   Graph parabolas with horizontal axes.
13.R.9   Solve quadratic inequalities.
13.R.10 Solve polynomial inequalities of degree 3 or greater.
13.R.11 Solve rational inequalities.

Key Terms

Use the vocabulary terms listed below to complete each statement in exercises 1–18.

system of linear equations	**solution of a system**
solution set of a system	**consistent system**
inconsistent system	**independent equations**
dependent equations	**addition property of equality**
elimination method	**substitution** **vertex**
system of linear inequalities	**solution set of a system of linear inequalities**
radical equation **proposed solution**	**extraneous solution**
quadratic inequality **rational inequality**	

1.   Equations of a system that have different graphs are called
     _____.

2.   A system of equations with at least one solution is a
     _____.

3.   The set of all ordered pairs that are solutions of a system is the
     _____.

4.   The _____ of linear equations is an ordered
     pair that makes all the equations of the system true at the same time.

5.   Equations of a system that have the same graph (because they are different forms of
     the same equation) are called _____.

13–1

6. A system with no solution is called a(n) _____.

7. A(n) _____ consists of two or more linear
    equations with the same variables.

8. Using the addition property to solve a system of equations is called the

    _____.

9. The _____ states that the same added quantity to each
    side of an equation results in equal sums.

10. _____ is being used when one expression is replaced by
    another.

11. All ordered pairs that make all inequalities of the system true at the same time is
    called the _____.

12. A _____ contains two or more linear inequalities
    (and no other kinds of inequalities).

13. A(n) _____ is a potential solution to an equation that does
    not satisfy the equation.

14. An equation with a variable in the radicand is a(n) _____.

15. A value of a variable that appears to be a solution of an equation is a(n)

    _____.

16. The maximum or minimum value of a quadratic function occurs at the
    _____ of its graph.

17. An inequality that involves a rational expression is a _____.

18. An inequality that can be written in the form $ax^2 + bx + c < 0$ or $ax^2 + bx + c > 0$,
    where a, b, and c are real numbers with $a \neq 0$ is called a

    _____.

Objective 13.R.1 Solve linear systems by graphing.

Video Examples

Review this example:

Solve the system of equation by graphing both equations on the same axes.

$$6x - 5y = 4$$

$$2x - 5y = 8$$

Graph these equations by plotting several points for each line. To find the x-intercept, let $y = 0$. To find the y-intercept, let $x = 0$.

The tables show the intercepts and a check point for each graph.

$6x - 5y = 4$

x	y
0	$-\dfrac{4}{5}$
$\dfrac{2}{3}$	0
4	4

$2x - 5y = 8$

x	y
0	$-\dfrac{8}{5}$
4	0
-6	-4

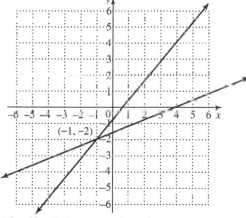

The lines suggest that the graphs intersect at the point $(-1, -2)$. We check by substituting -1 for x and -2 for y in both equations.

$$6x - 5y = 4$$
$$6(-1) - 5(-2) \overset{?}{=} 4$$
$$-6 + 10 \overset{?}{=} 4$$
$$4 = 4 \quad \text{True}$$

$$2x - 5y = 8$$
$$2(-1) - 5(-2) \overset{?}{=} 8$$
$$-2 + 10 \overset{?}{=} 8$$
$$\text{True} \quad 8 = 8$$

Because $(-1, -2)$ satisfies both equations, the solution set of this system is $\{(-1, -2)\}$.

Now Try:

Solve the system of equation by graphing both equations on the same axes.

$$3x - y = -7$$

$$2x + y = -3$$

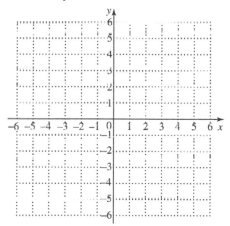

Practice Exercises

Solve each system by graphing both equations on the same axes.

1. $x - 2y = 6$
   $2x + y = 2$

1. _____

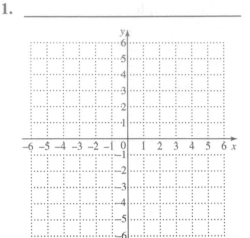

2. $2x = y$
   $5x + 3y = 0$

2. _____

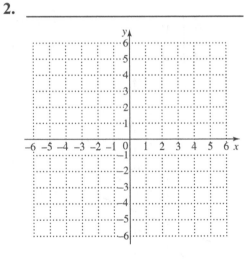

3. $3x + 2 = y$
   $2x - y = 0$

3. _____

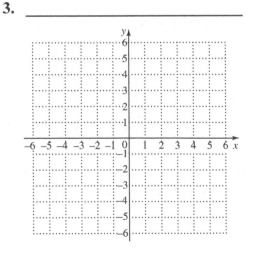

Objective 13.R.2 Solve linear systems by substitution.

Video Examples

Review these examples:

Solve the system by the substitution method.
$$2x + 5y = 22 \quad (1)$$
$$y = 4x \quad (2)$$

Equation (2) is already solved for y. We substitute $4x$ for y in equation (1).
$$2x + 5y = 22$$
$$2x + 5(4x) = 22$$
$$2x + 20x = 22$$
$$22x = 22$$
$$x = 1$$

Find the value of y by substituting 1 for x in either equation. We use equation (2).
$$y = 4x$$
$$y = 4(1) = 4$$

We check the solution $(1, 4)$ by substituting 1 for x and 4 for y in both equations.

$$2x + 5y = 22 \qquad\qquad y = 4x$$
$$2(1) + 5(4) \overset{?}{=} 22 \qquad\qquad 4 \overset{?}{=} 4(1)$$
$$2 + 20 \overset{?}{=} 22 \qquad\qquad \text{True} \;\; 4 - 4$$
$$22 - 22 \;\; \text{True}$$

Since $(1, 4)$ satisfies both equations, the solution set of the system is $\{(1, 4)\}$.

Now Try:

Solve the system by the substitution method.
$$x + y = 7$$
$$y = 6x$$

_____

Solve the system by the substitution method.
$$4x + 5y = 13 \qquad (1)$$
$$x = -y + 2 \quad (2)$$

Equation (2) gives x in terms of y. We substitute $-y + 2$ for x in equation (1).
$$4x + 5y = 13$$
$$4(-y + 2) + 5y = 13$$
$$-4y + 8 + 5y = 13$$
$$y + 8 = 13$$
$$y = 5$$

Find the value of x by substituting 5 for y in either equation. We use equation (2).
$$x = -y + 2$$
$$x = -5 + 2 = -3$$

Solve the system by the substitution method.
$$2x + 3y = 6$$
$$x = 5 - y$$

_____

We check the solution (–3, 5) by substituting –3 for x
and 5 for y in both equations.

$$4x + 5y = 13 \qquad\qquad x = -y + 2$$
$$4(-3) + 5(5) \overset{?}{=} 13 \qquad\qquad -3 \overset{?}{=} -5 + 2$$
$$-12 + 25 \overset{?}{=} 13 \qquad\quad \text{True} \;\; -3 = -3$$
$$13 = 13 \;\; \text{True}$$

Both results are true, so the solution set of the
system is {(–3, 5)}.

Solve the system by the substitution method.

$$4x = 5 - y \quad (1)$$
$$7x + 3y = 15 \qquad (2)$$

Step 1 Solve one of the equations for x or y.
Solve equation (1) for y to avoid fractions.

$$4x = 5 - y$$
$$y + 4x = 5$$
$$y = -4x + 5$$

Step 2 Now substitute $-4x + 5$ for y in
equation (2).

$$7x + 3y = 15$$
$$7x + 3(-4x + 5) = 15$$

Step 3 Solve the equation from Step 2.

$$7x - 12x + 15 = 15$$
$$-5x + 15 = 15$$
$$-5x = 0$$
$$x = 0$$

Step 4 Equation (1) solved for y is $y = -4x + 5$.
Substitute 0 for x.

$$y = -4(0) + 5 = 5$$

Step 5 Check that (0, 5) is the solution.

$$4x = 5 - y \qquad\quad 7x + 3y = 15$$
$$4(0) \overset{?}{=} 5 - 5 \qquad 7(0) + 3(5) \overset{?}{=} 15$$
$$0 = 0 \;\; \text{True} \qquad \text{True} \;\; 15 = 15$$

Since both results are true, the solution set of the
system is {(0, 5)}.

Solve the system by the
substitution method.

$$2x + 7y = 2$$
$$3y = 2 - x$$

_____

Use substitution to solve the system.
$$x = 7 - 3y \quad (1)$$
$$5x + 15y = 1 \quad\quad (2)$$

Because equation (1) is already solved for x, we substitute $7 - 3y$ for x in equation (2).
$$5x + 15y = 1$$
$$5(7 - 3y) + 15y = 1$$
$$35 - 15y + 15y = 1$$
$$35 = 1 \quad \text{False}$$
A false result, here $35 = 1$, means that the equations in the system have graphs that are parallel lines. The system is inconsistent and has no solution, so the solution set is $\varnothing$.

Use substitution to solve the system.
$$5x - 10y = 8$$
$$x = 2y + 5$$

_____ _____

Use substitution to solve the system.
$$14x - 7y = 21 \quad (1)$$
$$-2x + y = -3 \quad (2)$$

Begin by solving equation (2) for y to get $y = 2x - 3$. Substitute $2x - 3$ for y in equation (1).
$$14x - 7y = 21$$
$$14x - 7(2x - 3) = 21$$
$$14x - 14x + 21 = 21$$
$$0 = 0$$
This true result means that every solution of one equation is also a solution of the other, so the system has an infinite number of solutions. The solution set is $\{(x, y) | 14x - 7y = 21\}$.

Use substitution to solve the system.
$$5x + 4y = 20$$
$$-10x + 40 = 8y$$

_____

Practice Exercises

Solve each system by the substitution method. Use set-builder notation for dependent equations.

4. $3x + 2y = 14$
         $y = x + 2$

4.   _____

5. $x + y = 9$
          $5x - 2y = -4$

5. _____

6. $3x - 21 = y$
          $y + 2x = -1$

6. _____

7. $y = -\dfrac{1}{3}x + 5$
        $3y + x = -9$

7. _____

8. $\dfrac{1}{2}x + 3 = y$
          $6 = -x + 2y$

8. _____

9. $4x + 3y = 2$
          $8x + 6y = 6$

9. _____

Objective 13.R.3 Solve linear systems by elimination.

Video Examples

Review this example:

Use the elimination method to solve the system.
$$x + y = 6 \quad (1)$$
$$-x + y = 4 \quad (2)$$

Add the equations vertically.
$$x + y = 6 \quad (1)$$
$$\underline{-x + y = 4 \quad (2)}$$
$$2y = 10$$
$$y = 5$$

To find the x-value, substitute 5 for y in either of the two equations of the system. We choose equation (1).
$$x + y = 6$$
$$x + 5 = 6$$
$$x = 1$$

Check the solution $(1, 5)$, by substituting 1 for x and 5 for y in both equations of the given system.

$x + y = 6$	$-x + y = 4$
$1 + 5 \overset{?}{=} 6$	$-1 + 5 \overset{?}{=} 4$
$6 = 6$ True	True $4 = 4$

Since both results are true, the solution set of the system is $\{(1, 5)\}$.

Now Try:

Use the elimination method to solve the system.
$$x + y = 11$$
$$x - y = 5$$

_____

Practice Exercises

Solve each system by the elimination method. Check your answers.

10. $x - 4y = -4$
     $-x + y = -5$

10. _____

11. $2x - y = 10$

$3x + y = 10$

11. _____

12. $x - 3y = 5$

$-x + 4y = -5$

12. _____

Objective 13.R.4 Solve systems of linear inequalities by graphing.

Video Examples

Review these examples:

Graph the solution set of the system.

$$x + y \leq 3$$
$$5x - y \geq 5$$

To graph $x + y \leq 3$, graph the solid boundary line $x + y = 3$ using the intercepts (0, 3) and (3, 0). Determine the region to shade using (0, 0) as a test point.

$$x + y \leq 3$$
$$0 + 0 \overset{?}{\leq} 3$$
$$0 \leq 3 \quad \text{True}$$

Shade the region containing (0, 0).

To graph $5x - y \geq 5$, graph the solid boundary line $5x - y = 5$ using the intercepts (0, –5) and (1, 0). Determine the region to shade using (0, 0) as a test point.

$$5x \quad y \geq 5$$
$$5(0) - 0 \overset{?}{\geq} 5$$
$$0 \geq 5 \quad \text{False}$$

Shade the region that does not contain (0, 0).

The solution set of this system includes all points in the intersection (overlap) of the graph of the two inequalities. This intersection is the gray shaded region and portions of the two boundary lines that surround it.

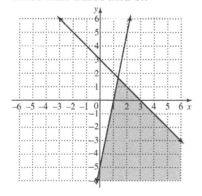

Now Try:

Graph the solution set of the system.

$$3x - y \leq 3$$
$$x + y \leq 0$$

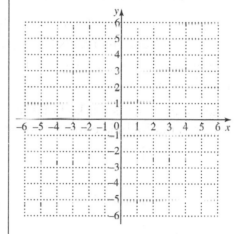

Graph the solution set of the system.
$$4x - y > 2$$
$$x + y > -2$$

Graph $4x - y > 2$ using a dashed line for

$4x - y = 2$ and intercepts (0, –2) and $\left(\frac{1}{2}, 0\right)$.

Using the test point (0, 0), we shade the region that does not contain the point (0, 0).

Graph $x + y > -2$ using a dashed line for $x + y = -2$ and intercepts (–2, 0) and (0, –2). Using the test point (0, 0), we shade the region that does contain the point (0, 0).

The solution set is marked in gray. The solution set does not include either boundary line.

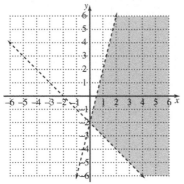

Graph the solution set of the system.
$$y \geq -1$$
$$2x - y > -1$$

Recall that $y = -1$ is a horizontal line through the point (0, –1). Use a solid line, and shade the region with the point (0, 0).

The graph of $2x - y > -1$ is created using a dashed line through the intercepts (0, 1) and $\left(-\frac{1}{2}, 0\right)$. Shade the region with the point (0, 0).

The solution set is marked in gray. The solution set includes the boundary line $y \geq -1$.

Graph the solution set of the system.
$$6x - y > 6$$
$$2x + 5y < 10$$

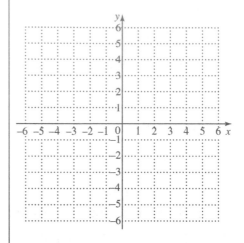

Graph the solution set of the system.
$$x - 3y \leq -7$$
$$x < 2$$

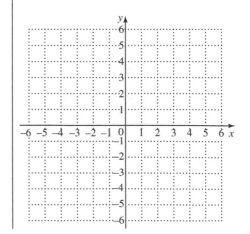

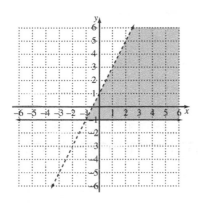

Practice Exercises

Graph the solution of each system of linear inequalities.

13. $4x + 5y \leq 20$

$y \leq x + 3$

13.

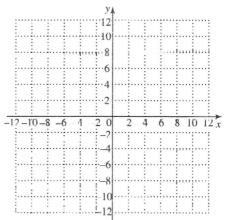

14. $x < 2y + 3$

$0 < x + y$

14.

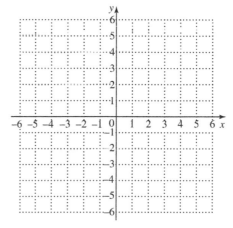

15. $y < 4$

$x \geq -3$

15.

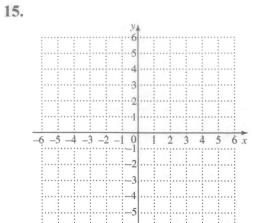

Objective 13R.5 Graph basic polynomial functions.

Video Examples

Review these examples:

Graph each function. Give the domain and range.

$$f(x) = -3x + 2$$

Plot the points and join them with a straight line. The domain and range are both $(-\infty, \infty)$.

x	$f(x) = -3x + 2$
-1	5
0	2
1	-1
2	-4
3	-7

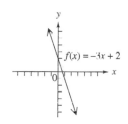

$$g(x) = -2x^2$$

The graph of $g(x)$ has the same shape as that of $f(x) = x^2$ but is narrower and opens downward. The domain is $(-\infty, \infty)$. The range is $(-\infty, 0]$.

x	$g(x) = -2x^2$
-2	-8
-1	-2
0	0
1	-2
2	-8

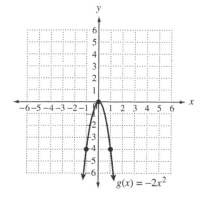

Now Try:

Graph each function. Give the domain and range.

$$f(x) = -2x - 3$$

_____

_____

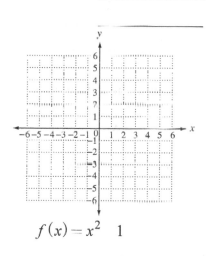

$$f(x) = x^2 \quad 1$$

_____

_____

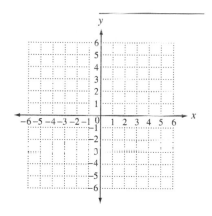

$f(x) = x^3 + 3.$

For this function, cube the input and add 3 to the result. The graph is the cubing function shifted 3 units up.

x	$f(x) = x^3 + 3$
-2	-5
-1	2
0	3
1	4
2	11

The domain and range is $(-\infty, \infty)$.

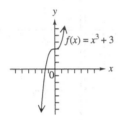

$f(x) = -x^3 + 1.$

_____

_____

Practice Exercises

Graph each function. Give the domain and range.

16. $f(x) = \dfrac{1}{2}x + \dfrac{1}{2}$

16. domain _____

range _____

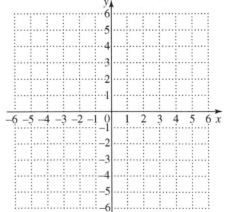

17. $f(x) = -2x^2$

17. domain _____

range _____

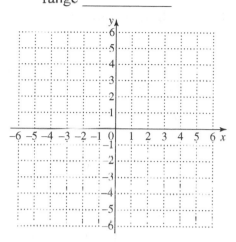

18. $f(x) = x^3 + 2$

18. domain _____

range _____

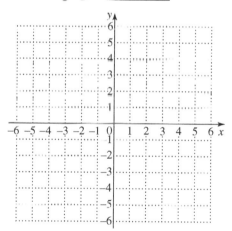

Objective 13.R.6 Solve radical equations that require additional steps.

Video Examples

Review these example:

Solve $\sqrt{x+3} = x-3$.

Step 1 The radical is isolated on the left side of the equation.

Step 2 Square each side.

$$\left(\sqrt{x+3}\right)^2 = (x-3)^2$$
$$x+3 = x^2 - 6x + 9$$

Step 3 Write the equation in standard form and solve.

$$0 = x^2 - 7x + 6$$
$$0 = (x-1)(x-6)$$
$$x-1=0 \quad \text{or} \quad x-6=0$$
$$x=1 \quad \text{or} \quad x=6$$

Step 4 Check each proposed solution in the original equation.

$$\sqrt{x+3} = x-3 \qquad \Big| \qquad \sqrt{x+3} = x-3$$
$$\sqrt{1+3} \overset{?}{=} 1-3 \qquad \Big| \qquad \sqrt{6+3} \overset{?}{=} 6-3$$
$$\sqrt{4} \overset{?}{=} -2 \qquad \Big| \qquad \sqrt{9} \overset{?}{=} 3$$
$$2 = -2 \quad \text{False} \qquad \Big| \qquad 3 = 3 \quad \text{True}$$

The solution set is $\{6\}$. The other proposed solution, 1, is extraneous.

Solve $\sqrt{3x} - 4 = \sqrt{x-2}$.

$$\left(\sqrt{3x}-4\right)^2 = \left(\sqrt{x-2}\right)^2$$
$$3x - 8\sqrt{3x} + 16 = x - 2$$
$$-8\sqrt{3x} = -2x - 18$$
$$\left(-8\sqrt{3x}\right)^2 = (-2x-18)^2$$
$$192x = 4x^2 + 72x + 324$$
$$0 = 4x^2 - 120x + 324$$
$$0 = 4\left(x^2 - 30x + 81\right)$$
$$0 = 4(x-3)(x-27)$$
$$x-3=0 \quad \text{or} \quad x-27=0$$
$$x=3 \quad \text{or} \quad x=27$$

Now Try:

Solve $\sqrt{x+11} = x-1$.

_____

Solve $\sqrt{3x+4} = \sqrt{9x} - 2$.

_____

Check

$$\sqrt{3x} - 4 = \sqrt{x-2}$$
$$\sqrt{3(3)} - 4 \overset{?}{=} \sqrt{3-2}$$
$$3 - 4 \overset{?}{=} \sqrt{1}$$
$$-1 = 1 \quad \text{False}$$

$$\sqrt{3x} - 4 = \sqrt{x-2}$$
$$\sqrt{3(27)} - 4 \overset{?}{=} \sqrt{27-2}$$
$$9 - 4 \overset{?}{=} \sqrt{25}$$
$$5 = 5 \quad \text{True}$$

The proposed solution, 27, is valid, but 3 is extraneous and must be rejected. The solution set is $\{27\}$.

Practice Exercises

Solve each equation.

19. $\sqrt{27 - 18v} = 2v - 3$

19. _____

20. $\sqrt{7r + 8} - \sqrt{r + 1} = 5$

20. _____

21. $\sqrt{k + 10} + \sqrt{2k + 19} = 2$

21. _____

Objective 13.R.7 Graph parabolas with vertical axes.

Video Examples

Review this example:

Graph the quadratic function defined by $f(x) = x^2 - 3x + 2$. Give the vertex, axis, domain, and range.

Step 1 From the equation, $a = 1$, so the graph opens up.

Step 2 The x-coordinate of the vertex is $\frac{3}{2}$.

The y-coordinate of the vertex is

$$f\left(\frac{3}{2}\right) = \left(\frac{3}{2}\right)^2 - 3\left(\frac{3}{2}\right) + 2 = -\frac{1}{4}.$$

The vertex is $\left(\frac{3}{2}, -\frac{1}{4}\right)$.

Step 3 Find any intercepts. Since the vertex is in quadrant IV and the graph opens up, there will be two x-intercepts. Let $f(x) = 0$ and solve.

$$x^2 - 3x + 2 = 0$$
$$(x - 1)(x - 2) = 0$$
$$x - 1 = 0 \quad \text{or} \quad x - 2 = 0$$
$$x = 1 \quad \text{or} \quad x = 2$$

The x-intercepts are (1, 0) and (2, 0). Find the y-intercept by evaluating $f(0)$.

$$f(0) = 0^2 - 3(0) + 2$$

The y-intercept is (0, 2).

Step 4 Plot the points found so far and additional points as needed using symmetry about the axis, $x = \frac{3}{2}$.

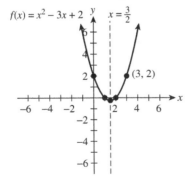

The domain is $(-\infty, \infty)$.

The range is $\left[-\frac{1}{4}, \infty\right)$.

Now Try:

Graph the quadratic function defined by $f(x) = x^2 + 4x + 5$. Give the vertex, axis, domain, and range.

Vertex _____

Axis _____

Domain _____

Range _____

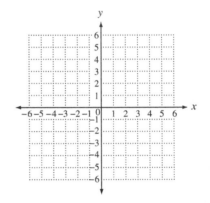

Practice Exercises

Sketch the graph of each parabola. Give the vertex, axis, domain, and range.

22. $f(x) = -x^2 + 8x - 10$

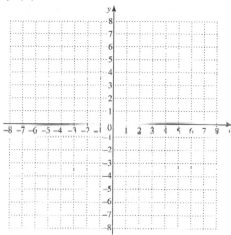

22. vertex _____

axis _____

domain _____

range _____

23. $f(x) - 3x^2 + 6x + 2$

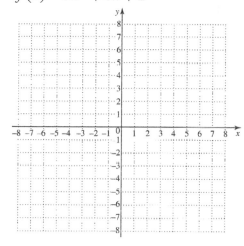

23. vertex _____

axis _____

domain _____

range _____

24. $f(x) = -2x^2 + 4x + 1$

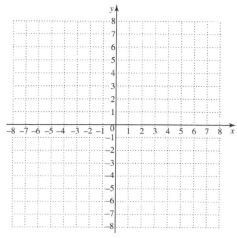

24. vertex _____

axis _____

domain _____

range _____

Objective 13.R.8   **Graph parabolas with horizontal axes.**

Video Examples

Review this example:

Graph $x = (y+1)^2 - 2$. Give the vertex, axis, domain, and range.

The graph has its vertex at $(-2,-1)$ since the roles of x and y are interchanged. It opens to the right since $a > 0$ and has the same shape as $y = x^2$ (but situated horizontally).

x	y
2	−3
−1	−2
−2	−1
−1	0
2	1

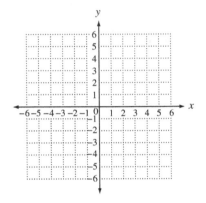

The axis is $y = -1$. The domain is $[-2, \infty)$.

The range is $(-\infty, \infty)$.

Now Try:

Graph $x = (y-2)^2 + 1$. Give the vertex, axis, domain, and range.

Vertex   _____

Axis   _____

Domain   _____

Range   _____

Practice Exercises

Sketch the graph of each parabola. Give the vertex, axis, domain, and range.

25. $x = -y^2 + 2$

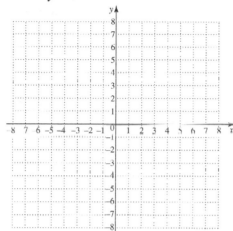

25. vertex _____

axis _____

domain _____

range _____

26. $x = y^2 - 3$

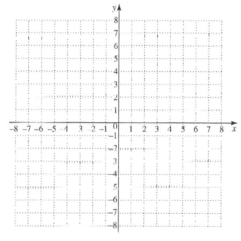

26. vertex _____

axis _____

domain _____

range _____

27. $x = -y^2 - 6y - 10$

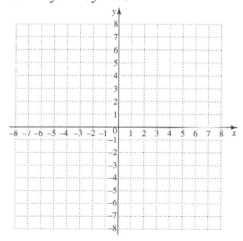

27. vertex _____

axis _____

domain _____

range _____

Objective 13.R.9 Solve quadratic inequalities.

Video Examples

Review these examples:

Solve and graph the solution set of
$x^2 + 5x + 4 \geq 0$.

Solve the quadratic equation by factoring.

$$(x+1)(x+4) = 0$$
$$x+1 = 0 \quad \text{or} \quad x+4 = 0$$
$$x = -1 \quad \text{or} \quad x = -4$$

The numbers –4 and –1 divide a number line into intervals A, B, and C, as shown below.

Since the numbers –4 and –1 are the only numbers that make the quadratic expression
$x^2 + 5x + 4$ equal to 0, all other numbers make the expression either positive or negative. If one number in an interval satisfies the inequality, then all the numbers in that interval will satisfy the inequality.

Choose any number in interval A as a test number; we will choose –5.

$$x^2 + 5x + 4 \geq 0$$
$$\overset{?}{(-5)^2 + 5(-5) + 4 \geq 0}$$
$$4 \geq 0 \quad \text{True}$$

Because –5 satisfies the inequality, all numbers from interval A are solutions.

Now try –2 from interval B.

$$x^2 + 5x + 4 \geq 0$$
$$\overset{?}{(-2)^2 + 5(-2) + 4 \geq 0}$$
$$-2 \geq 0 \quad \text{False}$$

The numbers in interval B are not solutions.
Finally, try 0 from interval C.

$$x^2 + 5x + 4 \geq 0$$
$$\overset{?}{0^2 + 5(0) + 4 \geq 0}$$
$$4 \geq 0 \quad \text{True}$$

Because 0 satisfies the inequality, all numbers from interval C are solutions.

Because the inequality is greater than or equal to zero, we include the endpoints of the intervals in

Now Try:

Solve and graph the solution set
of $x^2 + 7x + 12 > 0$.

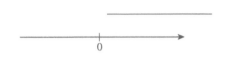

the solution set. Thus, the solution set is

$(-\infty, -4] \cup [-1, \infty)$.

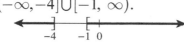

Solve and graph the solution set of $x^2 - 3x \leq 18$.

Step 1

$$x^2 - 3x = 18$$
$$x^2 - 3x - 18 - 0$$
$$(x+3)(x-6) = 0$$
$$x + 3 = 0 \quad \text{or} \quad x - 6 - 0$$
$$x = -3 \quad \text{or} \quad x = 6$$

Step 2

The numbers –3 and 6 divide a number line into intervals A, B, and C, as shown below.

Steps 3 and 4

Substitute a test value form each interval in the *original* inequality $x^2 - 3x \leq 18$ to determine which intervals satisfy the inequality.

Interval	Test Number	Test of inequality	True or False?
A	-4	$-28 \leq 18$	F
B	0	$8 \leq 18$	T
C	7	$-8 \leq 18$	F

The numbers in Interval B are solutions.
The solution set is the interval $[-3, 6]$.

Solve and graph the solution set of $x^2 - x < 2$.

Practice Exercises

Solve each inequality, and graph the solution set.

28. $2y^2 + 5y < 3$

28. _____

29. $8k^2 + 10k > 3$

29. _____

30. $(3x - 2)^2 < -1$

30. _____

Objective 13.R.10 Solve polynomial inequalities of degree 3 or greater.

Video Examples

Review this example:

Solve and graph the solution set of
$(x+1)(x-2)(x+4) \leq 0$.

Set the factored polynomial equal to 0, then use the zero-factor property.

$x+1=0$ or $x-2=0$ or $x+4=0$

    $x=-1$ or $x=2$ or $x=-4$

Locate –4, –1, and 2 on a number line to determine the intervals A, B, C, and D.

Substitute a test number from each interval in the original inequality to determine which intervals satisfy the inequality.

Interval	Test Number	Test of inequality	True or False?
A	-5	$-28 \leq 0$	T
B	-2	$8 \leq 0$	F
C	0	$8 \leq 0$	T
D	5	$162 \leq 0$	F

The numbers in intervals A and C are in the solution set. The three endpoints are included in the solution set since the inequality symbol, $\leq$, includes equality. Thus, the solution set is $(-\infty, -4] \cup [-1, \ 2]$.

Now Try:

Solve and graph the solution set of $(2x-1)(2x+3)(3x+1) \leq 0$.

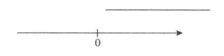

Objective 2 Practice Exercises

Solve each inequality, and graph the solution set.

31. $(y+2)(y-1)(y-2) < 0$

31. _____

32. $(k + 5)(k - 1)(k + 3) \leq 0$

32. _____

$\longleftarrow+\!+\!+\!+\!+\!+\!+\!+\!+\!+\!+\!+\!+\!+\!+\longrightarrow$

33. $(x - 1)(x - 3)(x + 2) \geq 0$

33. _____

$\longleftarrow+\!+\!+\!+\!+\!+\!+\!+\!+\!+\!+\!+\!+\!+\!+\longrightarrow$

Objective 13.R.11 Solve rational inequalities.

Video Examples

Review these examples:

Solve and graph the solution set of $\dfrac{x+1}{x-5} \geq 3$.

Write the inequality so that 0 is on one side.

$$\frac{x+1}{x-5} - 3 \geq 0$$

$$\frac{x+1}{x-5} - \frac{3(x-5)}{x-5} \geq 0$$

$$\frac{x+1}{x-5} - \frac{3x-15}{x-5} \geq 0$$

$$\frac{x+1-3x+15}{x-5} \geq 0$$

$$\frac{-2x+16}{x-5} \geq 0$$

The sign of $\dfrac{-2x+16}{x-5}$ will change from positive to negative or negative to positive only at those numbers that make the numerator or denominator 0. These two numbers, 5 and 8, divide a number line into three intervals.

Test a number in each interval using original inequality.

Interval	Test Number	Test of inequality	True or False?
A	0	$-\dfrac{1}{5} \geq 3$	F
B	6	$7 \geq 3$	T
C	10	$\dfrac{11}{5} \geq 3$	F

The solution set is $(5, \ 8]$. This interval does not include 5 because it would make the denominator of the original inequality 0. The number 8 is included because the inequality symbol, $\geq$, does includes equality.

Now Try:

Solve and graph the solution set of $\dfrac{z+2}{z-3} \leq 2$.

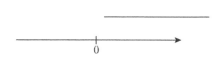

Practice Exercises

Solve each inequality, and graph the solution set.

34. $\dfrac{7}{x-1} \le 1$

34. _____

$\longleftrightarrow$

35. $\dfrac{2p-1}{3p+1} \le 1$

35. _____

$\longleftrightarrow$

36. $\dfrac{5}{x-3} \le -1$

36. _____

$\longleftrightarrow$

Chapter 1 THE REAL NUMBER SYSTEM

Key Terms

1. equivalent fractions
2. improper fraction
3. numerator
4. proper fraction
5. denominator
6. composite number
7. prime factorization
8. prime number
9. lowest terms
10. decimals
11. place value
12. percent

Objective 1.R.1 Multiply and divide fractions and mixed numbers.

Now Try

$$\frac{16}{21}$$

$$\frac{1}{10}$$

Practice Exercises

1. $\frac{15}{2}$ or $7\frac{1}{2}$

2. $\frac{7}{5}$ or $1\frac{2}{5}$

3. $\frac{45}{4}$ or $11\frac{1}{4}$

Objective 1.R.2 Add and subtract fractions and mixed numbers.

Now Try

$$\frac{23}{24}$$

$$\frac{41}{42}$$

Practice Exercises

4. $\frac{256}{225}$ or $1\frac{31}{225}$

5. $\frac{125}{12}$ or $10\frac{5}{12}$

6. $\frac{119}{24}$ or $4\frac{23}{24}$

Objective 1.R.3 Add and subtract decimals.

Now Try

7.024

Practice Exercises

7. 72.453

8. 755.098

9. 609.168

Objective 1.R.4 Multiply and divide decimals.

Now Try

251.116
32.4

5130.2
0.0986

Practice Exercises

10. 2.3424

11. 23.52

12. 0.4292

Chapter 2 EQUATIONS, INEQUALITIES, AND APPLICATIONS

Key Terms

1. decimals	2. percent	3. inequality
4. equation	5. variable	6. algebraic expression
7. solution	8. constant	9. whole numbers
10. integers	11. natural numbers	12. number line
13. irrational number	14. negative number	15. positive number
16. real numbers	17. signed numbers	18. rational number

19. identity element for addition   20. identity element for multiplication

Objective 2.R.1 Complete a table of fractions, decimals, and percents.

Now Try

0.35

$3.\overline{5}$; 3.556

0.0043

91%

$\dfrac{13}{25}$

Practice Exercises

1. 0.03; 3%   2. $\dfrac{7}{8}$; 87.5%   3. 0.87; $\dfrac{87}{100}$

Objective 2.R.2 Write statements that change the direction of inequality symbols.

Now Try

11 < 15

Practice Exercises

4. $\dfrac{2}{3} < \dfrac{3}{4}$ 5. $8 \le 12$ 6. 0.0002 < 0.002

Objective 2.R.3 Translate word phrases to algebraic expressions.

Now Try

$12 - x$

$5(x - 6)$

Practice Exercises

7. $10x + 21$ 8. $8x - 11$ 9. $\dfrac{2}{3}x - \dfrac{1}{2}x$

Objective 2.R.4 Evaluate algebraic expressions, given values for the variables.

Now Try

252, 567

64

20

Practice Exercises

10. 8

11. 4

12. $\dfrac{28}{13}$

Objective 2.R.5 Identify solutions of equations.
  Now Try
  yes
  no

Practice Exercises

13. no

14. no

15. yes

Objective 2.R.6 Translate sentences to equations.
  Now Try
  $3x - 7 = 12$

Practice Exercises

16. $\dfrac{10}{x} = 2 + x$

17. $6(5 + x) = 19$

18. $61 - 7x = 13 + x$

Objective 2.R.7 Classify numbers and graph them on number lines.
  Now Try

7

0, 7

−10, 0, 7

$-10, -\dfrac{5}{8}, 0, 0.\overline{4}, 5\dfrac{1}{2}, 7, 9.9$

$\sqrt{5}$

all of the numbers

Practice Exercises

19. 5

20. $\sqrt{2}$

21. $-8, -\dfrac{3}{5}, 1.\overline{6}, 3\dfrac{1}{4}, 5, 7.9$

22.

Objective 2.R.8 Identify properties of real numbers.
  Practice Exercises

23. identity

24. associative

25. distributive property

26. inverse

27. commutative

28. inverse

29. commutative

30. identity

31. associative

Objective 2.R.9 Simplify expressions.
  Now Try
  $71 + 18x$
  $11 - 7x$

Practice Exercises

32. $8x + 27$

33. $5 + s$

34. $10x + 3$

Chapter 3 GRAPHS OF LINEAR EQUATIONS AND INEQUALITIES IN TWO VARIABLES

Key Terms
1. contradiction
2. conditional equation
3. identity
4. formula
5. interval
6. linear inequality
7. inequalities
8. interval notation

Objective 3.R.1 Solve linear equations using more than one property of equality.
Now Try
{4}
{all real numbers}
$\varnothing$

Practice Exercises
1. $\left\{\dfrac{5}{2}\right\}$

2. {4}

3. $\left\{-\dfrac{1}{5}\right\}$

4. $\varnothing$

5. {all real numbers}

6. $\varnothing$

Objective 3.R.2 Solve a formula for one variable, given the values of the other variables.
Now Try
$W = 5.5$

Practice Exercises
7. $a = 36$

8. $p = 2400$

9. $h = 12$

Objective 3.R.3 Solve linear inequalities by using both properties of inequality.
Now Try
$[2, \infty)$

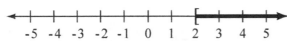

Practice Exercises

10. $(4, \infty);$

11. $(-\infty, 5];$

12. $[2, \infty);$

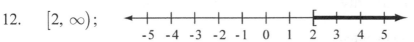

Chapter 4 SYSTEMS OF LINEAR EQUATIONS AND INEQUALITIES

Key Terms
1. line graph
2. linear equation in two variables
3. coordinates
4. *x*-axis
5. *y*-axis
6. ordered pair
7. rectangular (Cartesian) coordinate system
8. quadrants
9. origin
10. plot
11. scatter diagram
12. table of values
13. plane
14. *y*-intercept
15. *x*-intercept
16. graphing
17. graph
18. boundary line
19. linear inequality in two variables

Objective 4.R.1 Decide whether a given ordered pair is a solution of a given equation.
Now Try
yes

no

Practice Exercises
1. no, not a solution
2. no, not a solution
3. yes, a solution

Objective 4.R.2 Graph linear equations by plotting ordered pairs.
Now Try

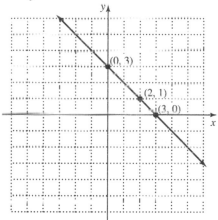

Answers

Practice Exercises

4. $(0, -2), \left(\frac{2}{3}, 0\right), (2, 4)$

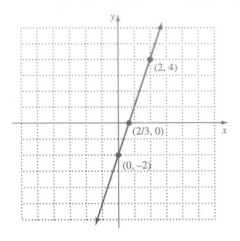

5. $(0, -4), (4, 0), (-2, -6)$

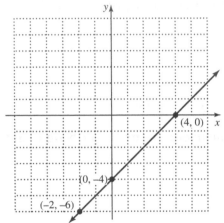

6. $\left(0, -\frac{1}{2}\right), (1, 0), (-3, -2)$

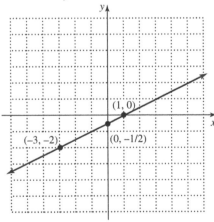

Objective 4.R.3 Use a linear equation to model data.
Now Try
a. 0 calculators, $45
5000 calculators, $42
20,000 calculators, $33
45,000 calculators, $18

b. (0, $45), (5, $42), (20, $33), (45, $18)

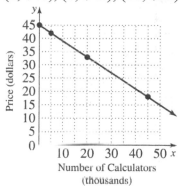

c. 30,000 calculators, $27

Practice Exercises

7. 2014, 4.9 million;
2015, 5.53 million;
2016, 6.16 million;
2017, 6.79 million

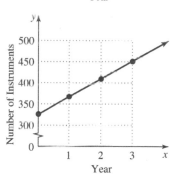

8. 2013, 325;
2014, 367;
2015, 409;
2016, 451

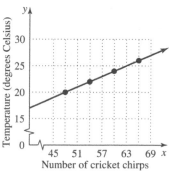

9. 48, 20°C;
54, 22°C;
60, 24°C;
66, 26°C

Objective 4.R.4 Graph linear inequalities in two variables.

Now Try

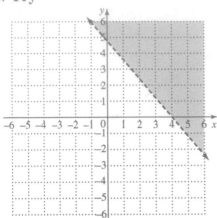
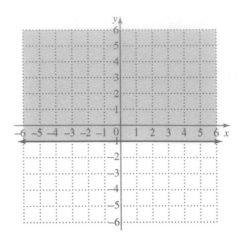

Practice Exercises

10.

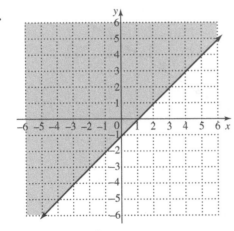

11.

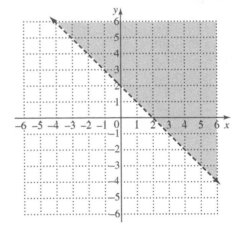

12.

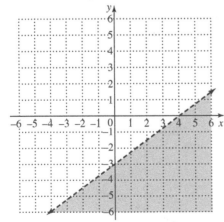

Chapter 5 EXPONENTS AND POLYNOMIALS

Key Terms
1. equation
2. variable
3. algebraic expression
4. solution
5. constant
6. quotient
7. reciprocals
8. product
9. dividend
10. divisor
11. numerical coefficient
12. term
13. like terms
14. formula

Objective 5.R.1 Evaluate algebraic expressions, given values for the variables.
Now Try
252, 567

64
20

Practice Exercises

1. 8

2. 4

3. $\dfrac{28}{13}$

Objective 5.R.2 Solve applied problems that involve real number operations.
Practice Exercises
4. 5

5. 1

6. 2

Objective 5.R.3 Simplify expressions.
Now Try
$71 + 18x$

$11 - 7x$

Practice Exercises
7. $8x + 27$

8. $5 + s$

9. $10x + 3$

Objective 5.R.4 Solve a formula for one variable, given the values of the other variables.
Now Try
$W = 5.5$

Practice Exercises
10. $a = 36$

11. $p = 2400$

12. $h = 12$

Chapter 6 FACTORING AND APPLICATIONS

Key Terms
1. composite number
2. prime factorization
3. prime number
4. quotient
5. reciprocals
6. dividend
7. divisor
8. formula
9. degree of a term
10. descending powers
11. term
12. trinomial
13. polynomial
14. monomial
15. degree of a polynomial
16. binomial
17. like terms
18. inner product
19. FOIL
20. outer product
21. binomial
22. conjugate

Objective 6.R.1 Write numbers in prime factored form.
  Now Try
    $5 \cdot 11$
    $2 \cdot 3 \cdot 5 \cdot 7$

Practice Exercises
1. $2 \cdot 7 \cdot 7$
2. $2 \cdot 2 \cdot 2 \cdot 2 \cdot 2 \cdot 2 \cdot 2 \cdot 2$
3. $2 \cdot 3 \cdot 7 \cdot 13$

Objective 6.R.2 Solve multiplication and division problems involving 0.
  Practice Exercises
4. 0
5. 0
6. undefined

Objective 6.R.3 Solve a formula for a specified variable.
  Now Try
    $t = \dfrac{d}{r}$
    $a = P - b - c$
    $y = 6x + 5$

Practice Exercises
7. $H = \dfrac{V}{LW}$
8. $n = \dfrac{S}{180} + 2$ or $n = \dfrac{S + 360}{180}$
9. $h = \dfrac{3V}{\pi r^2}$

Objective 6.R.4 Know the vocabulary for polynomials.
  Now Try
    $8x^3 + 4x^2 + 6$; degree 3; trinomial
    $4x^5$; degree 5; monomial

  Practice Exercises
10. $n^8 - n^2$; degree 8; binomial
11. $-5.7d^8 - 1.1d^5 + 3.2d^3 - d^2$; degree 8; none of these

12. $5c^5 + 3c^4 - 10c^2$; degree 5; trinomial

Objective 6.R.5 Multiply two polynomials.
  Now Try
  $$4x^7 - 2x^5 + 37x^4 - 18x^2 + 9x$$
  $$28x^4 - 33x^3 + 51x^2 + 17x - 15$$

  Practice Exercises
  13. $x^3 + 27$ 　　　　14. $6m^5 + 4m^4 - 5m^3 + 2m^2 - 4m$
  15. $6x^4 + 11x^3 - 9x^2 - 4x$

Objective 6.R.6 Multiply binomials using the FOIL method.
  Now Try
  $$x^2 + 3x - 54$$
  $$16xy + 72y - 14x - 63$$
  $$45p^2 + 11pq - 4q^2$$

  Practice Exercises
  16. $20a^2 + 11ab - 3b^2$　　17. $3 + 10a + 8a^2$　　　　18. 　$6m^2 - mn + 12n^2$

Objective 6.R.7 Square binomials.
  Now Try
  $$r^2 + 12x + 36$$

  $$64a^2 - 48ab + 9b^2$$
  $$4a^2 + 36ak + 81k^2$$

  Practice Exercises
  19. $49 + 14x + x^2$ 　　20. $4m^2 - 12mp + 9p^2$ 　　21. $16y^2 - 5.6y + 0.49$

Objective 6.R.8 Find the product of the sum and difference of two terms.
  Now Try
  $$x^2 - 81$$

  $$x^2 - \frac{9}{25}$$
  $$121x^2 - y^2$$
  $$4p^5 - 144p$$

  Practice Exercises
  22. $144 - x^2$ 　　　　23. $64k^2 - 25p^2$ 　　　　24. $\frac{16}{49}t^2 - 4u^2$

Answers

Objective 6.R.9 Divide a polynomial by a polynomial.

Now Try

$2x^2 - 2x - 2 + \dfrac{-5}{5x-1}$

$x^2 + 10x + 100$

$3x^2 + 5x + 5 + \dfrac{8x+29}{x^2-4}$

$x^2 + \dfrac{1}{8}x + \dfrac{5}{8} + \dfrac{6}{8x-8}$

Practice Exercises

25. $-3x + 4$

26. $3x^2 - 6x + 2 + \dfrac{13x-7}{2x^2+3}$

27. $a^2 + 1$

Chapter 7 FACTORING EXPRESSIONS AND APPLICATIONS

Key Terms

1. equivalent fractions   2. improper fraction   3. numerator

4. proper fraction   5. denominator   6. composite number

7. prime factorization   8. prime number   9. lowest terms

10. formula   11. power rule for exponents

12. base; exponent   13. product rule for exponents

14. factoring by grouping   15. FOIL

Objective 7.R.1 Write fractions in lowest terms.

Now Try

$\dfrac{3}{4}$

Practice Exercises

1. $\dfrac{7}{25}$ 2. $\dfrac{5}{6}$ 3. $\dfrac{33}{73}$

Objective 7.R.2 Multiply and divide fractions and mixed numbers.

Now Try

$\dfrac{16}{21}$

$\dfrac{1}{10}$

Practice Exercises

4. $\dfrac{15}{2}$ or $7\dfrac{1}{2}$ 5. $\dfrac{7}{5}$ or $1\dfrac{2}{5}$ 6. $\dfrac{45}{4}$ or $11\dfrac{1}{4}$

Objective 7.R.3 Add and subtract fractions.

Now Try

$\dfrac{23}{24}$

$\dfrac{41}{42}$

Practice Exercises

7. $\dfrac{256}{225}$ or $1\dfrac{31}{225}$ 8. $\dfrac{125}{12}$ or $10\dfrac{5}{12}$ 9. $\dfrac{119}{24}$ or $4\dfrac{23}{24}$

Objective 7.R.4 Use the distributive property.

Now Try

375

$4x + 5y - z$

Practice Exercises

10. $2an - 4bn + 6cn$ 11. $-10y + 18z$ 12. $2k - 7$

Objective 7.R.5 Solve equations with fractions or decimals as coefficients.
  Now Try
   $\{-18\}$
   $\{2\}$

  Practice Exercises
  13. $\{2\}$ 14. $\{-14\}$ 15. $\{10\}$

Objective 7.R.6 Solve problems involving unknown numbers.
  Now Try
   16

  Practice Exercises
  16. $4 + 3x = 7$; 1   17. $-2(4 - x) = 24$; 16   18. $4x + 7 = 6x - 5$; 6

Objective 7.R.7 Solve a formula for a specified variable.
  Now Try
   $t = \dfrac{d}{r}$
   $a = P - b - c$
   $y = 6x + 5$

  Practice Exercises
  19. $H = \dfrac{V}{LW}$ 20. $n = \dfrac{S}{180} + 2$ or $n = \dfrac{S + 360}{180}$

  21. $h = \dfrac{3V}{\pi r^2}$

Objective 7.R.8 Use combinations of rules for exponents that involve integer exponents.
  Now Try
   $\dfrac{243}{32p^{20}}$

   $\dfrac{x^{12}y^{15}}{8000}$

  Practice Exercises
  22. $\dfrac{1}{9xy}$ 23. $a^{16}b^{22}$ 24. $\dfrac{1}{k^4 t^{20}}$

Objective 7.R.9 Factor any polynomial.
  Now Try
   $(y + z)(7x - 5)$

   prime
   $(3t - 4w)(9t^2 + 12tw + 16w^2)$

   $(4k + 1)(k - 2)$
   $(c - 3)(m + x)$

Practice Exercises

25. $-6x(2x+1)$

26. $3ab(4ab+a-3b)$

27. $2(x+1)^2$

28. $2xy^2(xy+6)(xy-6)$

29. $2(4x-y)(16x^2+4xy+y^2)$

30. $(y^2+1)(y^4-y^2+1)$

31. $(2x-3y)^2$

32. $(a-6)(2a-5)$

33. $(2w-5x)(7w+3x)$

34. $(x-3)(x^2+7)$

35. $(a-3b+5)(a-3b-5)$

Chapter 8 EQUATIONS, INEQUALITIES, GRAPHS, AND SYSTEMS REVISITED

Key Terms

1. contradiction
2. conditional equation
3. identity
4. three-part inequality
5. interval
6. linear inequality
7. inequalities
8. interval notation
9. y-intercept
10. x-intercept
11. graphing
12. graph
13. $d = rt$
14. system of linear equations

Objective 8.R.1 Solve linear equations using more than one property of equality.

Now Try

$\{4\}$

{all real numbers}

$\varnothing$

Practice Exercises

1. $\left\{\dfrac{5}{2}\right\}$
2. $\{4\}$
3. $\left\{-\dfrac{1}{5}\right\}$
4. $\varnothing$
5. {all real numbers}
6. $\varnothing$

Objective 8.R.2 Solve linear inequalities by using both properties of inequality.

Now Try

$[2, \infty)$

Practice Exercises

7. $(4, \infty)$;

8. $(-\infty, 5]$;

9. $[2, \infty)$;

Objective 8.R.3 Use inequalities to solve applied problems.

Now Try

10 feet

Practice Exercises

10. 89
11. $38.84
12. all numbers greater than 5

Objective 8.R.4 Solve linear inequalities with three parts.

Now Try

$(-5, -1]$;

```
◄——●━━━━━━━━●——+——+——+——+——+——+——►
   -5  -4  -3  -2  -1   0   1   2   3   4   5
```

$[2, 4)$;

```
◄——+——+——+——+——+——+——+——●━━━━●——+——►
   -5  -4  -3  -2  -1   0   1   2   3   4   5
```

Practice Exercises

13. $(2, 5]$;

```
◄——+——+——+——+——+——+——+——●━━━━━━━━●——►
   -5  -4  -3  -2  -1   0   1   2   3   4   5
```

14. $[-5, -3)$;

```
◄——●━━━━●——+——+——+——+——+——+——+——+——►
    5  -4  -3  -2  -1   0   1   2   3   4   5
```

15. $(-1, 5)$

```
◄——+——+——+——+——●━━━━━━━━━━━━━━━●——►
   -5  -4  -3  -2  -1   0   1   2   3   4   5
```

Objective 8.R.5 Graph linear equations by plotting ordered pairs.
Now Try

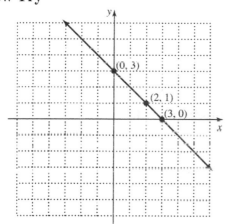

Practice Exercises

16. $(0, -2), \left(\dfrac{2}{3}, 0\right), (2, 4)$

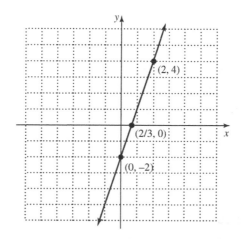

17. $(0,-4), (4, 0), (-2,-6)$

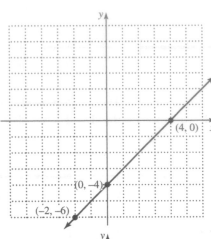

18. $\left(0,-\dfrac{1}{2}\right),(1,\ 0),(-3,-2)$

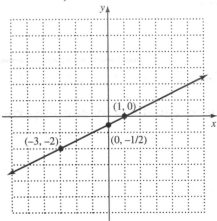

Objective 8.R.6 Solve problems about quantities and their costs.

   Now Try

   1500 general admission tickets; 750 reserved seats

   Practice Exercises

   19. 111 student tickets; 300 nonstudent tickets   20. 30 \$5 bills; 60 \$10 bills

   21. 8 \$14 ties; 2 \$25 ties

Objective 8.R.7 Solve problems about mixtures.

   Now Try

   32 oz of 65% solution; 48 oz of 85% solution

   Practice Exercises

   22. \$6 coffee: 100 lbs; \$12 coffee: 50 lb

   23. 50% solution: 375 mL; 10% solution: 125 mL

   24. \$1.60 candy: 20 lb; \$2.50 candy: 10 lb

Objective 8.R.8 Solve problems about distance, rate (or speed), and time.

   Now Try

   Enid: 44 mph; Jerry: 16 mph

   Practice Exercises

   25. boat speed: 12 mph; current speed: 4 mph

   26. plane A: 400 mph; plane B: 360 mph

   27. Steve: 6 mph; Vic: 4 mph

Chapter 9 RELATIONS AND FUNCTIONS

Key Terms
1. formula
2. line graph
3. linear equation in two variables
4. coordinates
5. x-axis
6. y-axis
7. ordered pair
8. rectangular (Cartesian) coordinate system
9. quadrants
10. origin
11. plot
12. table of values
13. plane
14. y-intercept
15. x-intercept
16. graphing
17. graph
18. degree of a term
19. descending powers
20. term
21. trinomial
22. polynomial
23. monomial
24. binomial
25. like terms
26. degree of a polynomial

Objective 9.R.1 Solve a formula for a specified variable.
Now Try

$$t = \frac{d}{r}$$
$$a = P - b - c$$
$$y = 6x + 5$$

Practice Exercises

1. $H = \dfrac{V}{LW}$

2. $n = \dfrac{S}{180} + 2$ or $n = \dfrac{S + 360}{180}$

3. $h = \dfrac{3V}{\pi r^2}$

Objective 9.R.2 Decide whether a given ordered pair is a solution of a given equation.
Now Try

yes

no

Practice Exercises

4. no, not a solution
5. no, not a solution
6. yes, a solution

Objective 9.R.3 Complete a table of values.
Now Try

x	y
1	−4
5	12
2	0
3	4

(1, –4), (5, 12), (2, 0), (3, 4)

Practice Exercises

7. $(1, -3)$, $(1, 0)$, $(1, 5)$

8. $(-4, 4), (0, 4), (6, 4)$

9. $(0, 4)$, $(3, 0)$, $\left(\frac{15}{4}, -1\right)$

Objective 9.R.4 Graph linear equations by plotting ordered pairs.
   Now Try

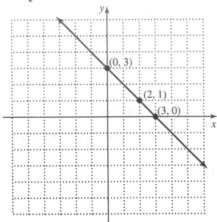

Practice Exercises

10. $(0, -2), \left(\frac{2}{3}, 0\right), (2, 4)$

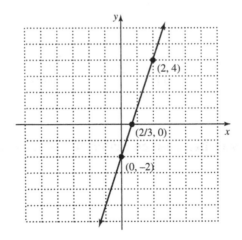

11. $(0, -4)$, $(4, 0)$, $(-2, -6)$

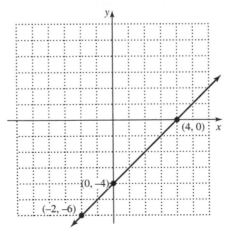

12. $\left(0, -\dfrac{1}{2}\right), (1, 0), (-3, -2)$

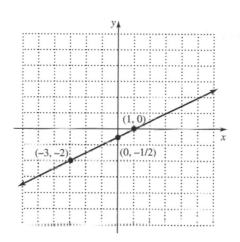

Objective 9.R.5 Graph linear equations of the form $y = b$ or $x = a$.

Now Try

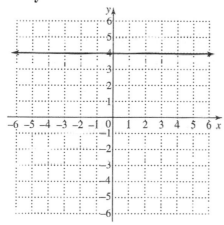

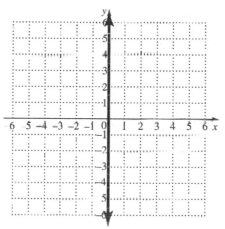

Practice Exercises

13. $x - 1 = 0$

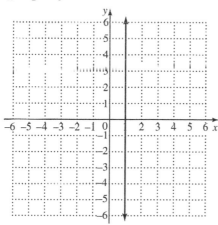

14. $y + 3 = 0$

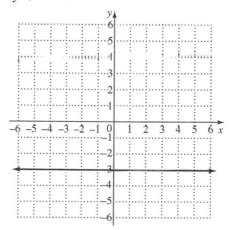

Objective 9.R.6 Evaluate polynomials.
  Now Try
    1285
    1357

  Practice Exercises
15. a. -51; b. 74   16. a. 71; b. -19 17. a. 29; b. 39

Objective 9.R.7 Perform operations with polynomials.
  Now Try
    $5x^3 - 5x^2 + 4x$
    $7x^3 + 8x^2 - 14x - 1$
    $-11x^3 - 7x + 1$
    $x^2y + 4xy$

  Practice Exercises
18. $5m^3 - 2m^2 - 4m + 4$ 19. $4x^2 + 6x - 18$

20. $3r^3 + 7r^2 - 5r - 2$ 21. $-8w^3 + 21w^2 - 15$

22. $8b^4 - 4b^3 - 2b^2 - b - 2$ 23. $3x^3 + 7x^2 + 2$

24. $-3a^6 + a^4b - 3b^2$ 25. $-5ab - 6ac$ 26. $-7x^2y + 3xy + 5xy^2$

Chapter 10 ROOTS, RADICALS, AND ROOT FUNCTIONS

Key Terms
1. *y*-intercept
2. *x*-intercept
3. graphing
4. graph
5. power rule for exponents
6. base; exponent
7. product rule for exponents
8. legs
9. hypotenuse
10. squaring function
11. polynomial function of degree *n*
12. cubing function
13. identity function

Objective 10.R.1 Use a linear equation to model data.

Now Try

a. 0 calculators, $45
   5000 calculators, $42
   20,000 calculators, $33
   45,000 calculators, $18

b. (0, $45), (5, $42), (20, $33), (45, $18)

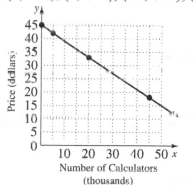

c. 30,000 calculators, $27

Practice Exercises

1. 2014, 4.9 million;
    2015, 5.53 million;
    2016, 6.16 million;
    2017, 6.79 million

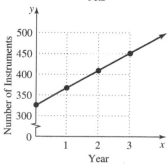

2. 2013, 325;
    2014, 367;
    2015, 409;
    2016, 451

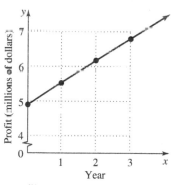

3. 48, 20°C;
   54, 22°C;
   60, 24°C;
   66, 26°C

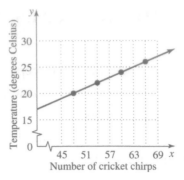

Objective 10.R.2 Use combinations of the rules for exponents.

Now Try

$$\frac{243}{32p^{20}}$$

$$\frac{x^{12}y^{15}}{8000}$$

Practice Exercises

4. $\dfrac{1}{9xy}$

5. $a^{16}b^{22}$

6. $\dfrac{1}{k^4 t^{20}}$

Objective 10.R.3 Solve problems by applying the Pythagorean theorem.

Now Try

16 ft

Practice Exercises

7. 45 m, 60 m, 75 m

8. car: 60 mi; train: 80 mi

9. 20 mi

Objective 10.R.4 Graph basic polynomial functions.

Now Try

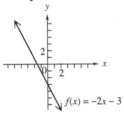

domain: $(-\infty, \infty)$;
range: $(-\infty, \infty)$

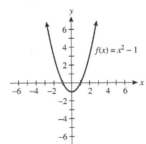

domain: $(-\infty, \infty)$;
range: $[-1, \infty)$

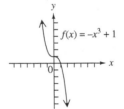

domain: $(-\infty, \infty)$;
range: $(-\infty, \infty)$

Practice Exercises

10.

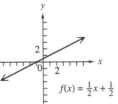

domain: $(-\infty, \infty)$;
range: $(-\infty, \infty)$

11.

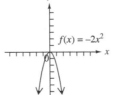

domain: $(-\infty, \infty)$;
range: $(-\infty, 0]$

12.

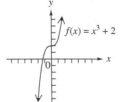

domain: $(-\infty, \infty)$;
range: $(-\infty, \infty)$

Chapter 11 QUADRATIC EQUATIONS, INEQUALITIES, AND FUNCTIONS

Key Terms
1. boundary line
2. linear inequality in two variables
3. factoring by grouping
4. FOIL
5. standard form
6. quadratic equation
7. squaring function
8. polynomial function of degree n
9. cubing function
10. identity function
11. extraneous solution
12. radical equation
13. proposed solution
14. complex number
15. complex conjugate
16. imaginary part
17. real part
18. standard form (of a complex number)
19. pure imaginary number
20. nonreal complex number

Objective 11.R.1 Graph linear inequalities in two variables.
Now Try

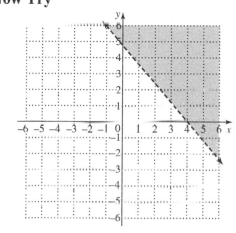

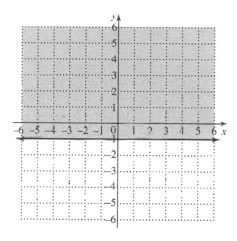

Practice Exercises

1.

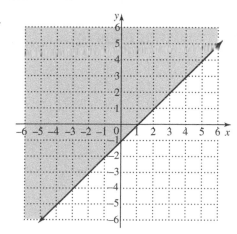

2.

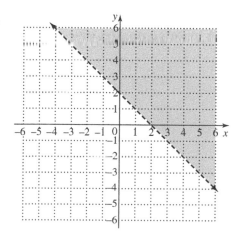

3.

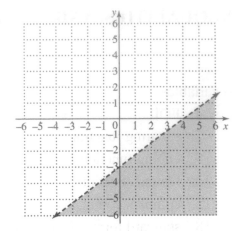

Objective 11.R.2 Factor any polynomial.
Now Try

$(y+z)(7x-5)$

prime

$(3t-4w)(9t^2+12tw+16w^2)$

$(4k+1)(k-2)$

$(c-3)(m+x)$

Practice Exercises

4. $-6x(2x+1)$ 5. $3ab(4ab+a-3b)$ 6. $2(x+1)^2$

7. $2xy^2(xy+6)(xy-6)$ 8. $2(4x-y)(16x^2+4xy+y^2)$

9. $(y^2+1)(y^4-y^2+1)$ 10. $(2x-3y)^2$ 11. $(a-6)(2a-5)$

12. $(2w-5x)(7w+3x)$ 13. $(x-3)(x^2+7)$ 14. $(a-3b+5)(a-3b-5)$

Objective 11.R.3 Solve quadratic equations using the zero-factor property.
Now Try

$\left\{-12, \dfrac{7}{4}\right\}$

$\{-3, 8\}$

$\left\{\dfrac{4}{5}, 3\right\}$

$\left\{-\dfrac{1}{4}, 6\right\}$

$\{0, 11\}$

Practice Exercises

15. $\left\{-\dfrac{5}{2}, 4\right\}$ 16. $\left\{0, \dfrac{4}{5}\right\}$ 17. $\left\{-4, \dfrac{3}{5}\right\}$

Objective 11.R.4 Solve application problems involving quadratic equations.
Practice Exercises

18.

d, in feet	0	1	2	3	5	7
S, in square feet	0	6	24	54	150	294

19. When the length is 0, there is no surface area.

Objective 11.R.5 Solve problems using given quadratic models.
Now Try
1 sec

Practice Exercises

20. a. $\frac{3}{2}$ or $1\frac{1}{2}$ sec; b. 1 or 2 sec

21. 40 items or 110 items

22. 236 ft

Objective 11.R.6 Solve application problems involving quadratic equations.
Practice Exercises

23. a. $P(x) = 9.49x - 40$; b $909

24. a. $P(x) = 16.8x - 85$; b. $755

Objective 11.R.7 Graph basic polynomial functions.
Now Try

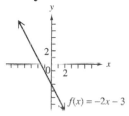
domain: $(-\infty, \infty)$;
range: $(-\infty, \infty)$

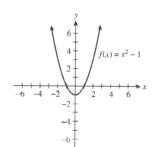
domain: $(-\infty, \infty)$;
range: $[-1, \infty)$

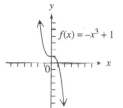
domain: $(-\infty, \infty)$;
range: $(-\infty, \infty)$

Practice Exercises

25.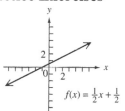
domain: $(-\infty, \infty)$;
range: $(-\infty, \infty)$

26.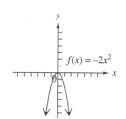
domain: $(-\infty, \infty)$;
range: $(-\infty, 0]$

27.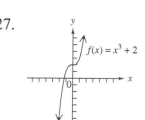
domain: $(-\infty, \infty)$;
range: $(-\infty, \infty)$

Objective 11.R.8 Solve applications involving radical expressions and graphs.
Practice Exercises

28. 64 m^2

29. 2.2 sec

Answers

Objective 11.R.9 Solve radical equations that require additional steps.
  Now Try
    $\{5\}$
    $\{4\}$
  Practice Exercises
30. $\left\{\dfrac{3}{2}\right\}$ 31. $\{8\}$ 32. $\{-9\}$

Objective 11.R.10 Solve applications and extensions involving complex numbers.
  Practice Exercises
33. $-i$ 34. $6 + 2i$

Chapter 12 INVERSE, EXPONENTIAL, AND LOGARITHMIC FUNCTIONS

Key Terms
1. power rule for exponents
2. base; exponent
3. product rule for exponents
4. range
5. relation
6. dependent variable
7. domain
8. function
9. independent variable
10. linear function
11. function notation
12. constant function
13. squaring function
14. polynomial function of degree n
15. cubing function
16. identity function
17. quadratic function
18. Pythagorean theorem

Objective 12.R.1 Use combinations of the rules for exponents.
Now Try

$$\frac{243}{32p^{20}}$$

$$\frac{x^{12}y^{15}}{8000}$$

Practice Exercises

1. $\dfrac{1}{9xy}$
2. $a^{16}b^{22}$
3. $\dfrac{1}{k^4 t^{20}}$

Objective 12.R.2 Identify functions defined by graphs and equations.
Now Try

function

function; domain: $(-\infty, \infty)$

Practice Exercises

4. function
5. not a function; $[-1, \infty)$

6. function; $(-\infty, -6) \cup (-6, \infty)$

Objective 12.R.3 Use function notation.
Now Try

$f(3) = 17$

$g(a-1) = 4a - 11$

$f(-6) = 21$

$f(x) = -\dfrac{2}{3}x + \dfrac{7}{3}, \ f(-3) = \dfrac{13}{3}$

Practice Exercises

7. (a) -13; (b) -7; (c) $-3x - 7$

8. (a) 1; (b) -5; (c) $2x^2 - x - 5$
9. (a) 9; (b) 9; (c) 9

Objective 12.R.4 Solve applications involving linear functions.
Practice Exercises
10. $f(x) = 0.20x + 20$
11. $30; The cost to drive 50 miles is $30.
12. 250; $f(250) = 70$; It costs $70 to drive a rental truck 250 miles.

Objective 12.R.5 Graph basic polynomial functions
Now Try

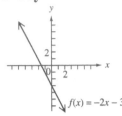

domain: $(-\infty, \infty)$;
range: $(-\infty, \infty)$

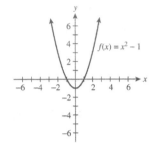

domain: $(-\infty, \infty)$;
range: $[-1, \infty)$

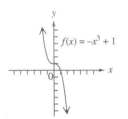

domain: $(-\infty, \infty)$;
range: $(-\infty, \infty)$

Practice Exercises
13.

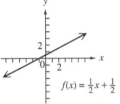

domain: $(-\infty, \infty)$;
range: $(-\infty, \infty)$

14.

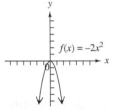

domain: $(-\infty, \infty)$;
range: $(-\infty, 0]$

15.

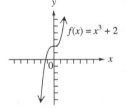

domain: $(-\infty, \infty)$;
range: $(-\infty, \infty)$

Objective 12.R.6 Solve applied problems using quadratic functions as models.
Now Try
4.1 sec
Practice Exercises
16. 17.1 hr

17. 1.2 sec

18. 2000 widgets

Chapter 13 NONLINEAR FUNCTIONS, CONIC SECTIONS, AND NONLINEAR SYSTEMS

Key Terms

1. independent equations
2. consistent system
3. solution set of a system
4. solution of a system
5. dependent equations
6. inconsistent system
7. system of linear equations
8. elimination method
9. addition property of equality
10. substitution
11. solution set of a system of linear inequalities
12. system of linear inequalities
13. extraneous solution   14. radical equation
15. proposed solution
16. vertex   17. rational inequality
18. quadratic inequality

Objective 13.R.1 Solve linear systems by graphing.

Now Try

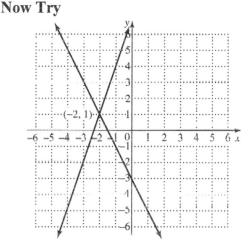

Practice Exercises

1.

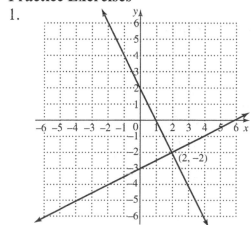

2.

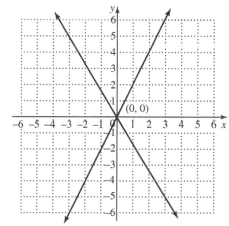

3.

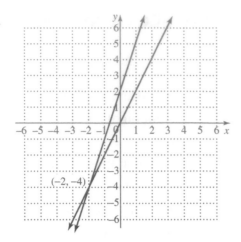

Objective 13.R.2 Solve linear systems by substitution.
  Now Try
    $\{(1, 6)\}$
    $\{(9, -4)\}$
    $\{(8, -2)\}$
    $\varnothing$
    $\{(x, y)\mid 5x + 4y = 20\}$
  Practice Exercises
  4. $\{(2, 4)\}$ 5. $\{(2, 7)\}$ 6. $\{(4, -9)\}$

  7. $\varnothing$ 8. $\{(x, y)\mid -x + 2y = 6\}$ 9. $\varnothing$

Objective 13.R.3 Solve linear systems by elimination.
  Now Try
    $\{(8, 3)\}$
  Practice Exercises
  10. $\{(8, 3)\}$ 11. $\{(4, -2)\}$ 12. $\{(5, 0)\}$

Objective 13.R.4 Solve systems of linear inequalities by graphing.
  Now Try

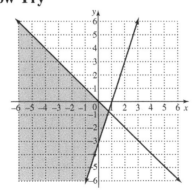   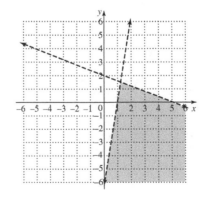

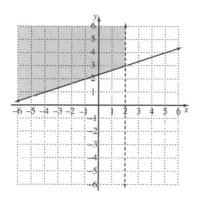

Practice Exercises

13.

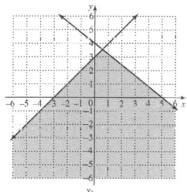

14.

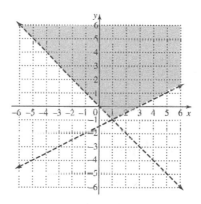

15.

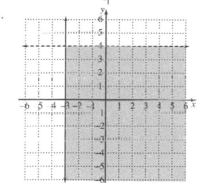

Objective 13.R.5 Graph basic polynomial functions.
  Now Try

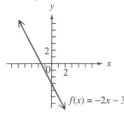

domain: $(-\infty, \infty)$;
range: $(-\infty, \infty)$

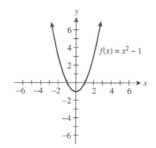

$f(x) = x^2 - 1$

domain: $(-\infty, \infty)$;
range: $[-1, \infty)$

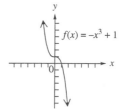

$f(x) = -x^3 + 1$

domain: $(-\infty, \infty)$;
range: $(-\infty, \infty)$

Practice Exercises

16.

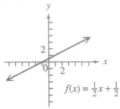

domain: $(-\infty, \infty)$;
range: $(-\infty, \infty)$

17.

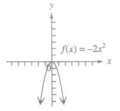

domain: $(-\infty, \infty)$;
range: $(-\infty, 0]$

18.

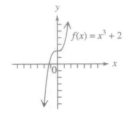

domain: $(-\infty, \infty)$;
range: $(-\infty, \infty)$

Objective 13.R.6 Solve radical equations that require additional steps.

  Now Try

    $\{5\}$

    $\{4\}$

  Practice Exercises

19. $\left\{\dfrac{3}{2}\right\}$ 20. $\{8\}$ 21. $\{-9\}$

Objective 13.R.7 Graph parabolas with vertical axes.

  Now Try

    vertex: $(-2, 1)$, axis: $x = -2$;

    domain: $(-\infty, \infty)$; range: $[1, \infty)$

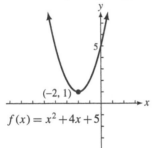

Practice Exercises

22. $f(x) = -x^2 + 8x - 10$

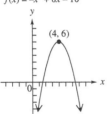

Vertex: $(4, 6)$
Axis: $x = 4$
Domain: $(-\infty, \infty)$
Range: $(-\infty, 6]$

23. $f(x) = 3x^2 + 6x + 2$

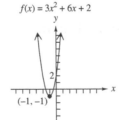

Vertex: $(-1, -1)$
Axis: $x = -1$
Domain:
$(-\infty, \infty)$
Range: $[-1, \infty)$

24. $f(x) = -2x^2 + 4x + 1$

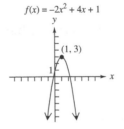

Vertex: $(1, 3)$
Axis: $x = 1$
Domain: $(-\infty, \infty)$
Range: $(-\infty, 3]$

Objective 13.R.8 Graph parabolas with horizontal axes.
  Now Try
    vertex: (1, 2), axis: $y = 2$;
    domain: $[1, \infty)$ range: $(-\infty, \infty)$;

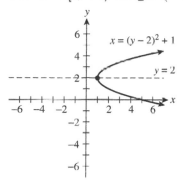

Practice Exercises

25.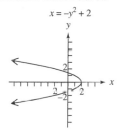

Vertex: (2, 0)
Axis: $y = 0$
Domain: $(-\infty, 2]$
Range: $(-\infty, \infty)$

26.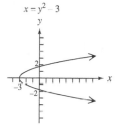

Vertex: (−3, 0)
Axis: $y = 0$
Domain: $[-3, \infty)$
Range: $(-\infty, \infty)$

27.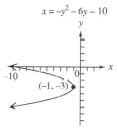

Vertex: (−1, −3)
Axis: $y = -3$
Domain: $(-\infty, -1]$
Range: $(-\infty, \infty)$

Objective 13.R.9 Solve quadratic inequalities.
  Now Try   This is 11.8.1
    $(-\infty - 4) \cup (-3, \infty)$

    $(-1, 2)$

Practice Exercises

28. $\left(-3, \frac{1}{2}\right)$

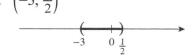

30. ∅

29. $\left(-\infty, -\frac{3}{2}\right) \cup \left(\frac{1}{4}, \infty\right)$

Answers

Objective 13.R.10 Solve polynomial inequalities of degree 3 or greater.

Now Try

$$\left(-\infty, -\frac{3}{2}\right] \cup \left[-\frac{1}{3}, \frac{1}{2}\right]$$

Practice Exercises

31. $(-\infty, -2) \cup (1, 2)$ 32. $(-\infty, -5] \cup [-3, 1]$ 33. $[-2, 1] \cup [3, \infty)$

Objective 13.R.11 Solve rational inequalities.

Now Try

$$(-\infty, 3) \cup [8, \infty)$$

Practice Exercises

34. $(-\infty, 1) \cup [8, \infty)$ 35. $(-\infty, -2] \cup \left(-\frac{1}{3}, \infty\right)$ 36. $[-2, 3)$

CONTENTS for LIAL VIDEO WORKBOOK

Chapter R PREALGEBRA REVIEW

R.1   Fractions

Learning Objectives
1   Identify prime numbers.
2   Write numbers in prime factored form.
3   Write fraction in lowest terms.
4   Convert between improper fractions and mixed numbers.
5   Multiply and divide fractions.
6   Add and subtract fractions.
7   Solve applied problems that involve fractions.
8   Interpret data from a circle graph.

Key Terms

Use the vocabulary terms listed below to complete each statement in exercises 1–9.

numerator	denominator	proper fraction
improper fraction	equivalent fractions	lowest terms
prime number	composite number	prime factorization

1. Two fractions are _____ when they represent the same portion of a whole.

2. A fraction whose numerator is larger than its denominator is called an
    _____.

3. In the fraction $\frac{2}{9}$, the 2 is the _____.

4. A fraction whose denominator is larger than its numerator is called a
    _____.

5. The _____ of a fraction shows the number of equal parts in a whole.

6. A _____ has at least one factor other than itself and 1.

7. In a _____ every factor is a prime number.

8. The factors of a _____ are itself and 1.

9. A fraction is written in _____ when its numerator and denominator have no common factor other than 1.

Objective 1   Identify prime numbers.

Objective 1 Practice Exercises

For extra help, see Example 1 on page 2 of your text.

Tell whether each number is prime, composite, *or* neither.

1.   29   1. _____

2.   35   2. _____

3.   1   3. _____

Objective 2   Write numbers in prime factored form.

Video Examples

Review these examples for Objective 2:   **Now Try:**
2.   Write each number in prime factored form.   2.   Write each number in prime factored form.

    a. 26   **a.** 55

We factor using prime factors 2 and 13, as
$26 = 2 \cdot 13$.

    b. 48   _____

We use a factor tree, as shown below. The prime
factors are boxed.   **b.** 210

Divide by the   54
least prime factor   / \
of 54, which is 2. $54 = 2 \cdot 27$ boxed{2} $\cdot$ 27
                                             / \   _____
Divide 27 by 3
to find two $54 = 2 \cdot 3 \cdot 9$ boxed{3} $\cdot$ 9
factors of 27.   / \

Now factor $54 = 2 \cdot 3 \cdot 3 \cdot 3$ boxed{3} $\cdot$ boxed{3}
9 as $3 \cdot 3$.

Objective 2 Practice Exercises

For extra help, see Example 2 on page 2 of your text.

Write each number in prime factored form.

4.   98   4. _____

5.   256   **5.** _____

6.   546   **6.** _____

Objective 3   **Write fractions in lowest terms.**

Video Examples

Review this example for Objective 3:

3c.   Write the fraction in lowest terms.

$$\frac{75}{100}$$

$$\frac{75}{100} = \frac{3 \cdot 25}{4 \cdot 25} = \frac{3}{4} \cdot 1 = \frac{3}{4}$$

Now Try:

3c.   Write the fraction in lowest terms.

$$\frac{81}{108}$$

_____

Objective 3 Practice Exercises

For extra help, see Example 3 on page 3 of your text.

Write each fraction in lowest terms.

7. $\dfrac{42}{150}$ **7.** _____

8. $\dfrac{180}{216}$ **8.** _____

9. $\dfrac{132}{292}$ **9.** _____

Name: Date:
Instructor: Section:

Objective 4 Convert between improper fractions and mixed numbers.

Video Examples

Review these examples for Objective 4:

4. Write $\frac{53}{6}$ as a mixed number.

We divide the numerator of the improper fraction by the denominator.

$$\begin{array}{r} 8 \\ 6\overline{)53} \\ \underline{48} \\ 5 \end{array} \qquad \frac{53}{6} = 8\frac{5}{6}$$

5. Write $5\frac{3}{8}$ as an improper fraction.

We multiply the denominator of the fraction by the whole number and add the numerator to get the numerator of the improper fraction.

$8 \cdot 5 + 3 = 40 + 3 = 43$

The denominator of the improper fraction is the same as the denominator in the mixed number, which is 8 here. Thus, $5\frac{3}{8} = \frac{43}{8}$

Now Try:

4. Write $\frac{74}{5}$ as a mixed number.

____

5. Write $12\frac{2}{7}$ as an improper fraction.

____

Objective 4 Practice Exercises

For extra help, see Examples 4–5 on page 4 of your text.

Write the improper fraction as a mixed number.

10. $\frac{321}{15}$

10. ____________

Write each mixed number as an improper fraction.

11. $13\frac{5}{9}$

11. ____________

12. $22\frac{2}{11}$

12. ____________

Objective 5   Multiply and divide fractions.

Video Examples

Review these examples for Objective 5:

7.   Find each quotient, and write it in lowest terms.

b. $\dfrac{2}{5} \div \dfrac{8}{7}$

$\dfrac{2}{5} \div \dfrac{8}{7} = \dfrac{2}{5} \cdot \dfrac{7}{8}$ Multiply by the reciprocal.

$= \dfrac{2 \cdot 7}{5 \cdot 4 \cdot 2}$ Multiply and factor.

$= \dfrac{7}{20}$

c. $\dfrac{7}{9} \div 14$

$\dfrac{7}{9} \div 14 = \dfrac{7}{9} \cdot \dfrac{1}{14}$ Multiply by the reciprocal.

$= \dfrac{7 \cdot 1}{9 \cdot 2 \cdot 7}$ Multiply and factor.

$= \dfrac{1}{18}$

Now Try:

7.   Find each quotient, and write it in lowest terms.

b. $\dfrac{6}{7} \div \dfrac{9}{8}$

_____

c. $\dfrac{4}{5} \div 8$

_____

Objective 5 Practice Exercises

For extra help, see Examples 6–7 on pages 4–6 of your text.

Find each product or quotient, and write it in lowest terms.

13. $\dfrac{25}{11} \cdot \dfrac{33}{10}$

13. _____

14. $\dfrac{5}{4} \div \dfrac{25}{28}$

14. _____

15. $4\dfrac{3}{8} \cdot 2\dfrac{4}{7}$

15. _____

Objective 6   Add and subtract fractions.

Video Examples

Review these examples for Objective 6:

9a. Add. Write sum in lowest terms.

$$\frac{5}{21} + \frac{3}{14}$$

Step 1 To find the LCD, factor the denominators to prime factored form.

$21 = 3 \cdot 7$ and $14 = 2 \cdot 7$

7 is a factor of both denominators.

$$\begin{array}{cc} 21 & 14 \\ \wedge & \wedge \end{array}$$

Step 2 LCD $= 3 \cdot 7 \cdot 2 = 42$

In this example, the LCD needs one factor of 3, one factor of 7 and one factor of 2.

Step 3 Now we can use the second property of 1 to write each fraction with 42 as the denominator.

$$\frac{5}{21} = \frac{5}{21} \cdot \frac{2}{2} = \frac{10}{42} \quad \text{and} \quad \frac{3}{14} = \frac{3}{14} \cdot \frac{3}{3} = \frac{9}{42}$$

Now add the two equivalent fractions to get the sum.

$$\frac{5}{21} + \frac{3}{14} = \frac{10}{42} + \frac{9}{42}$$

$$= \frac{19}{42}$$

10b.   Subtract. Write difference in lowest terms.

$$\frac{14}{15} - \frac{5}{8}$$

Since 15 and 8 have no common factors greater than 1, the LCD is $15 \cdot 8 = 120$.

$$\frac{14}{15} - \frac{5}{8} = \frac{14}{15} \cdot \frac{8}{8} - \frac{5}{8} \cdot \frac{15}{15}$$

$$= \frac{112}{120} - \frac{75}{120}$$

$$= \frac{37}{120}$$

Now Try:

9a. Add. Write sum in lowest terms.

$$\frac{7}{12} + \frac{3}{8}$$

_____

10b.   Subtract. Write difference in lowest terms.

$$\frac{17}{6} - \frac{13}{7}$$

_____

Objective 6 Practice Exercises

For extra help, see Example 8–10 on pages 7–10 of your text.

Find each sum or difference, and write it in lowest terms.

16. $\dfrac{23}{45} + \dfrac{47}{75}$ 16. _____

17. $2\dfrac{3}{4} + 7\dfrac{2}{3}$ 17. _____

18. $12\dfrac{5}{6} - 7\dfrac{7}{8}$ 18. _____

Objective 7   Solve applied problems that involve fractions.

Video Examples

Review this example for Objective 7: | **Now Try:**

12. A pie requires $3\dfrac{1}{3}$ cups of apples. How many pies can be made with $20\dfrac{1}{2}$ cups of apples?

To better understand the problem, we replace the fractions with whole numbers. Suppose each pie requires 3 cups apples and we have a total of 21 cups of apples. Dividing 21 by 3 gives 7, the number of pies that can be made. To solve the

12. A pumpkin pie requires $\dfrac{3}{4}$ cup sugar. A large container of sugar has $10\dfrac{2}{3}$ cups of sugar. How many pumpkin pies can be made with this sugar?

_____

original problem, we must divide $20\frac{1}{2}$ by $3\frac{1}{3}$.

Convert the mixed numbers to improper fractions. Then multiply by the reciprocal.

$$20\frac{1}{2} \div 3\frac{1}{3} = \frac{41}{2} \div \frac{10}{3}$$

$$= \frac{41}{2} \cdot \frac{3}{10}$$

$$= \frac{123}{20}, \text{ or } 6\frac{3}{20}$$

Thus, 6 pies can be made with some apples left over.

Objective 7 Practice Exercises

For extra help, see Examples 11–12 on pages 10–11 of your text.

Solve each applied problem. Write each answer in lowest terms.

19.   Arnette worked $24\frac{1}{2}$ hours and earned $9 per hour.   **19.** _____
How much did she earn?

20.   Debbie bought 15 yards of material at a sale. She   **20.** _____
made a shirt with $3\frac{1}{8}$ yards of the material, a dress
with $4\frac{7}{8}$ yards, and a jacket with $3\frac{3}{4}$ yards. How
many yards of material were left over?

21. Three sides of a parking lot are $35\frac{1}{4}$ yards, $42\frac{7}{8}$ yards, and $32\frac{3}{4}$ yards. If the total distance around the lot is $145\frac{1}{2}$ yards, find the length of the fourth side.

21. _____

Objective 8   Interpret data from a circle graph.

Video Examples

In August 2016, 1300 workers were surveyed on where they eat during their lunch time. The circle graph shows the approximate fractions of locations.

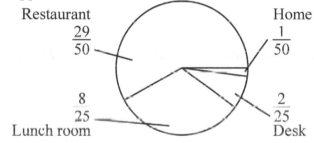

Restaurant $\frac{29}{50}$ Home $\frac{1}{50}$ $\frac{8}{25}$ Lunch room $\frac{2}{25}$ Desk

Review this example for Objective 8:
13c.

How many actual workers ate in the Lunch Room?

Multiply the actual fraction from the graph of Lunch Room by the number of workers surveyed.

$$\frac{8}{25}\cdot 1300 = \frac{8}{25}\cdot\frac{1300}{1}$$
$$= \frac{10,400}{25}$$
$$= 416$$

Thus, 416 workers ate in the Lunch Room.

Now Try:
13c.

How many actual workers ate at their Desk?

_____

Objective 8 Practice Exercises

For extra help, see Example 13 on page 11 of your text.

In August 2016, 1300 workers were surveyed on where they eat during their lunch time. The circle graph shows the approximate fractions of locations.

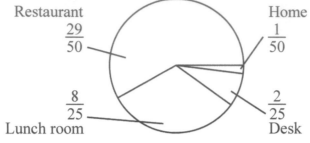

Restaurant $\frac{29}{50}$ Home $\frac{1}{50}$

$\frac{8}{25}$ $\frac{2}{25}$

Lunch room   Desk

22.   What location had the second-largest number of workers?

22. _____

23.   Estimate the number of workers who eat at a Restaurant.

23. _____

24.   How many actual workers ate at a Restaurant?

24. _____

Chapter R PREALGEBRA REVIEW

R.2   Decimals and Percents

Learning Objectives
1   Write decimals as fractions.
2   Add and subtract decimals.
3   Multiply and divide decimals.
4   Write fractions as decimals.
5   Write percents as decimals and decimals as percents.
6   Write percents as fractions and fractions as percents.
7   Solve applied problems that involve percents.

Key Terms

Use the vocabulary terms listed below to complete each statement in exercises 1–3.

decimals   place value   percent

1.   We use _____ to show parts of a whole.

2.   A _____ is assigned to each place to the left or right of the decimal point.

3.   _____ means per one hundred.

Objective 1   Write decimals as fractions.

Video Examples

Review this example for Objective 1:

1c. Write the decimal as a fraction. Do not write in lowest terms.

5.3084

Here we have 4 places.
$$5.3084 = 5 + 0.3084$$

$$= \frac{50,000}{10,000} + \frac{3084}{10,000} \quad \text{The LCD is } 10,000.$$

$$= \frac{53,084}{10,000}$$

       4 zeros

Now Try:

1c. Write the decimal as a fraction. Do not write in lowest terms.

3.7058

_____

Objective 1 Practice Exercises

For extra help, see Example 1 on pages 17–18 of your text.

Write each decimal as a fraction. Do not write in lowest terms.

1.   0.007   1. _____

2.   18.03   2. _____

3.   30.0005   3. _____

Objective 2   Add and subtract decimals.

Video Examples

Review this example for Objective 2:	**Now Try:**
2c. Add or subtract as indicated.	**2c.** Add or subtract as indicated.
$5 - 0.832$	$8 - 0.976$
A whole number is assumed to have the decimal point at the right of the number. Write 5 as 5.000.	_____

$$\begin{array}{r} 5.000 \\ -\ 0.832 \\ \hline 4.168 \end{array}$$

Objective 2 Practice Exercises

For extra help, see Example 2 on page 18 of your text.

Add or subtract as indicated.

4. $45.83 + 20.923 + 5.7$ 4. _____

5. $768.5 - 13.402$ 5. _____

6. $689 - 79.832$ 6. _____

Objective 3   Multiply and divide decimals.

Video Examples

Review these examples for Objective 3:

3a. Multiply.

37.4×5.26

There is 1 decimal place in the first number and 2 decimal places in the second number.
Therefore, there are $1 + 2 = 3$ decimal places in the answer.

$$\begin{array}{r} 37.4 \\ \times\ 5.26 \\ \hline 2244 \\ 748 \\ 1870 \\ \hline 196.724 \end{array}$$

4a. Divide.

$1191.45 \div 23.5$

Write the problem as follows.

$23.5\overline{)1191.45}$

To change 23.5 into a whole number, move the decimal point one place to the right. Move the decimal point in 1191.45 the same number of places to the right, to get 11,914.5.

$235\overline{)11,914.5}$

Move the decimal point straight up and divide as with whole numbers.

$$\begin{array}{r} 50.7 \\ 235\overline{)11,914.5} \\ \underline{1175} \\ 1645 \\ \underline{1645} \\ 0 \end{array}$$

5. Multiply or divide as indicated.

a. 97.648×100

Move the decimal point 2 places to the right because 100 has two zeros.
$97.648 \times 100 = 9764.8$

Now Try:

3a. Multiply.

26.8×9.37

_____

4a. Divide.

$2472.12 \div 76.3$

_____

5. Multiply or divide as indicated.

a. 51.302×100

_____

b. $85.1 \div 1000$

                                                   b. $98.6 \div 1000$

Move the decimal point three places to the left
because 1000 has three zeros. Insert a zero in
front of the 8 to do this.

    $85.1 \div 1000 = 0.0851$

                                                            _____

Objective 3 Practice Exercises

For extra help, see Examples 3–5 on pages 19–20 of your text.

Multiply or divide as indicated.

  7. 14.64×0.16 **7.** _____

  8. $498.624 \div 21.2$ **8.** _____

  9. $429.2 \div 1000$ **9.** _____

Objective 4   Write fractions as decimals.

Video Examples

Review these examples for Objective 4:

6.   Write each fraction as a decimal. For repeating decimals, write the answer by first using bar notation and then rounding to the nearest thousandth.

a. $\dfrac{27}{8}$

Divide 27 by 8. Add a decimal point and as many 0s as necessary.

$$
\begin{array}{r}
3.375 \\
8\overline{)27.000} \\
\underline{24} \\
30 \\
\underline{24} \\
60 \\
\underline{56} \\
40 \\
\underline{40} \\
0
\end{array}
$$

$\dfrac{27}{8} = 3.375$

b. $\dfrac{26}{9}$

$$
\begin{array}{r}
2.888... \\
9\overline{)26.000...} \\
\underline{18} \\
80 \\
\underline{72} \\
80 \\
\underline{72} \\
80 \\
\underline{72} \\
8
\end{array}
$$

$\dfrac{26}{9} = 2.888...$

The remainder is never 0. Because 8 is always left after the subtraction, this quotient is a repeating decimal. A convenient notation for a repeating decimal is a bar over the digit (or digits) that repeats.

$\dfrac{26}{9} = 2.\overline{8}$ or $\dfrac{26}{9} \approx 2.889$ rounded

Now Try:

6.   Write each fraction as a decimal. For repeating decimals, write the answer by first using bar notation and then rounding to the nearest thousandth.

a. $\dfrac{7}{20}$

_____

b. $\dfrac{32}{9}$

_____

Objective 4 Practice Exercises

For extra help, see Example 6 on page 21 of your text.

Write each fraction as a decimal. For repeating decimals, write the answer two ways: using the bar notation and rounding to the nearest thousandth.

10. $\dfrac{3}{7}$ 10. _____

11. $\dfrac{4}{9}$ 11. _____

12. $\dfrac{151}{200}$ 12. _____

Objective 5   Write percents as decimals and decimals as percents.

Video Examples

Review these examples for Objective 5:

9. Convert each percent to a decimal and each decimal to a percent.

    c. 0.29%

    0.29% = 0.0029

    d. 0.54

    0.54 = 54%

Now Try:

9. Convert each percent to a decimal and each decimal to a percent.

    c. 0.43%

    _____

    d. 0.91

    _____

Objective 5 Practice Exercises

For extra help, see Examples 7–9 on pages 21–22 of your text.

Convert each percent to a decimal and each decimal to a percent.

13. 362%   13. _____

14. 0.4%   14. _____

15. 0.0084   15. _____

Objective 6   Write percents as fractions and fractions as percents.

Video Examples

Review this example for Objective 6:

10a. Write the percent as a fraction. Give answers in lowest terms as needed.

65%

We use the fact that, $1\% = \dfrac{1}{100}$, and convert as follows.

$$65\% = 65 \cdot 1\% = 65 \cdot \frac{1}{100} = \frac{65}{100}$$

In lowest terms,

$$\frac{65}{100} = \frac{13 \cdot 5}{20 \cdot 5} = \frac{13}{20}.$$

Thus,

$$65\% = \frac{13}{20}.$$

Now Try:

10a. Write the percent as a fraction. Give answers in lowest terms as needed.

52%

_____

Objective 6 Practice Exercises

For extra help, see Examples 10–11 on page 23 of your text.

Write each percent as a fraction. Give answers in lowest terms as needed.

16. 42%   16. _____

17. 3.5%   17. _____

18.   160%   **18.** _____

Objective 7   **Solve applied problems that involve percents.**

Video Examples

Review this example for Objective 6:

12. Ranee bought a pair of shows with a regular price of $70, on sale at 20% off. How much money did she save? What is the sale price?

From the table in the text, $20\% = \frac{1}{5}$ and $\frac{1}{5}$ of $70 is $14, so she will save $14.

Original price – discount = sale price
    $70 – $14 = $56

Now Try:

12. At the end of the season, a swim suit is on sale at 75% off. The regular price is $56. What is the sale price? How much is saved?

              _____

Objective 6 Practice Exercises

For extra help, see Example 12 on page 24 of your text.

Solve each problem.

19.   A television set sells for $750 plus 8% sales tax. Find the price of the set including sales tax.   **19.** _____

20.   Geishe's Shoes sells shoes at $33\frac{1}{3}\%$ off the regular price. Find the price of a pair of shoes normally priced at $54, after the discount is given.   **20.** _____

21.   A house costs $225,000. The Lee's paid $45,000 as a down payment. What percent of the cost of the house is their down payment?   **21.** _____

Chapter 1 THE REAL NUMBER SYSTEM

1.1   Exponents, Order of Operations, and Inequality

Learning Objectives
1   Use exponents.
2   Use the rules for order of operations.
3   Use more than one grouping symbol.
4   Know the meanings of $\neq$, $<$, $>$, $\leq$, and $\geq$.
5   Translate word statements to symbols.
6   Write statements that change the direction of inequality symbols.

Key Terms

Use the vocabulary terms listed below to complete each statement in exercises 1–4.

> exponent   base   exponential expression   inequality

1.   A number written with an exponent is an _____   _____.

2.   The _____ is the number that is a repeated factor when written with an exponent.

3.   An _____ is a statement that two expressions may not be equal.

4.   An _____ is a number that indicates how many times a factor is repeated.

Objective 1   Use exponents.

Video Examples

Review this example for Objective 1:
1a. Find the value of the exponential expression.

6^2

6^2 means $6 \cdot 6$, which equals 36.

Now Try:
1a. Find the value of the exponential expression.
9^2

_____

Objective 1 Practice Exercises

For extra help, see Example 1 on page 30 of your text.

Find the value of each exponential expression.

1. 3^3

1.   _____

2. $\left(\dfrac{2}{3}\right)^4$

2.   _____

3. $(0.4)^2$

3.   _____

Objective 2   Use the rules for order of operations.

Video Examples

Review this example for Objective 2:
2a. Find the value of the expression.

$$56 - 32 \div 8$$

Divide, then subtract.
$$56 - 32 \div 8 = 56 - 4$$
$$= 52$$

Now Try:
2a. Find the value of the expression.

$$93 - 56 \div 7$$

_____

Objective 2 Practice Exercises

For extra help, see Example 2 on pages 31–32 of your text.

Find the value of each expression.

  4. $20 \div 5 - 3 \cdot 1$

4. _____

  5. $3 \cdot 5^2 - 3 \cdot 7 - 9$

5. _____

  6. $6^2 \div 3^2 - 4 \cdot 3 - 2 \cdot 5$

6. _____

Objective 3   Use more than one grouping symbol.

Video Examples

Review this example for Objective 3:
3a. Find the value of the expression.

$$3[9 + 4(7 + 8)]$$

Start by adding inside the parentheses.
$$3[9 + 4(7 + 8)] = 3[9 + 4(15)] \quad \text{Add.}$$
$$= 3[9 + 60] \quad \text{Multiply.}$$
$$= 3[69] \quad \text{Add.}$$
$$= 207 \quad \text{Multiply.}$$

Now Try:
3a. Find the value of the expression.

$$5[3 + 4(8 + 2)]$$

_____

Objective 3 Practice Exercises

For extra help, see Example 3 on page 32 of your text.

Find the value of each expression.

7. $\dfrac{10(5-3)-9(6-2)}{2(4-1)-2^2}$

7. _____

8. $19-3\big[8(5-2)+6\big]$

8. _____

9. $4\big[5+2(8-6)\big]+12$

9. _____

Objective 4 Know the meanings of $\neq$, $<$, $>$, $\leq$, and $\geq$.

Video Examples

Review this example for Objective 4:

4c. Determine whether the statement is true or false.

$14 \leq 50 \cdot 3$

The statement $14 \leq 50 \cdot 3$ is true, because $14 < 150$.

Now Try:

4c. Determine whether the statement is true or false.

$25 \leq 49 \cdot 2$

_____

Objective 4 Practice Exercises

For extra help, see Example 4 on page 33 of your text.

Tell whether each statement is true *or* false.

10. $3 \cdot 4 \div 2^2 \neq 3$

10. _____

11. $3.25 > 3.52$ **11.** _____

12. $2\left[7(4) - 3(5)\right] \leq 45$ **12.** _____

Objective 5   **Translate word statements to symbols.**

Video Examples

Review this example for Objective 5:
5d. Write the word statement in symbols.

Eight is greater than six.

$8 > 6$

Now Try:
5d. Write the word statement in symbols.
Five is greater than three.

_____

Objective 5 Practice Exercises

For extra help, see Example 5 on page 34 of your text.

Write each word statement in symbols.

13.   Seven equals thirteen minus six.   **13.** _____

14.   Five times the sum of two and nine is less than one   **14.** _____
          hundred six.

15.   Twenty is greater than or equal to the product of two **15.** _____
          and seven.

Objective 6   Write statements that change the direction of inequality symbols.

Video Examples

Review this example for Objective 6:

6a. Write the statement as another true statement with the inequality symbol reversed.

$9 > 7$

$7 < 9$

Now Try:

6a. Write the statement as another true statement with the inequality symbol reversed.
$15 > 11$

_____

Objective 6 Practice Exercises

For extra help, see Example 6 on page 34 of your text.

Write each statement with the inequality symbol reversed.

16. $\dfrac{3}{4} > \dfrac{2}{3}$

16. _____

17. $12 \geq 8$

17. _____

18. $0.002 > 0.0002$

18. _____

Chapter 1 THE REAL NUMBER SYSTEM

1.2 Variables, Expressions, and Equations

Learning Objectives
1   Evaluate algebraic expressions, given values for the variables.
2   Translate word phrases to algebraic expressions.
3   Identify solutions of equations.
4   Translate sentences to equations.
5   Distinguish between equations and expressions.

Key Terms

Use the vocabulary terms listed below to complete each statement in exercises 1−5.

variable	constant	algebraic expression
equation	solution	

1. A(n) _____ is a statement that says two expressions are equal.

2. A _____ is a symbol, usually a letter, used to represent an unknown number.

3. A collection of numbers, variables, operation symbols, and grouping symbols is an_____.

4. Any value of a variable that makes an equation true is a(n) _____ of the equation.

5. A _____ is a fixed, unchanging number.

Objective 1   Evaluate algebraic expressions, given values for the variables.

Video Examples

Review these examples for Objective 1:

1d. Find the value of each algebraic expression for $p = 4$ and then $p = 7$.

$$5p^2$$

For $p = 4$,

$\quad 5p^2 = 5 \cdot 4^2 \quad$ Let $p = 4$.

$\quad\quad = 5 \cdot 16 \quad$ Square 4.

$\quad\quad = 80 \quad\quad$ Multiply.

For $p = 7$,

$\quad 5p^2 = 5 \cdot 7^2 \quad$ Let $p = 7$.

$\quad\quad = 5 \cdot 49 \quad$ Square 7.

$\quad\quad = 245 \quad\quad$ Multiply.

Now Try:

1d. Find the value of each algebraic expression for $k = 6$ and then $k = 9$.

$$7k^2$$

_____

2. Find the value of each expression for $x = 7$ and $y = 6$.

    a. $3x + 4y$

    Replace x with 7 and y with 6.

$$3x + 4y = 3 \cdot 7 + 4 \cdot 6$$
$$= 21 + 24 \quad \text{Multiply.}$$
$$= 45 \quad \text{Add.}$$

    b. $\dfrac{8x - 6y}{4x - 3y}$

    Replace x with 7 and y with 6.

$$\frac{8x - 6y}{4x - 3y} = \frac{8 \cdot 7 - 6 \cdot 6}{4 \cdot 7 - 3 \cdot 6}$$
$$= \frac{56 - 36}{28 - 18} \quad \text{Multiply.}$$
$$= \frac{20}{10} \quad \text{Subtract.}$$
$$= 2 \quad \text{Divide.}$$

2. Find the value of each expression for $x = 8$ and $y = 4$.

    a. $5x + 6y$

    _____

    b. $\dfrac{9x + 2y}{3x - 5y}$

    _____

Objective 1 Practice Exercises

For extra help, see Examples 1–2 on pages 39–40 of your text.

Find the value of each expression if $x = 2$ and $y = 4$.

    1. $9x - 3y + 2$

    1. _____

    2. $\dfrac{2x + 3y}{3x - y + 2}$

    2. _____

    3. $\dfrac{3y^2 + 2x^2}{5x + y^2}$

    3. _____

Objective 2   Translate word phrases to algebraic expressions.

Video Examples

Review these examples for Objective 2:	**Now Try:**
3. Write each word phrase as an algebraic expression, using x as the variable.	3. Write each word phrase as an algebraic expression, using x as the variable.
c. A number subtracted from 13	**c.** A number subtracted from 12
$13 - x$	_____
f. The product of 3 and the difference between a number and 5	**f.** The product of 5 and the difference between a number and 6
$3 \cdot (x - 5)$, or $3(x - 5)$	_____

Objective 2 Practice Exercises

For extra help, see Example 3 on pages 40–41 of your text.

Write each word phrase as an algebraic expression. Use x as the variable.

4. Ten times a number, added to 21   4. _____

5. 11 fewer than eight times a number   5. _____

6. Half a number subtracted from two-thirds of the   6. _____
   number

Objective 3   Identify solutions of equations.

Video Examples

Review these examples for Objective 3:	**Now Try:**
4. Decide whether the given number is a solution of the equation.	4. Decide whether the given number is a solution of the equation.
a. $6p + 5 = 29$; 4	**a.** $7k + 5 = 26$; 3
$6p + 5 = 29$	_____

$$6 \cdot 4 + 5 \overset{?}{=} 29$$
$$24 + 5 \overset{?}{=} 29$$
$$29 = 29 \quad \text{True – the left side of the equation}$$
$$\text{equals the right side.}$$

The number 4 is a solution of the equation.

b. $8n - 7 = 41;\ \dfrac{9}{4}$

$8n - 7 = 41$

$8 \cdot \dfrac{9}{4} - 7 \overset{?}{=} 41$

$18 - 7 \overset{?}{=} 41$

$11 = 41$ False – the left side does not equal the right side.

The number $\dfrac{9}{4}$ is not a solution of the equation.

b. $9m - 8 = 41;\ \ 5$

_____

Objective 3 Practice Exercises

For extra help, see Example 4 on page 41 of your text.

Decide whether the given number is a solution of the equation.

7. $5 + 3x^2 = 19;\ 2$

7. _____

8. $\dfrac{m+2}{3m-10} = 1;\ 8$

8. _____

9. $3y + 5(y - 5) = 7;\ 4$

9. _____

Objective 4   Translate sentences to equations.

Video Examples

Review this example for Objective 4:

5c. Write the word sentence as an equation. Use x as the variable.

Nine less than five times a number is equal to twenty.

Five times
a number   less   nine   is equal to   twenty.
$\downarrow$ $\downarrow$ $\downarrow$ $\downarrow$ $\downarrow$
$5x$ $-$ 9 $=$ 20

$5x - 9 = 20$

Now Try:

5c. Write the word sentence as an equation. Use x as the variable.

Seven less than three times a number is equal to twelve.

_____

Objective 4 Practice Exercises

For extra help, see Example 5 on page 42 of your text.

Write each word sentence as an equation. Use x as the variable.

10. Ten divided by a number is two more than the   10. _____
     number.

11. The product of six and five more than a number is   11. _____
     nineteen.

12. Seven times a number subtracted from 61 is 13 plus   12. _____
     the number.

Objective 5   Distinguish between equations and expressions.

Video Examples

Review these examples for Objective 5:	**Now Try:**
6. Decide whether each is an equation or an expression.	6. Decide whether each is an equation or an expression.
a. $5x - 4$	**a.** $4y - 6 = 20$
Ask, "Is there an equality symbol?" The answer is no, so this is an expression.	_____
b. $12x - 11 = 7$	**b.** $\dfrac{3x - 15y}{2}$
Because there is an equality symbol with something on either side of it, this is an equation.	_____

Objective 5 Practice Exercises

For extra help, see Example 6 on page 42 of your text.

Identify each as an **expression** *or an* **equation**.

13. $y^2 - 4y - 3$ 13. _____

14. $\dfrac{x + 4}{5}$ 14. _____

15. $8x = 2y$ 15. _____

Chapter 1 THE REAL NUMBER SYSTEM

1.3   Real Numbers and the Number Line

Learning Objectives	
1	Classify numbers and graph them on number lines.
2	Tell which of two real numbers is less than the other.
3	Find the additive inverse of a real number.
4	Find the absolute value of a real number.

Key Terms

Use the vocabulary terms listed below to complete each statement in exercises 1−14.

natural numbers	whole numbers	number line	additive inverse
integers	negative number	positive number	signed numbers
rational number	set-builder notation		coordinate
irrational number	real numbers	absolute value	

1.   The set {0, 1, 2, 3, ...} is called the set of _____.

2.   The _____ of a number is the same distance from 0 on the number line as the original number, but located on the opposite side of 0.

3.   The whole numbers together with their opposites and 0 are called
      _____   _____.

4.   The set { 1, 2, 3, ...} is called the set of _____.

5.   The _____ of a number is the distance between 0 and the number on the number line.

6.   A _____ shows the ordering of the real numbers on a line.

7.   A real number that is not a rational number is called a(n) _____.

8.   The number that corresponds to a point on the number line is the
      _____ of that point.

9.   A number located to the left of 0 on a number line is a _____.

10.   A number located to the right of 0 on a number line is a _____.

11.   Numbers that can be represented by points on the number line are
      _____.

12.   _____ uses a variable and a description to describe a set.

13.   Positive numbers and negative numbers are _____.

14. A number that can be written as the quotient of two integers is a

_____.

Objective 1 Classify numbers and graph them on number lines.

Video Examples

Review these examples for Objective 1:

1a. Use an integer to express the boldface italic number in the application.

a. In August, 2012, the National Debt was approximately $*16* trillion.

Use –$16 trillion because "debt" indicates a negative number.

2. Graph each number on a number line.

$$-3\frac{1}{2}, \ -\frac{3}{2}, \ 0, \ \frac{7}{2}, \ 1$$

To locate the improper fractions on the number line, write them as mixed numbers or decimals.

```
        –3.5  –1.5    0  1       3.5
     ◄──┼──┼●┼──┼●┼──●──●──┼──┼●┼──┼──►
       –5 –4 –3 –2 –1  0  1  2  3  4  5
```

3. List the numbers in the following set that belong to each set of numbers.

$$\left\{-6, \ -\frac{5}{6}, \ 0, \ 0.\overline{3}, \ \sqrt{3}, \ 4\frac{1}{5}, \ 6, \ 6.7\right\}$$

a. Natural numbers

Answer: 6

b. Whole numbers

Answer: 0 and 6

c. Integers

Answer: –6, 0, and 6

d. Rational numbers

Answer: $-6, \ -\frac{5}{6}, \ 0, \ 0.\overline{3}, \ 4\frac{1}{5}, \ 6, \ 6.7$

e. Irrational numbers

Answer: $\sqrt{3}$

f. Real numbers

Answer: all the numbers in the set

Now Try:

1a. Use an integer to express the boldface italic number in the application.

a. Death Valley is *282* feet below sea level.

_____

2. Graph each number on a number line.

$$\frac{1}{2}, \ 0, \ -3, \ -\frac{5}{2}$$

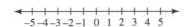

```
     ◄──┼──┼──┼──┼──┼──┼──┼──┼──┼──┼──►
       –5 –4 –3 –2 –1  0  1  2  3  4  5
```

3. List the numbers in the following set that belong to each set of numbers.

$$\left\{-10, -\frac{5}{8}, \ 0, \ 0.\overline{4}, \ \sqrt{5}, \ 5\frac{1}{2}, \ 7, \ 9.9\right\}$$

a. Natural numbers

_____

b. Whole numbers

_____

c. Integers

_____

d. Rational numbers

_____

e. Irrational numbers

_____

f. Real numbers

_____

Objective 1 Practice Exercises

For extra help, see Examples 1–3 on pages 48–50 of your text.

Use a real number to express each number in the following applications.

1. Last year Nina lost 75 pounds.   1. _____

2. Between 1970 and 1982, the population of Norway   2. _____
    increased by 279,867.

Graph the group of rational numbers on a number line.

3. –4.5, –2.3, 1.7, 4.2   3.
                                                          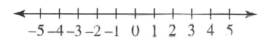
                                                          -5 -4 -3 -2 -1 0 1 2 3 4 5

Objective 2   Tell which of two real numbers is less than the other.

Video Examples

Review this example for Objective 2: | **Now Try:**
4. Is the statement –4 < –2 true or false? | 4. Is the statement –10 < –8 true or false.

    Because –4 is to the left of –2 on the number
    line, –4 is less than –2. The statement
    –4 < –2 is true.

    ← + + + + + + + + + + + →
    -5 -4 -3 -2 -1 0 1 2 3 4 5

                                                       _____

Objective 2 Practice Exercises

For extra help, see Example 4 on page 51 of your text.

*Decide whether each statement is **true** or **false**.*

4. $-76 < 45$ 4. _____

5. $-5 > -5$ 5. _____

6. $-12 > -10$ 6. _____

Objective 3   Find the additive inverse of a real number.

Objective 3 Practice Exercises

For extra help, see page 51 of your text.

Find the additive inverse of each number.

7.   −25   7. _____

8. $\dfrac{3}{8}$ 8. _____

9.   4.5   9. _____

Objective 4   Find the absolute value of a real number.

Video Examples

Review these examples for Objective 4:
5.   Simplify by finding the absolute value.

    b. $|16|$

    $|16| = 16$

    c. $|-16|$

    $|-16| = -(-16) = 16$

    e. $-|-16|$

    $-|-16| = -(16) = -16$

Now Try:
5.   Simplify by finding the absolute value.

    b. $|10|$

    _____

    c. $|-10|$

    _____

    e. $-|-10|$

    _____

Objective 4 Practice Exercises

For extra help, see Example 5 on page 52 of your text.

Simplify.

10. $-|49 - 39|$ 10. _____

11. $|-7.52 + 6.3|$ 11. _____

12. $|16 - 14|$ 12. _____

Chapter 1 THE REAL NUMBER SYSTEM

1.4 Adding Real Numbers

Learning Objectives	
1	Add two numbers with the same sign.
2	Add numbers with different signs.
3	Use the rules for order of operations when adding real numbers.
4	Translate words and phrases that indicate addition.

Key Terms

Use the vocabulary terms listed below to complete each statement in exercises 1–2.

sum   **addends**

1. The answer to an addition problem is called the _____.

2. In an addition problem, the numbers being added are the _____.

Objective 1   Add two numbers with the same sign.

Video Examples

Review these examples for Objective 1:

1. Use a number line to find each sum.

   a. $4 + 4$

   Step 1 Start at 0 and draw an arrow 4 units to the right.

   Step 2 From the right end of that arrow, draw another arrow 4 units to the right.

   The number below the end of this second arrow is 8, so $4 + 4 = 8$.

   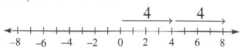

   b. $-3 + (-5)$

   Step 1 Start at 0 and draw an arrow 3 units to the left.

   Step 2 From the left end of that arrow, draw another arrow 5 units to the left.

   The number below the end of this second arrow is -5, so $-3 + (-5) = -8$.

   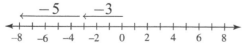

Now Try:

1. Use a number line to find each sum.

   a. $2 + 4$.

   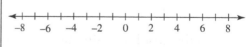

   _____

   b. $-4 + (-1)$

   _____

2. Find each sum.

     a. $-3 + (-7)$

     $-3 + (-7) = -10$

     b. $-5 + (-16)$

     $-5 + (-16) = -21$

     c. $-18 + (-7)$

     $-18 + (-7) = -25$

2. Find each sum.

     a. $-8 + (-4)$

     _____

     b. $-17 + (-14)$

     _____

     c. $-10 + (-30)$

     _____

Objective 1 Practice Exercises

For extra help, see Examples 1–2 on pages 57–58 of your text.

Find each sum.

1. $-7 + (-11)$ 1. _____

2. $-9 + (-9)$ 2. _____

3. $-2\frac{3}{8} + \left(-3\frac{1}{4}\right)$ 3. _____

Objective 2   Add numbers with different signs.

Video Examples

Review these examples for Objective 2:

4. Find each sum.

     a. $-10 + 6$

     Find the absolute value of each number.

       $\left|-10\right| = 10$ and $\left|6\right| = 6$

     Then find the difference between these absolute values: $10 - 6 = 4$. The sum will be negative since $\left|-10\right| > \left|6\right|$.

       $-10 + 6 = -4$

Now Try:

4. Find each sum.

     a. $-25 + 13$

     _____

b. $-16 + 9$

Find the absolute value of each number.
$$|-16| = 16 \quad \text{and} \quad |9| = 9$$
Then find the difference between these absolute values: $16 - 9 = 7$. The sum will be positive since $|-16| > |9|$.
$$-16 + 9 = -7$$

b. $-12 + 5$

_____

Objective 2 Practice Exercises

For extra help, see Examples 3–4 on pages 58–59 of your text.

Use a number line to find the sum.

4. $-8 + 5$

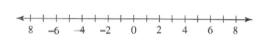

4. _____

Find each sum.

5. $\dfrac{7}{12} + \left(-\dfrac{3}{4}\right)$

5. _____

6. $-7.5 + 9.4$

6. _____

Objective 3   Use the rules for order of operations when adding real numbers.

Video Examples

Review this example for Objective 3:
5b. Find the sum.

    b. $9 + [(-3 + 7) + (-5)]$

$$9 + [(-3 + 7) + (-5)] = 9 + [4 + (-5)]$$
$$= 9 + (-1)$$
$$= 8$$

Now Try:
5b. Find the sum.

    b. $19 + [(-5 + 8) + (-7)]$

_____

Objective 3 Practice Exercises

For extra help, see Example 5 on pages 59–60 of your text.

Find each sum.

7. $-2+\left[4+(-18+13)\right]$ 7. _____

8. $\left[(-7)+14\right]+\left[(-16)+3\right]$ 8. _____

9. $-8.9+\left[6.8+(-4.7)\right]$ 9. _____

Objective 4   Translate words and phrases that indicate addition.

Video Examples

Review these examples for Objective 4:

6. Write a numerical expression for each phrase, and simplify the expression.

   a. The sum of –9 and 5 and 3

   –9 + 5 + 3 simplifies to –4 + 3, which equals –1.

   b. 4 more than –7, increased by 15

   (–7 + 4) + 15 simplifies to –3 + 15, which equals 12.

Now Try:

6. Write a numerical expression for each phrase, and simplify the expression.

   a. The sum of –10 and 11 and 2

   _____

   b. 15 more than –9, increased by 6

   _____

Objective 4 Practice Exercises

For extra help, see Examples 6–7 on page 60 of your text.

Write a numerical expression for each phrase, and then simplify the expression.

10.   The sum of –14 and –29, increased by 27   **10.** _____

_____

11.   –10 added to the sum of 20 and –4   **11.** _____

_____

Solve the problem.

12.   The temperature at dawn in Blackwood was 24°F.   **12.** _____
During the day the temperature decreased 30°. Then
it increased 11° by sunset. What was the temperature
at sunset?

Chapter 1 THE REAL NUMBER SYSTEM

1.5   Subtracting Real Numbers

Learning Objectives
1   Subtract two numbers on a number line.
2   Use the definition of subtraction.
3   Use the rules for order of operations when subtracting real numbers.
4   Translate words and phrases that indicate subtraction.

Key Terms

Use the vocabulary terms listed below to complete each statement in exercises 1–3.

minuend   **subtrahend**   **difference**

1.   The number from which another number is being subtracted is called the

_____.

2.   The _____ is the number being subtracted.

3.   The answer to a subtraction problem is called the _____.

Objective 1   Subtract two numbers on a number line.

Video Examples

Review this example for Objective 1:

1.   Use a number line to find the difference 5 – 3.

Step 1 Start at 0 and draw an arrow 5 units to the right.

Step 2 From the right end of that arrow, draw another arrow 3 units to the left.

The number below the end of this second arrow is 2, so 5 – 3 = 2.

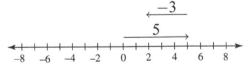

Now Try:

1.   Use a number line to find the difference 3 – 1.

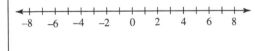

_____

Name: Date:
Instructor: Section:

Objective 1 Practice Exercises

For extra help, see Example 1 on page 64 of your text.

Use a number line to find the difference.

1. $8-5$ 1. _____

2. $7-10$ 2. _____

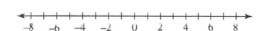

3. $-5-2$ 3. _____

Objective 2   Use the definition of subtraction.

Video Examples

Review these examples for Objective 2:
2. Find each difference.

    b. $8-11$

$$8-11=8+(-11)=-3$$

    c. $-9-16$

$$-9-16=-9+(-16)=-25$$

    d. $-6-(-9)$

$$-6-(-9)=-6+(9)=3$$

    e. $\dfrac{5}{6}-\left(-\dfrac{3}{7}\right)$

$$\frac{5}{6}-\left(-\frac{3}{7}\right)=\frac{35}{42}-\left(-\frac{18}{42}\right)$$
$$=\frac{35}{42}+\frac{18}{42}$$
$$=\frac{53}{42}$$

Now Try:
2. Find each difference.

    b. $13-17$

    _____

    c. $-11-27$

    _____

    d. $-5-(-7)$

    _____

    e. $\dfrac{5}{9}-\left(-\dfrac{4}{5}\right)$

    _____

Objective 2 Practice Exercises

For extra help, see Example 2 on page 65 of your text.

Find each difference.

4. $22-(-24)$

4. _____

5. $-3.2-(-7.6)$

5. _____

6. $\dfrac{15}{4}-\left(-\dfrac{17}{8}\right)$

6. _____

Objective 3   Use the rules for order of operations when subtracting real numbers.

Video Examples

Review these examples for Objective 3:
3.   Perform each operation.

   a. $-7-[3-(9+6)]$

   $$-7-[3-(9+6)]=-7-[3-15]$$
   $$=-7-[3+(-15)]$$
   $$=-7-(-12)$$
   $$=-7+12$$
   $$=5$$

   b. $4-\left[\left(-\dfrac{1}{5}-\dfrac{1}{4}\right)-(3-2)\right]$

   $4-\left[\left(-\dfrac{1}{5}-\dfrac{1}{4}\right)-(3-2)\right]$

   $=4-\left[\left(-\dfrac{1}{5}+\left(-\dfrac{1}{4}\right)\right)-1\right]$ $\quad -\dfrac{1}{5}+\left(-\dfrac{1}{4}\right)=-\dfrac{4}{20}+\left(-\dfrac{5}{20}\right)$
   $\qquad\qquad\qquad\qquad\qquad\qquad\qquad\qquad =-\dfrac{9}{20}$

   $=4-\left[\left(-\dfrac{9}{20}\right)-1\right]$

   $=4-\left[-\dfrac{9}{20}+(-1)\right]$

   $=4-\left[-\dfrac{9}{20}+\left(-\dfrac{20}{20}\right)\right]$

   $=4-\left(-\dfrac{29}{20}\right)=4+\dfrac{29}{20}$

   $=\dfrac{80}{20}+\dfrac{29}{20}=\dfrac{109}{20}$

Now Try:
3.   Perform each operation.

   a. $-10-[5-(7+3)]$

   _____

   b. $8-\left[\left(-\dfrac{1}{6}-\dfrac{1}{3}\right)-(5-4)\right]$

   _____

Objective 3 Practice Exercises

For extra help, see Example 3 on page 66 of your text.

Perform each operation.

7. $[8-(-12)]-2$ 7. _____

8. $3\left[4+(11-19)\right]$ 8. _____

9. $\dfrac{2}{9}-\left[\dfrac{5}{6}-\left(-\dfrac{2}{3}\right)\right]$ 9. _____

Objective 4   Translate words and phrases that indicate subtraction.

Video Examples

Review these examples for Objective 4:

4.   Write a numerical expression for each phrase, and simplify the expression.

   a. The difference between –10 and 7

   –10 –7 simplifies to $-10+(-7)$, which equals –17

   b. 3 subtracted from the sum of 9 and –4

   First, add 9 and –4. Next subtract 3 from this sum.
   $[9+(-4)]-3$ simplifies to $5-3$, which equals 2.

Now Try:

4.   Write a numerical expression for each phrase, and simplify the expression.

   a. The difference between –17 and 9

   _____

   b. 8 subtracted from the sum of 25 and –6

   _____

5. The early morning temperature on a mountain in California was –8°F. At noon the temperature was 38°F. What was the rise in temperature?

We must subtract the lowest temperature from the highest temperature.

$$38 - (-8) = 38 + 8 = 46$$

The rise was 46°F.

5. The floor of Death Valley is 282 ft below sea level. A nearby mountain has an elevation of 5182 ft above sea level. Find the difference between the highest and lowest elevations.

_____

Objective 4 Practice Exercises

For extra help, see Examples 4–5 on pages 67–68 of your text.

Write a numerical expression for each phrase, and then simplify the expression.

10.   4 less than –4

10. _____

11.   The sum of –4 and 12, decreased by 9

11. _____

Solve the problem.

12.   Dr. Somers runs an experiment at –43.3°C. He then lowers the temperature by 7.9°C. What is the new temperature for the experiment?

12. _____

Chapter 1 THE REAL NUMBER SYSTEM

1.6   Multiplying and Dividing Real Numbers

Learning Objectives
1   Find the product of a positive number and a negative number.
2   Find the product of two negative numbers.
3   Use the reciprocal of a number to apply the definition of division.
4   Use the rules for order of operations when multiplying and dividing real numbers.
5   Evaluate expressions involving variables.
6   Translate words and phrases that indicate multiplication and division.
7   Translate simple sentences into equations.

Key Terms

Use the vocabulary terms listed below to complete each statement in exercises 1–5.

product   **quotient**   **dividend**

divisor   **reciprocals**

1. The answer to a division problem is called the _____.

2. Pairs of numbers whose product is 1 are called _____.

3. The answer to a multiplication problem is called the _____.

4. In the division $x \div y$, x is called the _____.

5. In the division $x \div y$, y is called the _____.

Objective 1   Find the product of a positive number and a negative number.

Video Examples

Review these examples for Objective 1:

1. Find each product using the multiplication rule.

   a. $9(-6)$

   $9(-6) = -(9 \cdot 6) = -54$

   c. $-3.7(2.5)$

   $-3.7(2.5) = -9.25$

Now Try:

1. Find each product using the multiplication rule.

   a. $8(-7)$

   _____

   c. $-9.8(4.6)$

   _____

Objective 1 Practice Exercises

For extra help, see Example 1 on page 74 of your text.

Find each product.

1. $7(-4)$

                1. _____

2. $\left(\dfrac{1}{5}\right)\left(-\dfrac{2}{3}\right)$

                2. _____

3. $(-3.2)(4.1)$

                3. _____

Objective 2   Find the product of two negative numbers.

Video Examples

Review these examples for Objective 2:

2. Find each product using the multiplication rule.

    a. $-7(-3)$

    $-7(-3) = 21$

    b. $-8(-13)$

    $-8(-13) = 104$

Now Try:

2. Find each product using the multiplication rule.

    a. $-5(-6)$

    _____

    b. $-9(-15)$

    _____

Objective 2 Practice Exercises

For extra help, see Example 2 on page 75 of your text.

Find each product.

4. $(-4)(-10)$

                4. _____

5. $\left(-\dfrac{2}{7}\right)\left(-\dfrac{14}{5}\right)$

                5. _____

6. $(-0.4)(-3.4)$

                6. _____

Objective 3   Use the reciprocal of a number to apply the definition of division.

Video Examples

Review these examples for Objective 3:
3.   Find each quotient.

    a. $\dfrac{15}{-3}$

    $\dfrac{15}{-3} = -5$

    b. $-\dfrac{30}{6}$

    $-\dfrac{30}{6} = -5$

    c. $\dfrac{-7.5}{-0.03}$

    $\dfrac{-7.5}{-0.03} = 250$

    d. $-\dfrac{1}{9} \div \left(-\dfrac{2}{3}\right)$

    $\dfrac{1}{9} \div \left(-\dfrac{2}{3}\right) - -\dfrac{1}{9} \cdot \left(-\dfrac{3}{2}\right) = \dfrac{1}{6}$

Now Try:
3.   Find each quotient.

    a. $\dfrac{-18}{-6}$

    _____

    b. $\dfrac{16}{-8}$

    _____

    c. $\dfrac{-18.3}{-6.1}$

    _____

    d. $-\dfrac{2}{5} \div \left(-\dfrac{11}{10}\right)$

    _____

Objective 3 Practice Exercises

For extra help, see Example 3 on page 77 of your text.

Find each quotient.

7. $\dfrac{-120}{-20}$

8. $\dfrac{0}{-2}$

9. $\dfrac{10}{0}$

7.   _____

8.   _____

9.   _____

45

Objective 4   **Use the rules for order of operations when multiplying and dividing real numbers.**

Video Examples

Review this example for Objective 4:

4d. Simplify.

$$\frac{6(-4)-5(3)}{3(2-7)}$$

$$\frac{6(-4)-5(3)}{3(2-7)} = \frac{-24-15}{3(-5)}$$

$$= \frac{-39}{-15}$$

$$= \frac{13}{5}$$

Now Try:

4d. Simplify.

$$\frac{-9(-3)+4(-8)}{-4(5-6)}$$

_____

Objective 4 Practice Exercises

For extra help, see Example 4 on page 78 of your text.

Perform the indicated operations.

10. $-4\big[(-2)(7)-2\big]$

10. _____

11. $\dfrac{-7(2)-(-3)}{5+(-3)}$

11. _____

12. $\dfrac{-4\big[8-(-3+7)\big]}{-6\big[3-(-2)\big]-3(-3)}$

12. _____

Objective 5   Evaluate expressions involving variables.

Video Examples

Review these examples for Objective 5:

5.   Evaluate each expression for $x = -2$, $y = -4$, and $m = -5$.

   a. $(5x + 6y)(-3m)$

   Substitute the given values for the variables. Then simplify.

$$(5x + 6y)(-3m)$$
$$= [5(-2) + 6(-4)][-3(-5)]$$
$$= [-10 + (-24)][15]$$
$$= [-34]15$$
$$= -510$$

   b. $4x^2 - 5y^2$

$$4x^2 - 5y^2 = 4(-2)^2 - 5(-4)^2$$
$$= 4(4) - 5(16)$$
$$= 16 - 80$$
$$= -64$$

Now Try:

5.   Evaluate each expression for $x = -5$, $y = -3$, and $p = -4$.

   a. $(6x + 2y)(-3p)$

   _____

   b. $7x^2 - 8y^2$

   _____

Objective 5 Practice Exercises

For extra help, see Example 5 on page 79 of your text.

Evaluate the following expressions if x = −3, y = 2, and a = 4.

13. $-x + [(-a + y) - 2x]$ 13. _____

14. $(-4 + x)(-a) - |x|$ 14. _____

15. $\dfrac{4a - x}{y^2}$ 15. _____

Objective 6   Translate words and phrases that indicate multiplication and division.

Video Examples

Review these examples for Objective 6:

6.   Write a numerical expression for each phrase, and simplify the expression.

   a. The product of 15 and the sum of 4 and −7

   $15[4+(-7)]$ simplifies to $15[-3]$, which equals −45.

   d. 45% of the difference between 24 and −5

   $0.45[24-(-5)]$ simplifies to $0.45[29]$, which equals 13.05.

Now Try:

6.   Write a numerical expression for each phrase, and simplify the expression.
   a. The product of 16 and the sum of 5 and −7

   _____

   d. 8% of the difference between 18 and −4

   _____

Objective 6 Practice Exercises

For extra help, see Examples 6–7 on pages 80–81 of your text.

Write a numerical expression for each phrase and simplify.

16.   The product of −7 and 3, added to −7

16.   _____

17.   Three-tenths of the difference between 50 and −10, subtracted from 85

17.   _____

18.   The sum of −12 and the quotient of 49 and −7

18.   _____

Objective 7   Translate simple sentences into equations.

Video Examples

Review this example for Objective 7:

8d. Write the sentence in symbols, using x to represent the number.

   The quotient of 27 and a number is −3.

   $\dfrac{27}{x}=-3$

Now Try:

8d. Write the sentence in symbols, using x to represent the number.

   The quotient of 36 and a number is −4

   _____

Objective 7 Practice Exercises

For extra help, see Example 8 on page 81 of your text.

Write each statement in symbols, using x as the variable.

19. Two-thirds of a number is –7.

19. _____

20. –8 times a number is 72.

20. _____

21. When a number is divided by –4, the result is 1.

21. _____

Chapter 1 THE REAL NUMBER SYSTEM

1.7   Properties of Real Numbers

Learning Objectives	
1	Use the commutative properties.
2	Use the associative properties.
3	Use the identity properties.
4	Use the inverse properties.
5	Use the distributive property.

Key Terms

Use the vocabulary terms listed below to complete each statement in exercises 1−2.

identity element for addition

identity element for multiplication

1.   When the _____, which is 0, is added to a number, the number is unchanged.

2.   When a number is multiplied by the _____, which is 1, the number is unchanged.

Objective 1   Use the commutative properties.

Video Examples

Review these examples for Objective 1:

1.   Use a commutative property to complete each statement.

   a. $-7+6=6+$ _____

   Using the commutative property of addition,
   $-7+6=6+(-7)$

   b. $(-3)5=$ _____ (-3)

   Using the commutative property of multiplication,
   $(-3)5=5(-3)$

Now Try:

1.   Use a commutative property to complete each statement.

   a. $-12+8=8+$ _____

   _____

   b. $(-4)2=$ _____ (-4)

   _____

Name: _____   Date: _____

Instructor: _____   Section: _____

Objective 1 Practice Exercises

For extra help, see Example 1 on page 88 of your text.

Complete each statement. Use a commutative property.

1. $y + 4 = \underline{\hspace{1cm}} + y$ 1. _____

2. $5(2) = \underline{\hspace{1cm}}(5)$ 2. _____

3. $-4(4+z) - \underline{\hspace{1cm}}(\;4)$ 3. _____

Objective 2   Use the associative properties.

Video Examples

Review these examples for Objective 2:	Now Try:
2. Use an associative property to complete each statement.	**2.** Use an associative property to complete each statement.
a. $-5 + (3 + 7) = (-5 + \underline{\hspace{1cm}}) + 7$	**a.** $-8 + (4 + 6) = (-8 + \underline{\hspace{0.6cm}}) + 6$
Using the associative property of addition, $-5 + (3 + 7) = (\;5 + 3) + 7$	_____
b. $[4 \cdot (-9)] \cdot 2 = 4 \cdot \underline{\hspace{1cm}}$	**b.** $[8 \cdot (-3)] \cdot 4 = 8 \cdot \underline{\hspace{0.6cm}}$
Using the associative property of multiplication, $[4 \cdot (-9)] \cdot 2 = 4 \cdot [(-9) \cdot 2]$	_____
4. Find each sum or product.	**4.** Find each sum or product.
a. $54 + 21 + 3 + 17 + 29$	**a.** $48 + 15 + 12 + 24 + 8$
$54 + 21 + 3 + 17 + 29$	_____
$= 54 + (21 + 29) + (3 + 17)$	
$= 54 + 50 + 20$	
$= 124$	
b. $50(43)(4)$	**b.** $40(63)(5)$
$50(43)(4) = 50(4)(43)$	
$= 200(43)$	_____
$= 8600$	

Objective 2 Practice Exercises

For extra help, see Examples 2–4 on pages 88–89 of your text.

Complete each statement. Use an associative property.

4. $4(ab) = \underline{} \cdot b$ 4. _____

5. $[x + (-4)] + 3y = x + \underline{}$ 5. _____

6. $4r + (3s + 14t) = \underline{} + 14t$ 6. _____

Objective 3   Use the identity properties.

Video Examples

Review this example for Objective 3:   **Now Try:**

6a. Write $\dfrac{56}{35}$ in lowest terms.   **6a.** Write $\dfrac{49}{63}$ in lowest terms.

$$\frac{56}{35} = \frac{8 \cdot 7}{5 \cdot 7}$$ _____

$$= \frac{8}{5} \cdot \frac{7}{7}$$

$$= \frac{8}{5} \cdot 1$$

$$= \frac{8}{5}$$

Objective 3 Practice Exercises

For extra help, see Examples 5–6 on page 90 of your text.

Use an identity property to complete each statement.

7. $4 + 0 = \underline{}$ 7. _____

8. $\underline{} \cdot 1 = 12$ 8. _____

Use an identity property to simplify the expression.

9. $\dfrac{30}{35}$ 9. _____

Objective 4   Use the inverse properties.

Video Examples

Review these examples for Objective 4:	**Now Try:**

Review these examples for Objective 4:

7. Use an inverse property to complete each statement.

   b. $5 + \underline{\hspace{1cm}} = 0$

   Use the inverse property of addition.
      $5 + (-5) = 0$

   d. $\underline{\hspace{1cm}} \cdot \dfrac{6}{7} = 1$

   Use the inverse property of multiplication.
      $\dfrac{7}{6} \cdot \dfrac{6}{7} = 1$

Now Try:

7. Use an inverse property to complete each statement.

   b. $8 + \underline{\hspace{1cm}} = 0$

   _____

   d. $\dfrac{8}{5} \cdot \underline{\hspace{1cm}} = 1$

   _____

Objective 4 Practice Exercises

For extra help, see Example 7 on page 91 of your text.

Complete the statements so that they are examples of either an identity property or an inverse property. Identify which property is used.

10. $-4 + \underline{\hspace{1cm}} = 0$ 10. _____

11. $-9 + \underline{\hspace{1cm}} = -9$ 11. _____

12. $-\dfrac{3}{5} \cdot \underline{\hspace{1cm}} = 1$ 12. _____

Objective 5   Use the distributive property.

Video Examples

Review these examples for Objective 5:

8f. Use the distributive property to rewrite the expression.

   $4 \cdot 9 + 4 \cdot 2$

   Use the distributive property in reverse.
   $4 \cdot 9 + 4 \cdot 2 = 4(9 + 2)$
                $= 4(11)$
                $= 44$

Now Try:

8f. Use the distributive property to rewrite the expression.

   $25 \cdot 9 + 25 \cdot 6$

   _____

9c. Write the expression without parentheses.

$$-(-p-5r+9x)$$

$$-(-p-5r+9x)$$
$$=-1\cdot(-1p-5r+9x)$$
$$=-1\cdot(-1p)-1\cdot(-5r)-1\cdot(9x)$$
$$=p+5r-9x$$

9c. Write the expression without parentheses.
$$-(-4x-5y+z)$$

_____

Objective 5 Practice Exercises

For extra help, see Examples 8–9 on pages 92–93 of your text.

Use the distributive property to rewrite each expression. Simplify if possible.

13. $n(2a-4b+6c)$

13. _____

14. $-2(5y-9z)$

14. _____

15. $-(-2k+7)$

15. _____

Chapter 1 THE REAL NUMBER SYSTEM

1.8   Simplifying Expressions

Learning Objectives
1   Simplify expressions.
2   Identify terms and numerical coefficients.
3   Identify like terms.
4   Combine like terms.
5   Simplify expressions from word phrases.

Key Terms

Use the vocabulary terms listed below to complete each statement in exercises 1–3.

term   numerical coefficient   like terms

1.   In the term $4x^2$, "4" is the_____.

2.   A number, a variable, or a product or quotient of a number and one or more variables raised to powers is called a _____.

3.   Terms with exactly the same variables, including the same exponents, are called

    _____   _____.

Objective 1   Simplify expressions.

Video Examples

Review these examples for Objective 1:
1.   Simplify each expression.

c. $7+5(9k+6)$

$$7+5(9k+6) = 7+5(9k)+5(6)$$
$$= 7+45k+30$$
$$= 37+45k$$

d. $9-(4y-6)$

$$9-(4y-6) = 9-1(4y-6)$$
$$= 9-4y+6$$
$$= 15-4y$$

Now Try:
1.   Simplify each expression.

c. $8+9(2x+7)$

    _____

d. $8-(7x-3)$

    _____

Objective 1 Practice Exercises

For extra help, see Example 1 on page 98 of your text.

Simplify each expression.

1. $4(2x+5)+7$

1. _____

2. $-4+s-(12-21)$

2. _____

3. $-2(-5x+2)+7$

3. _____

Objective 2   Identify terms and numerical coefficients.

Objective 2 Practice Exercises

For extra help, see pages 98–99 of your text.

Give the numerical coefficient of each term.

4. $-2y^2$

4. _____

5. $\dfrac{7x}{9}$

5. _____

6. $5.6r^5$

6. _____

Objective 3   Identify like terms.

Objective 3 Practice Exercises

For extra help, see page 99 of your text.

Identify each group of terms as **like** *or* **unlike***.*

7. $4x^2, -7x^2$

7. _____

8. $-8m, -8m^2$

8. _____

9. $7xy, -6xy^2$

9. _____

Name: Date:
Instructor: Section:

Objective 4   Combine like terms.

Video Examples

Review these examples for Objective 4:	**Now Try:**
2c. Combine like terms in the expression.	**2c.** Combine like terms in the expression.

2c. Combine like terms in the expression.

$$9x + x$$

$$9x + x = 9x + 1x$$
$$= (9+1)x$$
$$-10x$$

Now Try:

2c. Combine like terms in the expression.
$$18x + x$$

_____

3a. Simplify the expression.

$$15y + 3(5 + 4y)$$

$$15y + 3(5 + 4y) = 15y + 3(5) + 3(4y)$$
$$= 15y + 15 + 12y$$
$$= 27y + 15$$

3a. Simplify each expression.

$$9y + 5(3 + 8y)$$

_____

Objective 4 Practice Exercises

For extra help, see Examples 2–3 on pages 99–101 of your text.

Simplify.

10. $12y - 7y^2 + 4y - 3y^2$ **10.** _____

11. $-4(x+4) + 2(3x+1)$ **11.** _____

12. $2.5(3y+1) - 4.5(2y-3)$ **12.** _____

Objective 5   Simplify expressions from word phrases.

Video Examples

Review this example for Objective 5:	**Now Try:**
4. Translate the phrase into a mathematical expression and simplify. The sum of 8, three times a number, nine times a number, and seven times a number. Use x for the number. $8 + 3x + 9x + 7x$ simplifies to $8 + 19x$.	4. Translate the phrase into a mathematical expression and simplify. The sum of 11, ten times a number, eight times a number, and four times a number _____

Objective 5 Practice Exercises

For extra help, see Example 4 on page 101 of your text.

Write each phrase as a mathematical expression and simplify by combining like terms. Use x as the variable.

13. The sum of six times a number and 12, added to four times the number.

13. _____

14. The sum of seven times a number and 2, subtracted from three times the number.

14. _____

15. Four times the difference between twice a number and six times the number, added to six times the sum of the number and 9.

15. _____

Chapter 2 EQUATIONS, INEQUALITIES, AND APPLICATIONS

2.1   The Addition Property of Equality

Learning Objectives
1   Identify linear equations.
2   Use the addition property of equality.
3   Simplify, and then use the addition property of equality.

Key Terms

Use the vocabulary terms listed below to complete each statement in exercises 1–3.

linear equation   solution set   equivalent equations

1.   Equations that have exactly the same solutions sets are called

   _____.

2.   An equation that can be written in the form $Ax + B = C$, where A, B, and C are real numbers and $A \neq 0$, is called a _____.

3.   The set of all numbers that satisfy an equation is called its _____.

Objective 1   Identify linear equations.

Objective 1 Practice Exercises

For extra help, see page 118 of your text.

Tell whether each of the following is a linear equation.

1. $3x^2 + 4x + 3 = 0$ 1. _____

2. $\dfrac{5}{x} - \dfrac{3}{2} = 0$ 2. _____

3. $4x - 2 = 12x + 9$ 3. _____

Objective 2   Use the addition property of equality.

Video Examples

Review these examples for Objective 2:

Now Try:

1.   Solve $x - 15 = 8$.

$$x - 15 = 8$$
$$x - 15 + 15 = 8 + 15$$
$$x = 23$$

Check $x - 15 = 8$
$$23 - 15 \overset{?}{=} 8$$
$$8 = 8 \quad \text{True}$$

The solution is 23, and the solution set is $\{23\}$.

1.   Solve $x - 12 = 9$.

_____

3.   Solve $-5 = x + 17$.

$$-5 = x + 17$$
$$-5 - 17 = x + 17 - 17$$
$$-22 = x$$

Check $-5 = x + 17$
$$-5 \overset{?}{=} -22 + 17$$
$$-5 = -5 \qquad \text{True}$$

The solution set is $\{-22\}$.

3.   Solve $-10 = x + 9$.

_____

4.   Solve $-5p = -6p + 3$.

$$-5p = -6p + 3$$
$$-5p + 6p = -6p + 3 + 6p$$
$$p = 3$$

Check by substituting 3 in the original equation. The solution set is $\{3\}$.

4.   Solve $-8p = -9p + 7$.

_____

Objective 2 Practice Exercises

For extra help, see Examples 1–6 on pages 119–121 of your text.

Solve each equation by using the addition property of equality. Check each solution.

4. $y - 4 = 16$

4. _____

5. $\quad \dfrac{9}{8}p - \dfrac{1}{2} = \dfrac{1}{8}p$

5. _____

6. $\quad 9.5y - 2.4 - 10.5y$

6. _____

Objective 3   Simplify, and then use the addition property of equality.

Video Examples

Review these examples for Objective 3:	Now Try:
7. Solve $5t - 16 + t + 4 = 9 + 5t + 6$.	**7.** Solve $8t - 9 + t + 7 = 12 + 8t + 15$.

$5t - 16 + t + 4 = 9 + 5t + 6$

$6t - 12 = 15 + 5t$

$6t - 12 - 5t = 15 + 5t - 5t$

$t - 12 = 15$

$t - 12 + 12 = 15 + 12$

$t = 27$

Check by substituting 27 in the original equation. The solution set is {27}.

8. Solve $4(3 + 6x) - (5 + 23x) = 19$.

$4(3 + 6x) - (5 + 23x) = 19$

$4(3) + 4(6x) - 1(5) - 1(23x) = 19$

$12 + 24x - 5 - 23x = 19$

$x + 7 = 19$

$x + 7 - 7 = 19 - 7$

$x = 12$

Check by substituting 12 in the original equation. The solution set is {12}.

7. Solve
$8t - 9 + t + 7 = 12 + 8t + 15$.

_____

8. Solve
$5(7 + 8x) - (29 + 39x) = 14$.

_____

Objective 3 Practice Exercises

For extra help, see Examples 7–8 on page 122 of your text.

Solve each equation. First simplify each side of the equation as much as possible. Check each solution.

7. $3(t+3)-(2t+7)=9$

7. _____

8. $-4(5g-7)+3(8g-3)=15-4+3g$

8. _____

9. $3.6p+4.8+4.0p=8.6p-3.1+0.7$

9. _____

Chapter 2 EQUATIONS, INEQUALITIES, AND APPLICATIONS

2.2   The Multiplication Property of Equality

Learning Objectives
1   Use the multiplication property of equality.
2   Simplify, and then use the multiplication property of equality.

Key Terms

Use the vocabulary terms listed below to complete each statement in exercises 1–2.

multiplication property of equality   addition property of equality

1.   The _____ states that multiplying both sides of an equation by the same nonzero number will not change the solution.

2.   When the same quantity is added to both sides of an equation, the _____ is being applied.

Objective 1   Use the multiplication property of equality.

Video Examples

Review these example for Objective 1:

1.   Solve $6x = 78$.

$$6x = 78$$

$$\frac{6x}{6} = \frac{78}{6}$$

$$x - 13$$

Check $6x = 78$

$$6(13) \overset{?}{=} 78$$

$$78 = 78 \quad \text{True}$$

The solution set is $\{13\}$.

6.   Solve $-k = -22$.

$$-k = -22$$

$$-1 \cdot k = -22$$

$$-1(-1 \cdot k) = -1(-22)$$

$$[-1(-1)] \cdot k = 22$$

$$1 \cdot k = 22$$

$$k = 22$$

Check by substituting 22 in the original equation. The solution set is $\{22\}$.

Now Try:

1.   Solve $4x = 56$.

_____

6.   Solve $-y = -39$.

_____

4. Solve $\frac{x}{7} = 5$.

$$\frac{x}{7} = 5$$

$$\frac{1}{7}x = 5$$

$$7 \cdot \frac{1}{7}x = 7 \cdot 5$$

$$x = 35$$

Check by substituting 35 in the original equation. The solution set is $\{35\}$.

5. Solve $\frac{5}{6}x = 15$.

$$\frac{5}{6}x = 15$$

$$\frac{6}{5} \cdot \frac{5}{6}x = \frac{6}{5} \cdot 15$$

$$1 \cdot x = \frac{6}{5} \cdot \frac{15}{1}$$

$$x = 18$$

Check by substituting 18 in the original equation. The solution set is $\{18\}$.

4. Solve $\frac{x}{8} = 3$.

_____

5. Solve $\frac{7}{9}h = 28$.

_____

Objective 1 Practice Exercises

For extra help, see Examples 1–6 on pages 127–129 of your text.

Solve each equation and check your solution.

1. $-3w = 51$

1. _____

2. $\frac{3p}{7} = -6$

2. _____

3. $-2.7v = -17.28$ **3.** _____

Objective 2   Simplify, and then use the multiplication property of equality.

Video Examples

Review this example for Objective 2:   **Now Try:**

7. Solve $9m + 4m - 39$.   **7.** Solve $12m + 8m = 80$.

$$9m + 4m = 39$$

$$13m = 39$$

$$\frac{13m}{13} = \frac{39}{13}$$ _____

$$m = 3$$

Check by substituting 3 in the original equation.
The solution set is {3}.

Objective 2 Practice Exercises

For extra help, see Examples 7–8 on pages 129–130 of your text.

Solve each equation and check your solution.

4. $-7b + 12b - 125$ **4.** _____

5. $3w - 7w = 20$ **5.** _____

6. $-11h - 6h + 14h - -21$ **6.** _____

Chapter 2 EQUATIONS, INEQUALITIES, AND APPLICATIONS

2.3   More on Solving Linear Equations

Learning Objectives
1   Learn and use the four steps for solving a linear equation.
2   Solve equations that have no solution or infinitely many solutions.
3   Solve equations with fractions or decimals as coefficients.
4   Write expressions for two related unknown quantities.

Key Terms

Use the vocabulary terms listed below to complete each statement in exercises 1–3.

   conditional equation   identity   contradiction

1.   An equation with no solution is called a(n) _____.

2.   A(n) _____ is an equation that is true for some values
      of the variable and false for other values.

3.   An equation that is true for all values of the variable is called a(n) _____.

Objective 1   Learn and use the four steps for solving a linear equation.

Video Examples

Review this example for Objective 1:

3.   Solve $5(k-4)-k=k-2$.

   Step 1 Clear parentheses using the distributive
   property.

$$5(k-4)-k=k-2$$
$$5(k)+5(-4)-k=k-2$$
$$5k-20-k=k-2$$
$$4k-20=k-2$$

   Step 2 $4k-20-k=k-2-k$
$$3k-20=-2$$
$$3k-20+20=-2+20$$
$$3k=18$$

   Step 3 $\dfrac{3k}{3}=\dfrac{18}{3}$
$$k=6$$

Now Try:

3.   Solve $9(k-2)-k=k+10$.

   _____

Name: Date:
Instructor: Section:

Step 4 Check by substituting 6 for *k* in the original equation.

$$5(k-4)-k = k-2$$
$$5(6-4)-6 \overset{?}{=} 6-2$$
$$5(2)-6 \overset{?}{=} 4$$
$$10-6 \overset{?}{=} 4$$
$$4 = 4 \quad \text{True}$$

The solution, 6, checks, so the solution set is {6}.

Objective 1 Practice Exercises

For extra help, see Examples 1–5 on pages 133–136 of your text.

Solve each equation and check your solution.

1. $7t + 6 = 11t - 4$

1. _____

2. $3a - 6a + 4(a - 4) = -2(a+2)$

2. _____

3. $3(t+5) = 6 - 2(t-4)$

3. _____

Objective 2   Solve equations that have no solution or infinitely many solutions.

Video Examples

Review these examples for Objective 2:

6.   Solve $6x - 18 = 6(x - 3)$.

$$6x - 18 = 6(x - 3)$$
$$6x - 18 = 6x - 18$$
$$6x - 18 - 6x = 6x - 18 - 6x$$
$$-18 = -18$$
$$-18 + 18 = -18 + 18$$
$$0 = 0$$

The solution set is {all real numbers}.

7.   Solve $3x + 4(x - 5) = 7x + 5$.

$$3x + 4(x - 5) = 7x + 5$$
$$3x + 4x - 20 = 7x + 5$$
$$7x - 20 = 7x + 5$$
$$7x - 20 - 7x = 7x + 5 - 7x$$
$$-20 = 5 \quad \text{False}$$

There is no solution. The solution set is $\varnothing$.

Now Try:

6.   Solve $3x + 4(x - 5) = 7x - 20$.

_____

7.   Solve $-5x + 17 = x - 6(x + 3)$.

_____

Objective 2 Practice Exercises

For extra help, see Examples 6–7 on page 137 of your text.

Solve each equation and check your solution.

4. $3(6x - 7) = 2(9x - 6)$

4. _____

5. $6y - 3(y + 2) = 3(y - 2)$

5. _____

6. $3(r-2)-r+4=2r+6$ 6. _____

Objective 3   Solve equations with fractions or decimals as coefficients.

Video Examples

Review these examples for Objective 3:

8.   Solve $\frac{3}{4}x-\frac{1}{2}x=-\frac{1}{8}x-6$.

Multiply each side by 8, the LCD.

$$\frac{3}{4}x-\frac{1}{2}x=-\frac{1}{8}x-6$$

Step 1 $8\left(\frac{3}{4}x-\frac{1}{2}x\right)=8\left(-\frac{1}{8}x-6\right)$

$$8\left(\frac{3}{4}x\right)+8\left(-\frac{1}{2}x\right)=8\left(-\frac{1}{8}x\right)+8(-6)$$

$$6x-4x=-x-48$$

$$2x=-x-48$$

Step 2 $2x+x=-x-48+x$

$$3x=-48$$

Step 3 $\frac{3x}{3}=-\frac{48}{3}$

$$x=-16$$

Step 4 $\frac{3}{4}x-\frac{1}{2}x=-\frac{1}{8}x-6$

$$\frac{3}{4}(-16)-\frac{1}{2}(-16)\overset{?}{=}-\frac{1}{8}(-16)-6$$

$$-12+8\overset{?}{=}2-6$$

$$-4=-4 \quad \text{True}$$

The solution set is $\{-16\}$.

Now Try:

8.   Solve $\frac{2}{9}x-\frac{1}{6}x=\frac{2}{3}x+11$.

10. Solve $0.2x + 0.04(10 - x) = 0.06(4)$.

To clear decimals, multiply by 100.

$$0.2x + 0.04(10 - x) = 0.06(4)$$

Step 1 $100[0.2x + 0.04(10 - x)] = 100[0.06(4)]$

$$100(0.2x) + 100[0.04(10 - x)] = 100[0.06(4)]$$

$$20x + 4(10) + 4(-x) = 24$$

$$20x + 40 - 4x = 24$$

$$16x + 40 = 24$$

Step 2 $16x + 40 - 40 = 24 - 40$

$$16x = -16$$

Step 3 $\dfrac{16x}{16} = \dfrac{-16}{16}$

$$x = -1$$

Step 4 Check to confirm that $\{-1\}$ is the solution set.

10. Solve
$$0.5x + 0.04(5 - 8x) = 0.07(8).$$

_____

Objective 3 Practice Exercises

For extra help, see Examples 8–10 on pages 138–139 of your text.

Solve each equation and check your solution.

7. $\dfrac{3}{8}x - \dfrac{1}{3}x = \dfrac{1}{12}$

7. _____

8. $\dfrac{1}{3}(2m - 1) - \dfrac{3}{4}m = \dfrac{5}{6}$

8. _____

Copyright © 2018 Pearson Education, Inc.

9. $0.45a - 0.35(20 - a) = 0.02(50)$ 9. _____

Objective 4   Write expressions for two related unknown quantities.

Video Examples

Review this example for Objective 4:
11a. Perform the translation.

> Two numbers have a sum of 51. If one of the numbers is represented by x, find an expression for the other number.
>
> If one number is x, then the other number is obtained by subtracting x from 51.
> $51 - x$.
> To check, we find the sum of the two numbers.
> $x + (51 - x) = 51$

Now Try:
11a. Perform the translation.

> Two numbers have a sum of 67. If one of the numbers is represented by t, find an expression for the other number.

_____

Objective 4 Practice Exercises

For extra help, see Example 11 on page 140 of your text.

Write an expression for the two related unknown quantities.

10.   Two numbers have a sum of 36. One is m. Find the other number.   **10.** _____

11. The product of two numbers is 17. One number is p. What is the other number?

11. _____

12. Admission to the circus costs x dollars for an adult and y dollars for a child. Find the total cost of 6 adults and 4 children.

12. _____

Chapter 2 EQUATIONS, INEQUALITIES, AND APPLICATIONS

2.4   An Introduction to Applications of Linear Equations

Learning Objectives
1   Learn the six steps for solving applied problems.
2   Solve problems involving unknown numbers.
3   Solve problems involving sums of quantities.
4   Solve problems involving consecutive integers.
5   Solve problems involving complementary and supplementary angles.

Key Terms

Use the vocabulary terms listed below to complete each statement in exercises 1–5.

complementary angles   right angle   supplementary angles

straight angle   consecutive integers

1.   Two angles whose measures sum to 180° are _____.

2.   Two angles whose measures sum to 90° are _____   _____.

3.   An angle whose measure is exactly 90° is a _____.

4.   An angle whose measure is exactly 180° is a _____.

5.   Two integers that differ by 1 are _____   _____.

Objective 1   Learn the six steps for solving applied problems.

Objective 1 Practice Exercises

For extra help, see page 147 of your text.

1.   Write the six problem-solving steps.   1.   _____

_____

_____

_____

_____

_____

Objective 2   Solve problems involving unknown numbers.

Video Examples

Review this example for Objective 2:

1. The product of 5, and a number decreased by 8, is 150. What is the number?

    Step 1 Read the problem carefully. We are asked to find a number.

    Step 2 Assign a variable to represent the unknown quantity.
        Let x = the number.

    Step 3 Write an equation.

    The product   a   decreased
        of 5,   and number   by   8, is 150.
         ↓   ↓   ↓   ↓ ↓ ↓
         $5 \cdot$ $(x$ $-$ $8) = 150$

    Step 4 Solve the equation.
    $$5(x-8) = 150$$
    $$5x - 40 = 150$$
    $$5x - 40 + 40 = 150 + 40$$
    $$5x = 190$$
    $$\frac{5x}{5} = \frac{190}{5}$$
    $$x = 38$$

    Step 5 State the answer. The number is 38.

    Step 6 Check. The number 38 decreased by 8 is 30. The product of 5 and 30 is 150. The answer, 38, is correct.

Now Try:

1. The product of 8, and a number decreased by 11, is 40. What is the number?

    _____

Objective 2 Practice Exercises

For extra help, see Example 1 on page 147 of your text.

Write an equation for each of the following and then solve the problem. Use x as the variable.

2. If 4 is added to 3 times a number, the result is 7. Find the number.

2. _____

    _____

3. If −2 is multiplied by the difference between 4 and a number, the result is 24. Find the number.

3. _____

_____

4. If four times a number is added to 7, the result is five less than six times the number. Find the number.

4. _____

_____

Objective 3   Solve problems involving sums of quantities.

Video Examples

Review this example for Objective 3:

2. George and Al were opposing candidates in the school board election. George received 21 more votes than Al, with 439 votes cast. How many votes did Al receive?

Step 1 Read the problem carefully. We are given total votes and asked to find the number of votes Al received.

Step 2 Assign a variable.
Let x = the number of votes Al received.
Then $x + 21$ = the number of votes George received.

Step 3 Write an equation.

$$\begin{array}{ccccc} \text{The} & & \text{votes} & & \text{votes for} \\ \text{total} & \text{is} & \text{for Al} & \text{plus} & \text{George} \\ \downarrow & \downarrow & \downarrow & \downarrow & \downarrow \\ 439 & = & x & + & (x+21) \end{array}$$

Now Try:

2. On a psychology test, the highest grade was 38 points more than the lowest grade. The sum of the two grades was 142. Find the lowest grade.

_____

Step 4 Solve the equation.

$$439 = x + (x + 21)$$
$$439 = 2x + 21$$
$$439 - 21 = 2x + 21 - 21$$
$$418 = 2x$$
$$\frac{418}{2} = \frac{2x}{2}$$
$$209 = x \quad \text{or} \quad x = 209$$

Step 5 State the answer. Al received 209 votes.

Step 6 Check. George won $209 + 21 = 230$ votes. The total number of votes is $209 + 230 = 439$. The answer checks.

Objective 3 Practice Exercises

For extra help, see Examples 2–4 on pages 148–150 of your text.

Write an equation for each of the following and then solve the problem. Use x as the variable.

5. Mount McKinley in Alaska is 5910 feet higher than Mount Rainier in Washington. Together, their heights total 34,730 feet. How high is each mountain?

5. _____

   Mt. Rainier _____

   Mt. McKinley_____

6. Charles bought five general admission tickets and four student tickets for a movie. He paid $35.25. If each student ticket cost $3.50, how much did each general admission ticket cost?

6. _____

   _____

7.   Pablo, Faustino, and Mark swim at a public pool   7. _____
     each day for exercise. One day Pablo swam five
     more than three times as many laps as Mark, and   Mark _____
     Faustino swam four times as many laps as Mark. If
     the men swam 29 laps altogether, how many laps did   Pablo_____
     each one swim?
                                                            Faustino _____

Objective 4   Solve problems involving consecutive integers.

Video Examples

Review these examples for Objective 4:

5.   Two pages that face each other in this book have
     337 as the sum of their page numbers. What are
     the page numbers?

     Step 1 Read the problem. Because the two pages
     face each other, they must have page numbers
     that are consecutive integers.

     Step 2 Assign a variable.
        Let x = the lesser page number.
     Then $x + 1$ = the greater page number.

     Step 3 Write an equation. The sum of the page
     numbers is 337.
        $$x + (x + 1) = 337$$

     Step 4 Solve the equation.
        $$2x + 1 = 337$$
        $$2x = 336$$
        $$x = 168$$

     Step 5 State the answer. The lesser page number
     is 168, and the greater is $168 + 1 = 169$.

     Step 6 Check. The sum of 168 and 169 is 337.
     The answer is correct.

Now Try:

5.   Two pages that face each other
     in this book have 705 as the sum
     of their page numbers. What are
     the page numbers?

     _____

6. Find two consecutive odd integers such that if three times the smaller is added to twice the larger, the sum is 69.

Step 1 Read the problem. We must find two consecutive odd integers.

Step 2 Assign a variable.
   Let x = the lesser consecutive odd integer.
Then $x + 2$ = the greater consecutive odd integer.

Step 3 Write an equation.

Three times   is added   twice the
the smaller   to   larger   is   69.
   ↓   ↓   ↓   ↓   ↓
   $3x$ $+$ $2(x+2)$ $=$ 69

Step 4 Solve the equation.
$$3x + 2x + 4 = 69$$
$$5x + 4 = 69$$
$$5x = 65$$
$$x = 13$$

Step 5 State the answer. The lesser integer is 13. The greater is $13 + 2 = 15$.

Step 6 Check. Three times the smaller is 39, added to twice the larger, 30, is a sum of 69. The answers check.

6. The sum of four consecutive even integers is 4. Find the integers.

_____

Objective 4 Practice Exercises

For extra help, see Examples 5–6 on pages 151–152 of your text.

Solve each problem.

8. Find two consecutive even integers such that the smaller, added to twice the larger, is 292.

8. _____

9.   Find two consecutive integers such that the larger,   **9.** _____
added to three times the smaller, is 109.

10.   Find three consecutive odd integers whose sum is   **10.** _____
363.

Objective 5   Solve problems involving complementary and supplementary angles.

Video Examples

Review this example for Objective 5:

7.   Find the measure of an angle such that the difference between the measures of an angle and its complement is 20°.

Step 1 Read the problem. We must find the measure of an angle.

Step 2 Assign a variable.
    Let x = the degree measure of the angle
Then $90 - x$ = the degree measure of its complement.

Step 3 Write an equation.

The angle	minus	Measure of the complement	is	20
↓	↓	↓	↓	↓
x	$-$	$(90 - x)$	$=$	20

Now Try:

7.   Find the measure of an angle whose complement is 4 times its measure.

_____

Step 4 Solve the equation.

$$x - 90 + x = 20$$
$$2x - 90 = 20$$
$$2x = 110$$
$$\frac{2x}{2} = \frac{110}{2}$$
$$x = 55$$

Step 5 State the

answer. The angle is 55°.

Step 6 Check. If the angle measures 55°, then its complement measures 90° − 55° = 35°. The difference between 55° and 35° is 20°. The answer is correct.

Objective 5 Practice Exercises

For extra help, see Examples 7–8 on pages 153–154 of your text.

Solve each problem.

11. Find the measure of an angle if the measure of the angle is 8° less than three times the measure of its supplement.

11. _____

12. Find the measure of an angle whose supplement measures 20° more than twice its complement.

12. _____

13.   Find the measure of an angle whose complement is   13.   _____
      9° more than twice its measure.

Chapter 2 EQUATIONS, INEQUALITIES, AND APPLICATIONS

2.5   Formulas and Additional Applications from Geometry

Learning Objectives

1   Solve a formula for one variable, given the values of the other variables.
2   Use a formula to solve an applied problem.
3   Solve problems involving vertical angles and straight angles.
4   Solve a formula for a specified variable.

Key Terms

Use the vocabulary terms listed below to complete each statement in exercises 1−4.

       formula   **area**   **perimeter**   **vertical angles**

1.   The nonadjacent angles formed by two intersecting lines are called

      _____.

2.   An equation in which variables are used to describe a relationship is called a(n)

      _____.

3.   The distance around a figure is called its _____.

4.   A measure of the surface covered by a figure is called its _____.

Objective 1   Solve a formula for one variable, given the values of the other variables.

Video Examples

Review this example for Objective 1:

1a. Find the value of the remaining variable in the formula.

$$A = LW; \quad A = 54, L = 8$$

Substitute the given values for A and L into the formula.

$$A = LW$$
$$54 = 8W$$
$$\frac{54}{8} = \frac{8W}{8}$$
$$6.75 = W$$

The width is 6.75. Since $8(6.75) = 54$, the answer checks.

Now Try:

1a. Find the value of the remaining variable in the formula.

$$A = LW; \quad A = 88, L = 16$$

      _____

Objective 1 Practice Exercises

For extra help, see Example 1 on page 161 of your text.

In the following exercises, a formula is given, along with the values of all but one of the variables in the formula. Find the value of the variable that is not given.

1. $S = \dfrac{a}{1-r}$; $S = 60$, $r = 0.4$

    1. _____

2. $I = prt$; $I = 288$, $r = 0.04$, $t = 3$

    2. _____

3. $A = \frac{1}{2}(b + B)h$; $b = 6$, $B = 16$, $A = 132$

    3. _____

Objective 2   Use a formula to solve an applied problem.

Video Examples

Review these examples for Objective 2:

2. Find the dimensions of a rectangle. The length is 4 m less than three times the width. The perimeter is 96 m.

    Step 1 Read the problem. We must find the dimensions of the rectangle.

    Step 2 Assign a variable.
    Let W = the width of the rectangle, in meters.
    Then $L = 3W - 4$ is the length, in meters.

    Step 3 Write an equation. Use the formula for the perimeter of a rectangle. Substitute $3W - 4$ for the length.
    $$P = 2L + 2W$$
    $$96 = 2(3W - 4) + 2W$$

Now Try:

2. Ruth has 42 feet of binding for a rectangular rug that she is weaving. If the rug is 9 feet wide, how long can she make the rug if she wishes to use all the binding on the perimeter of the rug?

    _____

Step 4 Solve.
$$96 = 6W - 8 + 2W$$
$$96 = 8W - 8$$
$$96 + 8 = 8W - 8 + 8$$
$$104 = 8W$$
$$\frac{104}{8} = \frac{8W}{8}$$
$$13 = W$$

Step 5 State the answer. The width is 13 m. The length is $3(13) - 4 = 35$ m.

Step 6 Check. The perimeter is $2(13) + 2(35) = 96$ m. The answer checks.

3. The longest side of a triangle is 4 feet longer than the shortest side. The medium side is 2 feet longer than the shortest side. If the perimeter is 36 feet, what are the lengths of the three sides?

Step 1 Read the problem. We must find the lengths of the sides.

Step 2 Assign a variable.
Let s = the length of the shortest side, in feet.
Then $s + 2$ = the length of the medium side, in feet,
and $s + 4$ = the length of the longest side, in feet.

Step 3 Write an equation. Use the formula for the perimeter of a triangle.
$$P = a + b + c$$
$$36 = s + (s + 2) + (s + 4)$$

Step 4 Solve.
$$36 = 3s + 6$$
$$30 = 3s$$
$$10 = s$$

Step 5 State the answer. Since s represents the length of the shortest side, its measure is 10 ft.
$s + 2 = 10 + 2 = 12$ ft is the length of the medium side.
$s + 4 = 10 + 4 = 14$ ft is the length of the longest side.

Step 6 Check. The perimeter is $10 + 12 + 14 = 36$ ft, as required.

3. The longest side of a triangle is twice as long as the shortest side. The medium side is 5 feet longer than the shortest side. If the perimeter is 65 feet, what are the lengths of the three sides?

_____

Name: _____ Date: _____
Instructor: _____ Section: _____

Objective 2 Practice Exercises

For extra help, see Examples 2–4 on pages 162–163 of your text.

Use a formula to write an equation for each of the following applications; then solve the application. (Use 3.14 as an approximation for π.)

4. Find the height of a triangular banner whose area is 48 square inches and base is 12 inches.

    4. _____

5. Linda invests $5000 at 6% simple interest and earns $450. How long did Linda invest her money?

    5. _____

6. The circumference of a circular garden is 628 feet. Find the area of the garden. (Hint: First find the radius of the garden.)

    6. _____

Objective 3   Solve problems involving vertical angles and straight angles.

Video Examples

Review this example for Objective 3:

5b. Find the measure of the marked angles in the figure below.

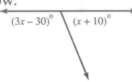

The measures of the marked angles must add to 180° because together they form a straight angle. The angles are supplements of each other.

$$(3x - 30) + (x + 10) = 180$$
$$4x - 20 = 180$$
$$4x = 200$$
$$x = 50$$

Replace x with 50 in the measure of each marked angle.

$$3x - 30 = 3(50) - 30 = 150 - 30 = 120$$
$$x + 10 = 50 + 10 = 60$$

The two angles measure 120° and 60°.

Now Try:

5b. Find the measure of the marked angles in the figure below.

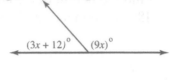

_____

Objective 3 Practice Exercises

For extra help, see Example 5 on page 164 of your text.

Find the measure of each marked angle.

7.

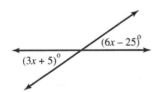

7. _____

8.

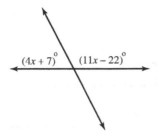

8. _____

9.

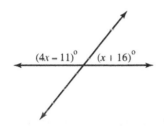

9. _____

Objective 4   Solve a formula for a specified variable.

Video Examples

Review these examples for Objective 4:

6. Solve $A = \frac{1}{2}bh$ for h.

$$A = \frac{1}{2}bh$$

$$2A = bh$$

$$\frac{2A}{b} = \frac{bh}{b}$$

$$\frac{2A}{b} = h \quad \text{or} \quad h = \frac{2A}{b}$$

7. Solve $A = p + prt$ for r.

$$A = p + prt$$

$$A - p = p + prt - p$$

$$A - p = prt$$

$$\frac{A - p}{pt} = \frac{prt}{pt}$$

$$\frac{A - p}{pt} = r \quad \text{or} \quad r = \frac{A - p}{pt}$$

9b. Solve the equation for y.

$$-4x + 5y = 15$$

$$-4x + 5y = 15$$

$$-4x + 5y + 4x = 15 + 4x$$

$$5y = 4x + 15$$

$$\frac{5y}{5} = \frac{4x + 15}{5}$$

$$y = \frac{4x}{5} + \frac{15}{5}$$

$$y = \frac{4}{5}x + 3$$

Now Try:

6. Solve $d = rt$ for t.

_____

7. Solve $P = a + b + c$ for a.

_____

9b. Solve the equation for y.

$$-18x + 3y = 15$$

_____

Objective 4 Practice Exercises

For extra help, see Examples 6–9 on pages 165–166 of your text.

Solve each formula for the specified variable.

10. $V = LWH$ for H **10.** _____

11. $S = (n-2)180$ for n **11.** _____

12. $V = \frac{1}{3}\pi r^2 h$ for h **12.** _____

Chapter 2 EQUATIONS, INEQUALITIES, AND APPLICATIONS

2.6   Ratio, Proportion, and Percent

Learning Objectives
1   Write ratios.
2   Solve proportions.
3   Solve applied problems using proportions.
4   Find percents and percentages.

Key Terms

Use the vocabulary terms listed below to complete each statement in exercises 1−4.

      ratio   proportion   cross products   terms

1. A _____ is a statement that two ratios are equal.

2. A _____ is a comparison of two quantities using a quotient.

3. In the proportion, $\dfrac{a}{b} = \dfrac{c}{d}$, a, b, c, and d are called the _____.

4. To see whether a proportion is true, determine if the _____ are equal.

Objective 1   Write ratios.

Video Examples

Review these examples for Objective 1:

1. Write a ratio for each word phrase.

    a. 7 hr to 9 hr

$$\frac{7 \text{ hr}}{9 \text{ hr}} = \frac{7}{9}$$

    b. 15 hr to 4 days

First convert 4 days to hours.
    4 days $= 4 \cdot 24 = 96$ hr
Now write the ratio using the common unit of measure, hours.

$$\frac{15 \text{ hr}}{4 \text{ days}} = \frac{15 \text{ hr}}{96 \text{ hr}} = \frac{15}{96}, \quad \text{or} \quad \frac{5}{32}$$

Now Try:

1. Write a ratio for each word phrase.

    a. 11 hr to 17 hr

    _____

    b. 32 hr to 5 days

    _____

2. The local grocery store charges the following prices for a bottle of olive oil.

  16-ounce bottle: $6.99
  25.5-ounce bottle: $9.99
  32-ounce bottle: $12.99
  44-ounce bottle: $14.99

Which size is the best buy? That is, which size has the lowest unit price?

To find the best buy, write ratios comparing the price for each size bottle to the number of units (ounces) per bottle.

Size	Unit price (dollars per ounce)
16 oz	$\dfrac{\$6.99}{16} = \$0.437$
25.5 oz	$\dfrac{\$9.99}{25.5} = \$0.392$
32 oz	$\dfrac{\$12.99}{32} = \$0.406$
44 oz	$\dfrac{\$14.99}{44} = \$0.341$

Because the 44-oz size has the lowest unit price, $0.341, it is the best buy.

2. The local grocery store charges the following prices for a jar of applesauce.

  16-ounce jar: $1.19
  24-ounce jar: $1.29
  48-ounce jar: $2.69
  64-ounce jar: $3.49

Which size is the best buy? That is, which size has the lowest unit price?

_____

Objective 1 Practice Exercises

For extra help, see Examples 1–2 on pages 174–175 of your text.

Write a ratio for each word phrase. Write fractions in lowest terms.

1. 8 men to 3 men

1. _____

2. 9 dollars to 48 quarters

2. _____

A supermarket was surveyed and the following prices were charged for items in various sizes. Find the best buy (based on price per unit) for each of the following items.

3.   Trash bags   3. _____
     10-count box: $2.89
     20-count box: $5.29
     45-count box: $6.69
     85-count box: $13.99

Objective 2   Solve proportions.

Video Examples

Review this example for Objective 2:

4.   Solve the equation $\dfrac{n-1}{3} = \dfrac{2n+1}{4}$.

$$\frac{n-1}{3} = \frac{2n+1}{4}$$

$$3(2n+1) = 4(n-1)$$

$$6n+3 = 4n-4$$

$$2n+3 = -4$$

$$2n = -7$$

$$n = -\frac{7}{2}$$

A check confirms that the solution is $-\dfrac{7}{2}$, so the solution set is $\left\{-\dfrac{7}{2}\right\}$.

Now Try:

4.   Solve the equation $\dfrac{2x+1}{2} = \dfrac{7x+3}{9}$.

_____

Objective 2 Practice Exercises

For extra help, see Examples 3–4 on page 176 of your text.

Solve each equation.

4. $\dfrac{z}{20} = \dfrac{25}{125}$

4. _____

5. $\dfrac{m}{5} - \dfrac{m-2}{2}$

5. _____

6. $\dfrac{z+1}{4} = \dfrac{z+7}{2}$

6. _____

Objective 3   Solve applied problems using proportions.

Video Examples

Review this example for Objective 3:

5. If four pounds of fertilizer will cover 50 square feet of garden, how many pounds would be needed for 125 square feet?

To solve this problem, set up a proportion, with pounds in the numerator and square feet in the denominator.

Now Try:

5. Margie earns $168.48 in 26 hours. How much does she earn in 40 hours?

    _____

$$\frac{4}{50} = \frac{x}{125}$$

$$4(125) = 50x$$

$$500 = 50x$$

$$10 = x$$

10 lb of fertilizer are needed.

Objective 3 Practice Exercises

For extra help, see Example 5 on page 177 of your text.

Solve each problem.

7. On a road map, 6 inches represents 50 miles. How many inches would represent 125 miles?

7. _____

8. If 12 rolls of tape cost $4.60, how much will 15 rolls cost?

8. _____

9. A garden service charges $30 to install 50 square feet of sod. Find the charge to install 225 square feet.

9. _____

Objective 4   Find percents and percentages.

Video Examples

Review these examples for Objective 4:	Now Try:
6. Solve each problem.	6. Solve each problem.

a. What is 18% of 700?

Let n = the number. The word of indicates multiplication.

What is 18% of 700
$\downarrow$ $\downarrow$ $\downarrow$ $\downarrow$ $\downarrow$
n = 0.18 · 700
$n = 126$

Thus, 126 is 18% of 700.

a. What is 35% of 400?

_____

b. 54% of what number is 162?

54% of what number is 162
$\downarrow$ $\downarrow$ $\downarrow$ $\downarrow$ $\downarrow$
0.54 · n = 162

$$n = \frac{162}{0.54}$$
$$n = 300$$

54% of 300 is 162.

b. 42% of what number is 399?

_____

Objective 4 Practice Exercises

For extra help, see Examples 6–7 on page 178 of your text.

Answer each question about percent.

10.   What is 2.5% of 3500?   10. _____

11.   What percent of 5200 is 104?   11. _____

Solve the problem.

12.   Paul recently bought a duplex for $144,000. He   12. _____
       expects to earn $6120 per year on this investment.
       What percent of the purchase price will he earn?

Chapter 2 EQUATIONS, INEQUALITIES, AND APPLICATIONS

2.7   Solving Linear Inequalities

Learning Objectives	
1	Graph intervals on a number line.
2	Use the addition property of inequality.
3	Use the multiplication property of inequality.
4	Solve linear inequalities.
5	Solve applied problems using inequalities.
6	Solve linear inequalities with three parts.

Key Terms

Use the vocabulary terms listed below to complete each statement in exercises 1–5.

inequalities   **interval**   **interval notation**

linear inequality   **three-part inequality**

1. An inequality that says that one number is between two other numbers is a(n)_____.

2. A portion of a number line is called a(n) _____.

3. A(n) _____ can be written in the form $Ax + B < C$, $Ax + B \leq C$, $Ax + B > C$, or $Ax + B \geq C$, where A, B, and C are real numbers with $A \neq 0$.

4. Algebraic expressions related by $<, \leq, >,$ or $\geq$ are called _____.

5. The _____ for $a \leq x < b$ is $[a, b)$.

Objective 1   Graph intervals on a number line.

Video Examples

Review this example for Objective 1:

1. Graph $x > -3$.

   The statement $x > -3$ says that x can represent any value greater than –3, but cannot equal –3, written $(-3, \infty)$. We graph this interval by placing a parenthesis at –3 and drawing an arrow to the right. The parenthesis indicates that –3 is not part of the graph.

Now Try:

1. Graph $x > -1$.

   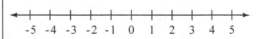

Name: Date:

Instructor: Section:

Objective 1 Practice Exercises

For extra help, see Examples 1–2 on page 187 of your text.

Write each inequality in interval notation and graph the interval.

1. $3 < a$

1. _____

2. $y \geq -2$

2. _____

3. $x < -4$

3. _____
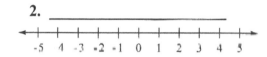

Objective 2   Use the addition property of inequality.

Video Examples

Review this example for Objective 2:

3. Solve $8 + 4k \geq 3k + 3$ and graph the solution set.

$$8 + 4k \geq 3k + 3$$
$$8 + 4k - 3k \geq 3k + 3 \quad 3k$$
$$8 + k \geq 3$$
$$8 + k - 8 \geq 3 \quad 8$$
$$k \geq -5$$

The solution set, $[-5, \infty)$ is graphed below.

Now Try:

3. Solve $5 + 9k \geq 8k + 2$ and graph the solution set.

Objective 2 Practice Exercises

For extra help, see Example 3 on page 188 of your text.

Solve each inequality. Write the solution set in interval notation and then graph it.

4. $5a + 3 \leq 6a$

4. _____

5. $6 + 3x < 4x + 4$

5. _____

6. $3 + 5p \leq 4p + 3$

6. _____

Objective 3   Use the multiplication property of inequality.

Video Examples

Review these examples for Objective 3:

4.   Solve each inequality, and graph the solution set.

    a. $6x < -24$

We divide each side by 6.
$$6x < -24$$
$$\frac{6x}{6} < \frac{-24}{6}$$
$$x < -4$$

The graph of the solution set $(-\infty, -4)$, is shown below.

    b. $-6x \geq 30$

Here each side of the inequality must be divided by –6, a negative number, which does require changing the direction of the inequality symbol.
$$-6x \geq 30$$
$$\frac{-6x}{-6} \leq \frac{30}{-6}$$
$$x \leq -5$$

The solution set, $(-\infty, -5]$, is graphed below.

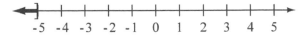

Now Try:

4.   Solve each inequality, and graph the solution set.

    a. $8x \leq -40$

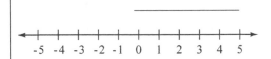

    b. $-9t > 36$

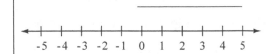

Name: Date:

Instructor: Section:

Objective 3 Practice Exercises

For extra help, see Example 4 on page 190 of your text.

Solve each inequality. Write the solution set in interval notation and then graph it.

7. $-2s < 4$

7. _____

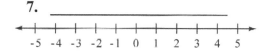

8. $4k \geq -16$

8. _____

9. $-9m \geq -36$

9. _____

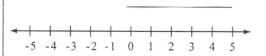

Objective 4   Solve linear inequalities.

Video Examples

Review this example for Objective 4:

5.   Solve $4x + 3 - 7 > -2x + 8 + 3x$. Graph the solution set.

     Step 1 Combine like terms and simplify.

$$4x + 3 - 7 > -2x + 8 + 3x$$

$$4x - 4 > x + 8$$

     Step 2 Use the addition property of inequality.

$$4x - 4 - x > x + 8 - x$$

$$3x - 4 > 8$$

$$3x - 4 + 4 > 8 + 4$$

$$3x > 12$$

     Step 3 Use the multiplication property of inequality.

$$\frac{3x}{3} > \frac{12}{3}$$

$$x > 4$$

     The solution set is $(4, \infty)$. The graph is shown below.

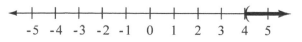

Now Try:

5.   Solve $8x - 5 + 4 \geq 6x - 3x + 9$. Graph the solution set.

Objective 4 Practice Exercises

For extra help, see Example 5–7 on pages 191–192 of your text.

Solve each inequality. Write the solution set in interval notation and then graph it.

10. $4(y-3)+2>3(y-2)$

10. _____

11. $-3(m+2)+3\le-4(m-2)-6$

11. _____

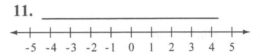

12. $7(2-x)\le-2(x-3)-x$

12. _____

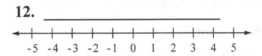

Objective 5   Solve applied problems using inequalities.

Video Examples

Review this example for Objective 5:

8. Ruth tutors mathematics in the evenings in an office for which she pays $600 per month rent. If rent is her only expense and she charges each student $40 per month, how many students must she teach to make a profit of at least $1600 per month?

Step 1 Read the problem again.

Step 2 Assign a variable.
    Let x = the number of students.

Step 3 Write an inequality.
    $40x-600\ge1600$

Now Try:

8. Two sides of a triangle are equal in length, with the third side 8 feet longer than one of the equal sides. The perimeter of the triangle cannot be more than 38 feet. Find the largest possible value for the length of the equal sides.

    _____

Step 4 Solve.

$$40x - 600 + 600 \geq 1600 + 600$$

$$40x \geq 2200$$

$$\frac{40x}{40} \geq \frac{2200}{40}$$

$$x \geq 55$$

Step 5 State the answer. Ruth must have 55 or more students to have at least $1600 profit.

Step 6 Check. $40(55) - 600 = 1600$ Also, any number greater than 55 makes the profit greater than $1600.

Objective 5 Practice Exercises

For extra help, see Example 8 on page 193 of your text.

Solve each problem.

13. Lauren has grades of 98 and 86 on her first two chemistry quizzes. What must she score on her third quiz to have an average of at least 91 on the three quizzes?

13. _____

14. Nina has a budget of $230 for gifts for this year. So far she has bought gifts costing $47.52, $38.98, and $26.98. If she has three more gifts to buy, find the average amount she can spend on each gift and still stay within her budget.

14. _____

Name: Date:

Instructor: Section:

15. If twice the sum of a number and 7 is subtracted from three times the number, the result is more than −9. Find all such numbers.

15. _____

Objective 6 Solve linear inequalities with three parts.

Video Examples

Review these examples for Objective 6:

9. Write the inequality $-4 \le x < 3$ in interval notation, and graph the interval.

$-4 \le x < 3$

x is between −4 and 3 (excluding 3).

In interval notation, we write $[-4, 3)$.

10a. Solve the inequality, and graph the solution set.

$3 \le 4x - 5 < 7$

$3 \le 4x - 5 < 7$

$3 + 5 \le 4x - 5 + 5 < 7 + 5$

$8 \le 4x < 12$

$\dfrac{8}{4} \le \dfrac{4x}{4} < \dfrac{12}{4}$

$2 \le x < 3$

The solution set is $[2, 3)$. The graph is shown below.

Now Try:

9. Write the inequality $-5 < x \le -1$ in interval notation, and graph the interval.

_____

10a. Solve the inequality, and graph the solution set.

$8 \le 6x - 4 < 20$

_____

Name: _____ Date: _____

Instructor: _____ Section: _____

Objective 6 Practice Exercises

For extra help, see Examples 9–10 on pages 193–194 of your text.

Solve each inequality. Write the solution set in interval notation and then graph it.

16. $7 < 2x + 3 \le 13$

16. _____

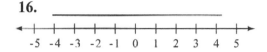

17. $-17 \le 3x - 2 < -11$

17. _____

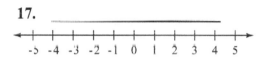

18. $1 < 3z + 4 < 19$

18. _____

Chapter 3 GRAPHS OF LINEAR EQUATIONS AND INEQUALITIES IN TWO VARIABLES

3.1   Linear Equations and Rectangular Coordinates

Learning Objectives
1   Interpret line graphs.
2   Write a solution as an ordered pair.
3   Decide whether a given ordered pair is a solution of a given equation.
4   Complete ordered pairs for a given equation.
5   Complete a table of values.
6   Plot ordered pairs.

Key Terms

Use the vocabulary terms listed below to complete each statement in exercises 1−13.

line graph   linear equation in two variables

ordered pair   table of values   *x*-axis

y-axis   rectangular (Cartesian) coordinate system

origin   quadrants   plane   coordinates

plot   scatter diagram

1.   A _____ _____ uses dots connected by lines to show trends.

2.   An equation that can be written in the form $Ax + By = C$, where A, B, and C are real numbers and A, $B \neq 0$, is called a _____.

3.   _____ are the numbers in the ordered pair that specify the location of a point on a rectangular coordinate system.

4.   In a coordinate system, the horizontal axis is called the _____.

5.   In a coordinate system, the vertical axis is called the _____.

6.   A pair of numbers written between parentheses in which order is important is called a(n) _____.

7.   Together, the *x*-axis and the *y*-axis form a _____.

8.   A coordinate system divides the plane into four regions called _____.

9.   The axis lines in a coordinate system intersect at the _____.

10.   To _____ an ordered pair is to find the corresponding point on a coordinate system.

11.   A graph of ordered pairs is called a _____.

12.   A table showing selected ordered pairs of numbers that satisfy an equation is called a
_____.

13.   A flat surface determined by two intersecting lines is a _____.

Objective 1   Interpret line graphs.

Video Examples

*The line graph shows the number of degrees awarded by a university for the years
2010–2015.*

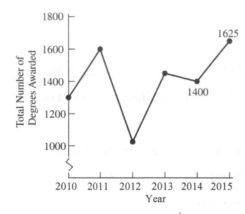

Review these examples for Objective 1:

1.

a. Between which years did the number of
degrees awarded decrease?

The line between 2011 and 2012 and between
2013 and 2014 falls, so the number of degrees
awarded decreased between 2011-2012 and
2013-2014.

b. Between which two years did the total
number of degrees awarded show the smallest
change?

The line between 2013 and 2014 falls the least,
so the total number of degrees awarded declined
the least from 2013 to 2014.

c. Estimate the total number of degrees awarded
in 2012 and in 2015. About how many more
degrees were awarded in 2015?

Move up from 2012 on the horizontal scale to
the point plotted for 2012. This point is about
1000. So about 1000 degrees were awarded in
2012.

Now Try:

1.

a. Between which years did the
number of degrees awarded
increase?

_____

b. Between which two years did
the total number of degrees
awarded show the greatest
decline?

_____

c. Estimate the total number of
degrees awarded in 2010 and in
2011. About how many more
degrees were awarded in 2011?

_____

Similarly, locate the point plotted for 2015. Moving across to the vertical scale, the graph indicates that the number of degrees awarded in 2015 was 1625.

Between 2012 and 2015, the increase was

$$1625 - 1000 = 625.$$

Objective 1 Practice Exercises

For extra help, see Example 1 on page 214 of your text.

The line graph shows the number of degrees awarded by a university for the years 2010 2015. Use this graph to answer exercises 1–3.

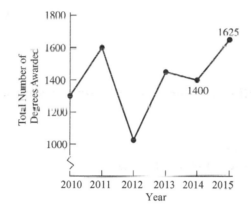

1. Between what years did the number of degrees awarded have the least change?

1. _____

2. Over which years was the number of degrees awarded more than 1400?

2. _____

3. Estimate the average number of degrees awarded in 2011 and 2012. About how much did the number of degrees awarded decrease between 2011 and 2012?

3. _____

Objective 2   Write a solution as an ordered pair.

Objective 2 Practice Exercises

For extra help, see page 215 of your text.

Write each solution as an ordered pair.

4. $x = 4$ and $y = 7$ 4. _____

5. $y = \frac{1}{3}$ and $x = 0$ 5. _____

6. $x = 0.2$ and $y = 0.3$ 6. _____

Objective 3   Decide whether a given ordered pair is a solution of a given equation.

Video Examples

Review these examples for Objective 3:

2.   Decide whether each ordered pair is a solution of the equation $4x + 5y = 40$.

a. $(5, \ 4)$

Substitute 5 for x and 4 for y in the given equation.
$$4x + 5y = 40$$
$$4(5) + 5(4) \overset{?}{=} 40$$
$$20 + 20 \overset{?}{=} 40$$
$$40 = 40 \quad \text{True}$$
This result is true, so $(5, \ 4)$ is a solution of $4x + 5y = 40$.

b. $(-3, \ 6)$

Substitute -3 for x and 6 for y in the given equation.
$$4x + 5y = 40$$
$$4(-3) + 5(6) \overset{?}{=} 40$$
$$-12 + 30 \overset{?}{=} 40$$
$$18 = 40 \quad \text{False}$$
This result is false, so $(-3, \ 6)$ is not a solution of $4x + 5y = 40$.

Now Try:

2.   Decide whether each ordered pair is a solution of the equation $3x - 4y = 12$.

a. $(8, \ 3)$

_____

b. $(5, -4)$

_____

Objective 3 Practice Exercises

For extra help, see Example 2 on page 216 of your text.

Decide whether the given ordered pair is a solution of the given equation.

7. $4x - 3y = 10$; $(1, 2)$ 7. _____

8. $2x - 3y = 1$; $\left(0, \frac{1}{3}\right)$ 8. _____

9. $x = -7$; $(-7, 9)$ 9. _____

Objective 4   Complete ordered pairs for a given equation.

Video Examples

Review this example for Objective 4:

3a. Complete each ordered pair for the equation $y = 5x + 8$.

$(3, \underline{\hspace{1cm}})$

Replace x with 3.
$$y = 5x + 8$$
$$y = 5(3) + 8$$
$$y = 15 + 8$$
$$y = 23$$
The ordered pair is (3, 23).

Now Try:

3a. Complete each ordered pair for the equation $y = 4x - 7$.

$(5, \underline{\hspace{1cm}})$

_____

Objective 4 Practice Exercises

For extra help, see Example 3 on page 216 of your text.

For each of the given equations, complete the ordered pairs beneath it.

10. $y = 2x - 5$

    (a)   (2,)

    (b)   (0,)

    (c)   (,3)

    (d)   (,−7)

    (e)   (,9)

10.

(a) _____

(b) _____

(c) _____

(d) _____

(e) _____

11. $y = 3 + 2x$

    (a)   (−4,)

    (b)   (2,)

    (c)   (,0)

    (d)   (−2,)

    (e)   (,−7)

11.

(a) _____

(b) _____

(c) _____

(d) _____

(e) _____

Objective 5   Complete a table of values.

Video Examples

Review this example for Objective 5:

4a. Complete the table of values for the equation. Then write the results as ordered pairs.

$2x - 3y = 6$

x	y
9	
6	
	2
	8

From the table, we can write the ordered pairs:
$(9, \underline{\quad}), (6, \underline{\quad}), (\underline{\quad}, -2), (\underline{\quad}, 8)$.
From the first row of the table, let $x = 9$ in the equation. From the second row of the table, let $x = 6$.

If $x = 9$,
$$2x - 3y = 6$$
$$2(9) - 3y = 6$$
$$18 - 3y = 6$$
$$-3y = -12$$
$$y = 4$$

If $x = 6$,
$$2x - 3y = 6$$
$$2(6) - 3y = 6$$
$$12 - 3y = 6$$
$$-3y = -6$$
$$y = 2$$

The first two ordered pairs are $(9, 4)$ and $(6, 2)$.

From the third and fourth rows of the table, let $y = -2$ and $y = 8$, respectively.

If $y = -2$,
$$2x - 3y = 6$$
$$2x - 3(-2) = 6$$
$$2x + 6 = 6$$
$$2x = 0$$
$$x = 0$$

If $y = 8$,
$$2x - 3y = 6$$
$$2x - 3(8) = 6$$
$$2x - 24 = 6$$
$$2x = 30$$
$$x = 15$$

The last two ordered pairs are $(0, -2)$ and $(15, 8)$. The completed table and corresponding ordered pairs follow.

x	y	Ordered pairs
9	4	→ (9, 4)
6	2	→ (6, 2)
0	−2	→ (0, −2)
15	8	→ (15, 8)

Now Try:

4a. Complete the table of values for the equation. Then write the results as ordered pairs.

$4x - y = 8$

x	y
1	
5	
	0
	4

_____

Objective 5 Practice Exercises

For extra help, see Example 4 on pages 217–218 of your text.

Complete each table of values. Write the results as ordered pairs.

12. $2x + 5 = 7$ **12.** _____

x	y
	-3
	0
	5

13. $y - 4 = 0$ **13.** _____

x	y
-4	
0	
6	

14. $4x + 3y = 12$ **14.** _____

x	y
0	
	0
	-1

Name: Date:
Instructor: Section:

Objective 6   Plot ordered pairs.

Video Examples

Review these examples for Objective 6:

5.   Plot the given points in a coordinate system.
   (5, 4)   (−2,−1)   (−3, 5)   (4,−2)
   (1,−2.5)   (3, 0)   (0, 4)

Step 1 Move right or left the number of units that correspond to the x-coordinate in the ordered pair—right if the x-coordinate is positive and left if it is negative.

Step 2 Then turn and move up or down the number of units that corresponds to the y-coordinate—up if the y-coordinate is positive or down if it is negative.

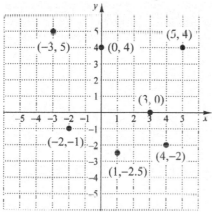

Now Try:

5.   Plot the given points in a coordinate system.
   (2, 4)   (−5, 1)   (−4,−2)
   (3,−5)   (2,−1.5)   (6, 0)
   (0,−6)

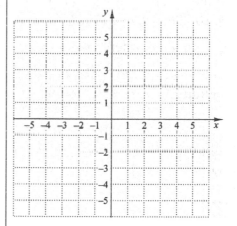

An accountant keeps track of the balance for a business loan given by the linear equation
$$y = 137.24x + 9879,$$
where x is the number of months, and y is the balance due. For this loan, payment is expected in full at the end of the term.

6.

   a. Complete the table of values for the given linear equation.

x (months)	y (balance due)
1	
3	
10	

   To find y for x = 1, substitute into the equation.

6.

   a. Complete the table of values for the given linear equation.

x (months)	y (balance due)
2	
5	
11	

   _____

$$y = 137.24x + 9879$$

$$y = 137.24(1) + 9879$$

$$y = 10,016.24$$

x (months)	y (balance due)
1	10,016.24
3	10,290.72
10	11,251.40

So the ordered pairs are (1, 10,016.24),
(3, 10,290.72), and (10, 11,251.40).

b. Graph the ordered pairs found in part (a).

A graph of ordered pairs of data is a scatter diagram.

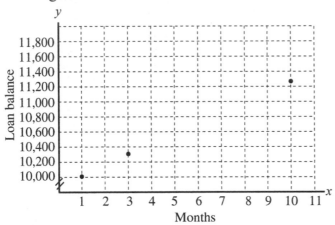

b. Graph the ordered pairs found in part (a).

Objective 6 Practice Exercises

For extra help, see Examples 5–6 on pages 219–220 of your text.

Plot the each ordered pair on a coordinate system.

15. $(0, -2)$

15. _____

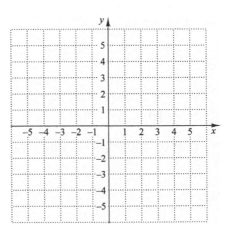

16. $(-3,\ 4)$

16. _____

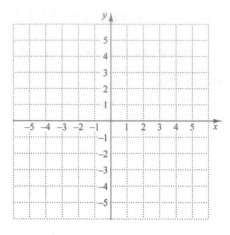

17. $(2,-5)$

17. _____

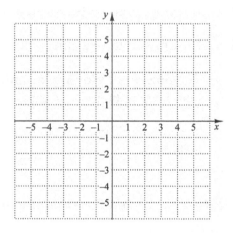

Chapter 3 GRAPHS OF LINEAR EQUATIONS AND INEQUALITIES IN TWO VARIABLES

3.2   Graphing Linear Equations in Two Variables

Learning Objectives
1   Graph linear equations by plotting ordered pairs.
2   Find intercepts.
3   Graph linear equations of the form $Ax + By = 0$.
4   Graph linear equations of the form $y = b$ or $x = a$.
5   Use a linear equation to model data.

Key Terms

Use the vocabulary terms listed below to complete each statement in exercises 1–4.

graph   graphing   *y*-intercept   *x*-intercept

1.   If a graph intersects the *y*-axis at *k*, then the _____ is $(0, k)$

2.   If a graph intersects the *x*-axis at *k*, then the _____ is $(k, 0)$.

3.   The process of plotting the ordered pairs that satisfy a linear equation and drawing a line through them is called _____.

4.   The set of all points that correspond to the ordered pairs that satisfy the equation is called the _____ of the equation.

Objective 1   Graph linear equations by plotting ordered pairs.

Video Examples

Review this example for Objective 1:
2.   Graph $2x + 3y = 6$.

First let $x = 0$ and then let $y = 0$ to determine two ordered pairs.

$$
\begin{array}{c|c}
2(0)+3y=6 & 2x+3(0)=6 \\
0+3y=6 & 2x+0=6 \\
3y=6 & 2x=6 \\
y=2 & x=3
\end{array}
$$

The ordered pairs are $(0, 2)$ and $(3, 0)$. Find a third ordered pair by choosing a number other than 0 for *x* or *y*. We choose $y = 4$.

Now Try:
2.   Graph $x + y = 3$.

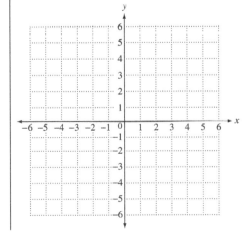

$$2x + 3(4) = 6$$
$$2x + 12 = 6$$
$$2x = -6$$
$$x = -3$$

This gives the ordered pair (−3, 4). We plot the three ordered pairs (0, 2), (3, 0), and (−3, 4) and draw a line through them.

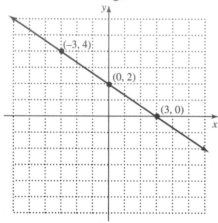

Objective 1 Practice Exercises

For extra help, see Examples 1–2 on pages 229–230 of your text.

Complete the ordered pairs for each equation. Then graph the equation by plotting the points and drawing a line through them.

1. $y = 3x - 2$

(0,)

(,0)

(2,)

1.

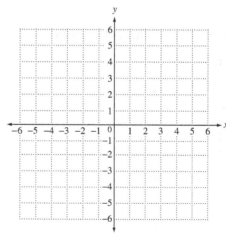

2. $x - y = 4$

$(0, \quad)$

$(\quad , 0)$

$(-2, \quad)$

2.

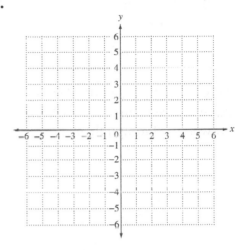

3. $x = 2y + 1$

$(0, \quad)$

$(\quad , 0)$

$(\quad , -2)$

3.

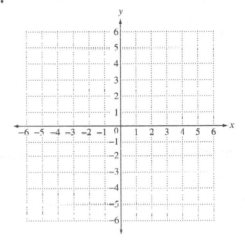

Objective 2   Find intercepts.

Video Examples

Review this example for Objective 2:

3.   Graph $3x + y = 6$ using intercepts.

To find the y-intercept, let $x = 0$.
To find the x-intercept, let $y = 0$.

$$3(0) + y = 6 \quad \bigm| \quad 3x + 0 = 6$$
$$0 + y = 6 \quad \bigm| \quad 3x = 6$$
$$y = 6 \quad \bigm| \quad x = 2$$

The intercepts are $(0, 6)$ and $(2, 0)$. To find a third point, as a check, we let $x = 1$.

$$3(1) + y = 6$$
$$3 + y = 6$$
$$y = 3$$

Now Try:

3.   Graph $5x - 2y = -10$ using intercepts.

Name: Date:
Instructor: Section:

This gives the ordered pair (1, 3).

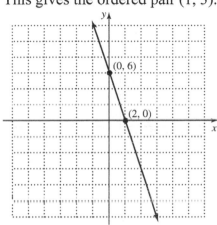

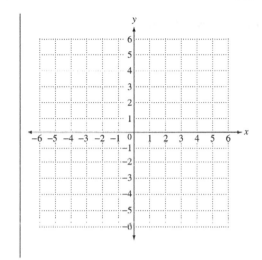

Objective 2 Practice Exercises

For extra help, see Examples 3–4 on pages 230–231 of your text.

Find the intercepts for each equation. Then graph the equation.

4. $y = \dfrac{2}{3}x - 2$ 4.

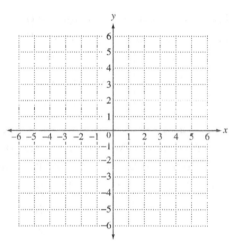

5. $4x - 7y = 8$ 5.

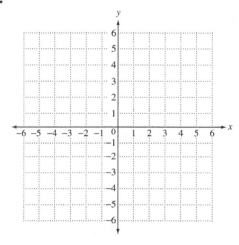

6. $y = 4x - 4$

6.

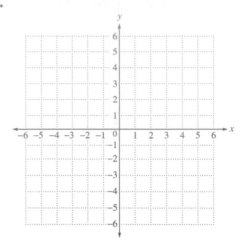

Objective 3   Graph linear equations of the form $Ax + By = 0$.

Video Examples

Review this example for Objective 3:

5.   Graph $x + 5y = 0$.

To find the y-intercept, let $x = 0$.
To find the x-intercept, let $y = 0$.

$$
\begin{array}{c|c}
0 + 5y = 0 & x + 5(0) = 0 \\
5y = 0 & x + 0 = 0 \\
y = 0 & x = 0
\end{array}
$$

The x- and y-intercepts are the same point $(0, 0)$. We must select two other values for x or y to find two other points. We choose $y = 1$ and $y = -1$.

$$
\begin{array}{c|c}
x + 5(1) = 0 & x + 5(-1) = 0 \\
x + 5 = 0 & x - 5 = 0 \\
x = -5 & x = 5
\end{array}
$$

We use $(-5, 1)$, $(0, 0)$, and $(5, -1)$ to draw the graph.

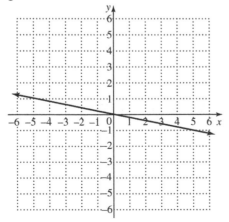

Now Try:

5.   Graph $3x - y = 0$.

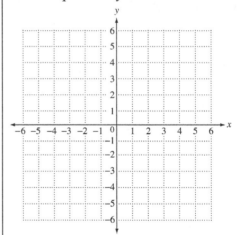

Copyright © 2018 Pearson Education, Inc.

Name: Date:
Instructor: Section:

Objective 3 Practice Exercises

For extra help, see Example 5 on page 232 of your text.

Graph each equation.

7. $-3x - 2y = 0$ 7.

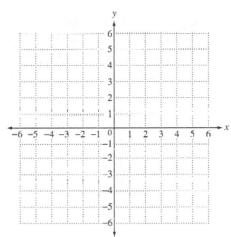

8. $x + y = 0$ 8.

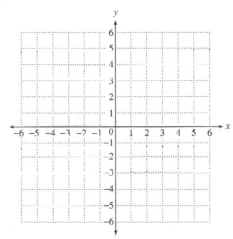

9. $y = 2x$ 9.

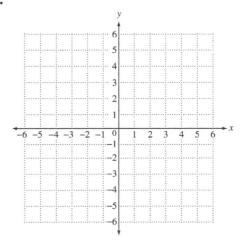

Objective 4 Graph linear equations of the form $y = b$ or $x = a$.

Video Examples

Review these examples for Objective 4:

6. Graph $y = -2$.

For any value of x, y is always -2. Three ordered pairs that satisfy the equation are $(-4, -2)$, $(0, -2)$ and $(2, -2)$. Drawing a line through these points gives the horizontal line. The y-intercept is $(0, -2)$. There is no x-intercept.

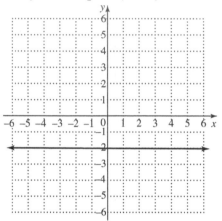

Now Try:

6. Graph $y = 4$.

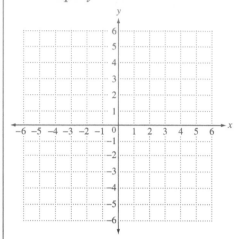

7. Graph $x + 4 = 0$.

First we subtract 4 from each side of the equation to get the equivalent equation $x = -4$. All ordered-pair solutions of this equation have x-coordinate -4.

Three ordered pairs that satisfy the equation are $(-4, -1)$, $(-4, 0)$, and $(-4, 3)$. The graph is a vertical line. The x-intercept is $(-4, 0)$. There is no y-intercept.

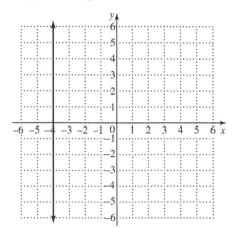

7. Graph $x = 0$.

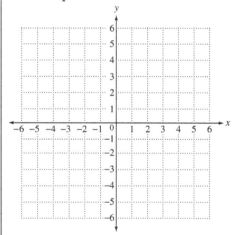

Name: Date:

Instructor: Section:

Objective 4 Practice Exercises

For extra help, see Examples 6–7 on page 233 of your text.

Graph each equation.

10. $x - 1 = 0$ **10.**

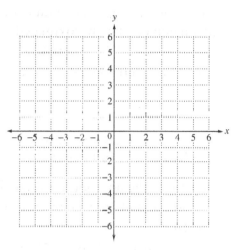

11. $y + 3 = 0$ **11.**

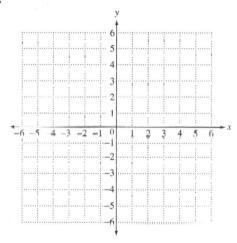

Name: _____   Date: _____

Instructor: _____   Section: _____

Objective 5   Use a linear equation to model data.

Video Examples

Review these examples for Objective 5:

8. Every year sea turtles return to a certain group of islands to lay eggs. The number of turtle eggs that hatch can be approximated by the equation $y = -70x + 3260$, where y is the number of eggs that hatch and $x = 0$ representing 1990.

a. Use this equation to find the number of eggs that hatched in 1995, 2000, and 2005, and 2015.

Substitute the appropriate value for each year x to find the number of eggs hatched in that year.

For 1995:

$y = -70(5) + 3260$ $1995 - 1990 = 5$

$y = 2910$ eggs   Replace x with 5.

For 2000:

$y = -70(10) + 3260$ $2000 - 1990 = 10$

$y = 2560$ eggs   Replace x with 10.

For 2005:

$y = -70(15) + 3260$ $2005 - 1990 = 15$

$y = 2210$ eggs   Replace x with 15.

For 2015:

$y = -70(25) + 3260$ $2015 - 1990 = 25$

$y = 1510$ eggs   Replace x with 25.

b. Write the information from part (a) as four ordered pairs, and use them to graph the given linear equation.

Since x represents the year and y represents the number of eggs, the ordered pairs are (5, 2910), (10, 2560), (15, 2210), and (25, 1510).

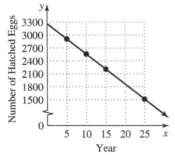

Now Try:

8. Suppose that the demand and price for a certain model of calculator are related by the equation $y = 45 - \frac{3}{5}x$, where y is the price (in dollars) and x is the demand (in thousands of calculators).

a. Assuming that this model is valid for a demand up to 50,000 calculators, use this equation to find the price of calculators at each level of demand.

0 calculators _____

5000 calculators _____

20,000 calculators _____

45,000 calculators _____

b. Write the information from part (a) as four ordered pairs, and use them to graph the given linear equation.

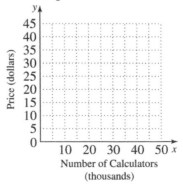

c. Use this graph and the equation to estimate the number of eggs that will hatch in 2010.

For 2010, $x = 20$. On the graph, find 20 on the horizontal axis, move up to the graphed line and then across to the vertical axis. It appears that in 2010, there were about 1900 eggs.

To use the equation, substitute 20 for x.

$$y = -70(20) + 3260$$

$$y = 1860 \text{ eggs}$$

This result for 2020 is close to our estimate of 1900 eggs from the graph.

c. Use this graph and the equation to estimate the price of 30,000 calculators.

_____

Objective 5 Practice Exercises

For extra help, see Example 8 on pages 234–235 of your text.

Solve each problem. Then graph the equation

12. The profit y in millions of dollars earned by a small computer company can be approximated by the linear equation $y = 0.63x + 4.9$, where $x = 0$ corresponds to 2014, $x = 1$ corresponds to 2015, and so on. Use this equation to approximate the profit in each year from 2014 through 2017.

12. 2014 _____

2015 _____

2016 _____

2017 _____

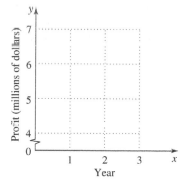

13. The number of band instruments sold by Elmer's Music Shop can be approximated by the equation $y = 325 + 42x$, where y is the number of instruments sold and x is the time in years, with $x = 0$ representing 2013. Use this equation to approximate the number of instruments sold in each year from 2013 through 2016.

13. 2013 _____

2014 _____

2015 _____

2016 _____

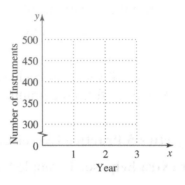

14. According to *The Old Farmer's Almanac*, the temperature in degrees Celsius can be determined by the equation $y = \frac{1}{3}x + 4$, where x is the number of cricket chirps in 25 seconds and y is the temperature in degrees Celsius. Use this equation to find the temperature when there are 48 chirps, 54 chirps, 60 chirps, and 66 chirps.

14. 48 _____

54 _____

60 _____

66 _____

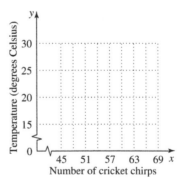

Chapter 3 GRAPHS OF LINEAR EQUATIONS AND INEQUALITIES IN TWO VARIABLES

3.3   The Slope of a Line

Learning Objectives
1   Find the slope of a line given two points.
2   Find the slope from the equation of a line.
3   Use slope to determine whether two lines are parallel, perpendicular, or neither.
4   Solve problems involving average rate of change.

Key Terms

Use the vocabulary terms listed below to complete each statement in exercises 1−5.

rise   run   slope   parallel lines

perpendicular lines

1.   Two lines that intersect in a 90° angle are called _____.

2.   The _____ of a line is the ratio of the change in *y* compared to the change in *x* when moving along the line from one point to another.

3.   The vertical change between two different points on a line is called the
_____.

4.   Two lines in a plane that never intersect are called _____.

5.   The horizontal change between two different points on a line is called the
_____.

Objective 1   Find the slope of a line given two points.

Video Examples

Review these examples for Objective 1:	Now Try:
1. Find the slope of the line.	1. Find the slope of the line.

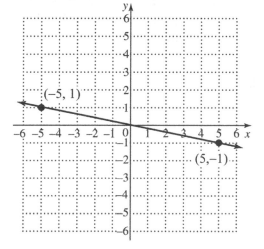

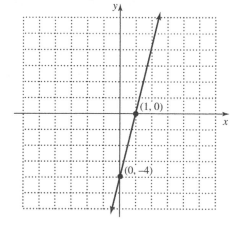

_____

Name: Date:
Instructor: Section:

We use the two points shown on the line. The vertical change is the difference in the y-values, or $-1 - 1 = -2$, and the horizontal change is the difference in the x-values or $5 - (-5) = 10$. Thus, the line has

$$\text{slope} = \frac{-2}{10}, \text{ or } -\frac{1}{5}.$$

2. Find the slope of each line.

a. The line passing through $(-5, 4)$ and $(2, -6)$

Apply the slope formula.

$(x_1, y_1) = (-5, 4)$ and $(x_2, y_2) = (2, -6)$

$$\text{slope } m = \frac{y_2 - y_1}{x_2 - x_1} = \frac{-6 - 4}{2 - (-5)}$$

$$= \frac{-10}{7}, \text{ or } -\frac{10}{7}$$

b. The line passing through $(-7, -3)$ and $(11, 8)$

Apply the slope formula.

$(x_1, y_1) = (-7, -3)$ and $(x_2, y_2) = (11, 8)$

$$\text{slope } m = \frac{y_2 - y_1}{x_2 - x_1} = \frac{8 - (-3)}{11 - (-7)} = \frac{11}{18}$$

3. Find the slope of the line passing through $(-9, 3)$ and $(4, 3)$.

$(x_1, y_1) = (-9, 3)$ and $(x_2, y_2) = (4, 3)$

$$m = \frac{y_2 - y_1}{x_2 - x_1} = \frac{3 - 3}{4 - (-9)} = \frac{0}{13} = 0$$

4. Find the slope of the line passing through $(-5, 3)$ and $(-5, 8)$.

$(x_1, y_1) = (-5, 3)$ and $(x_2, y_2) = (-5, 8)$

$$m = \frac{y_2 - y_1}{x_2 - x_1} = \frac{8 - 3}{-5 - (-5)} = \frac{5}{0} \quad \text{undefined slope}$$

2. Find the slope of each line.

a. The line passing through $(-6, 7)$ and $(3, -9)$

_____

b. The line passing through $(-5, -2)$ and $(12, 7)$

_____

3. Find the slope of the line passing through $(8, -5)$ and $(-7, -5)$.

_____

4. Find the slope of the line passing through $(9, 11)$ and $(9, -7)$.

_____

Name: Date:
Instructor: Section:

Objective 1 Practice Exercises

For extra help, see Examples 1–4 on pages 243–246 of your text.

Find the slope of the line through the given points.

1.   (4, 3) and (3, 5)   1. _____

2.   (−4, 6) and (−4, −1)   2. _____

3.   (−3, 3) and (6, 3)   3. _____

Objective 2   Find the slope from the equation of a line.

Video Examples

Review these examples for Objective 2:

5.   Find the slope of each line.

c. $5y + x = 5$

$5y + x = 5$

$5y = -x + 5$

$y = -\dfrac{x}{5} + 1$

$y = -\dfrac{1}{5}x + 1$

The coefficient of x is $-\dfrac{1}{5}$, so the slope of the

line is $-\dfrac{1}{5}$.

Now Try:

5.   Find the slope of each line.

c. $8y + x = 16$

_____

a. $4x - 3y = 7$

Step 1 Solve the equation for y.
$$4x - 3y = 7$$
$$-3y = -4x + 7$$
$$y = \frac{4}{3}x - \frac{7}{3}$$

Step 2 The slope is given by the coefficient of x, so the slope is $\frac{4}{3}$.

a. $7x - 4y = 8$

_____

Objective 2 Practice Exercises

For extra help, see Example 5 on page 247 of your text.

Find the slope of each line.

4. $7y - 4x = 11$

4. _____

5. $3y = 2x - 1$

5. _____

6. $y = -\frac{2}{5}x - 4$

6. _____

Copyright © 2018 Pearson Education, Inc.

Objective 3   **Use slope to determine whether two lines are parallel, perpendicular, or neither.**

Video Examples

Review this example for Objective 3:	**Now Try:**
6b. Decide whether the pair of lines is parallel, perpendicular, or neither.	**6b.** Decide whether the pair of lines is parallel, perpendicular, or neither.

Review this example for Objective 3:

6b. Decide whether the pair of lines is parallel, perpendicular, or neither.

$$5x - y = 3$$
$$15x - 3y = 12$$

Solve each equation for y.
$$y = 5x - 3$$
$$y = 5x - 4$$

Both lines have slope 5, so the lines are parallel.

Now Try:

6b. Decide whether the pair of lines is parallel, perpendicular, or neither.

$$2x - 4y = 7$$
$$3x - 6y = 8$$

_____

Objective 3 Practice Exercises

For extra help, see Example 6 on page 249 of your text.

In each pair of equations, give the slope of each line, and then determine whether the two lines are **parallel,** **perpendicular,** *or* **neither.**

7. $-x + y = -7$
      $x - y = -3$

7. _____

8. $4x + 2y = 8$
      $x + 4y = 3$

8. _____

9. $9x + 3y = 2$
      $x - 3y = 5$

9. _____

Objective 4   Solve problems involving average rate of change.

Video Examples

Review this example for Objective 4:

7. Enrollment in a college was 11,500 two years ago, 10,975 last year, and 10,800 this year. What is the average rate of change in enrollment per year for this 3-year period?

We use the ordered pairs (1, 11,500) and (3, 10,800).

average rate of change $= \dfrac{10,800 - 11,500}{3 - 1}$

$= \dfrac{-700}{2}$

$= -350$

The enrollment decreases at a rate of 350 students per year.

Now Try:

7. A company had 44 employees during the first year of operation. During their eighth year, the company had 79 employees. What was the average rate of change in the number of employees per year?

_____

Objective 4 Practice Exercises

For extra help, see Examples 7–8 on pages 250–251 of your text.

Solve each problem.

10. Suppose in 2010, the sales of a company was $1,625,000. In 2015, the company had sales of $2,250,000. Find the average rate of change in the sales per year.

10. _____

11. A state had a population of 755,000 in 2000 and a population of 809,000 in 2012. Find the average rate of change in population per year.

11. _____

12. Suppose a man's salary was $55,250 in 2012 and $60,000 in 2017. Find the average rate of change in the salary per year.

12. _____

Chapter 3 GRAPHS OF LINEAR EQUATIONS AND INEQUALITIES IN TWO VARIABLES

3.4   Slope-Intercept Form of a Linear Equation

Learning Objectives
1   Use slope-intercept form of the equation of a line.
2   Graph a line by using its slope and a point on the line.
3   Write an equation of a line by using its slope and any point on the line.
4   Graph and write equations of horizontal and vertical lines.
5   Write an equation of a line that models real data.

Key Terms

Use the vocabulary terms listed below to complete each statement in exercises 1–4.

slope   *y*-intercept   slope-intercept form   standard form

1.   A linear equation in the form $y - mx + b$ is written in

      _____.

2.   A linear equation in the form $Ax + By = C$ is written in

      _____.

3.   In the linear equation $y = mx + b$, the variable b represents the

      _____ of the line.

4.   In the linear equation $y = mx + b$, the variable m represents the

      _____ of the line.

Objective 1   Use the slope-intercept form of the equation of a line.

Video Examples

Review these examples for Objective 1:

1.   Identify the slope and *y*-intercept of the line with each equation.

   a. $y = -8x + 7$

   The slope is –8, and the *y*-intercept is (0, 7).

   c. $y = 11x$

   The equation can be written as
      $y = 11x + 0$.
   The slope is 11, and the *y*-intercept is (0, 0).

Now Try:

1.   Identify the slope and *y*-intercept of the line with each equation.

   a. $y = -12x + 6$

   _____

   c. $y = 23x$

   _____

Name: Date:

Instructor: Section:

Objective 1 Practice Exercises

For extra help, see Example 1 on page 258 of your text.

Identify the slope and y-intercept of the line with each equation.

1. $y = \dfrac{3}{2}x - \dfrac{2}{3}$ 1. _____

                                                   _____

2. $y = -4x$ 2. _____

                                                     _____

Write an equation of the line with the given slope and y-intercept.

3.   slope -3; y-intercept $(0, 3)$ 3. _____

Objective 2   Graph a line using its slope and a point on the line.

Video Examples

Review these examples for Objective 2:

2b. Graph the equation by using the slope and *y*-intercept.

$$2x - 3y = 6$$

Step 1 Solve for *y* to write the equation in slope-intercept form.

$$2x - 3y = 6$$
$$-3y = -2x + 6$$
$$y = \frac{2}{3}x - 2$$

Step 2 The *y*-intercept is $(0, -2)$. Graph this point.

Step 3 The slope is $\dfrac{2}{3}$. By definition,

$$\text{slope } m = \frac{\text{change in } y \text{ (rise)}}{\text{change in } x \text{ (run)}} = \frac{2}{3}$$

From the *y*-intercept, count up 2 units and to the right 3 units to obtain the point $(3, 0)$.

Now Try:

2b. Graph the equation by using the slope and *y*-intercept.

$$y = \frac{2}{3}x$$

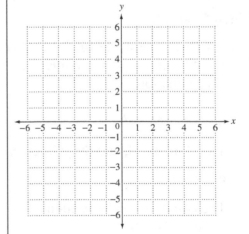

Step 4 Draw the line through the points
(0,–2) and (3, 0) to obtain the graph.

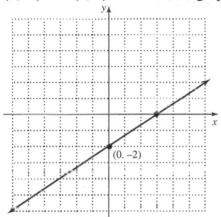

3. Graph the line passing through the point (1,–3), with slope $-\dfrac{5}{2}$.

First, locate the point (1,–3). Then write the slope $-\dfrac{5}{2}$ as

$$\text{slope } m = \frac{\text{change in } y \text{ (rise)}}{\text{change in } x \text{ (run)}} = \frac{5}{-2}.$$

Locate another point on the line by counting up 5 units from (1, –3), and then to the left 2 units. Finally, draw the line through this new point, (–1, 2).

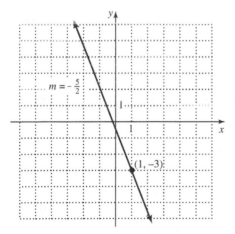

3. Graph the line passing through the point (2, 2), with slope $\dfrac{1}{3}$.

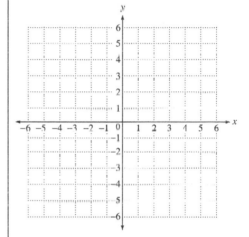

Objective 2 Practice Exercises

For extra help, see Examples 2–3 on pages 259–260 of your text.

Graph each equation by using the slope and y-intercept.

4. $4x - y = 4$ **4.**

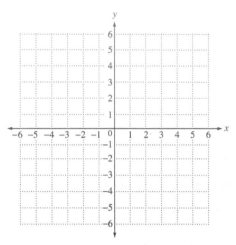

5. $y = -3x + 6$ **5.**

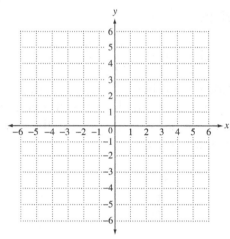

Objective 3   Write an equation of a line using its slope and any point on the line.

Video Examples

Review these examples for Objective 3:	Now Try:
4. Write an equation, in slope-intercept form of the line passing through the given point and having the given slope.	4. Write an equation, in slope-intercept form of the line passing through the given point and having the given slope.
a. $(0,-3)$; $m = \dfrac{2}{5}$	**a.** $(0,-6)$; $m = \dfrac{3}{4}$
Because the point $(0,-3)$ is the *y*-intercept, $b = -3$. We can substitute this value for *b* and the given	

slope $m = \dfrac{2}{5}$ directly into slope-intercept form

$y = mx + b$ to write an equation.

$$y = mx + b$$

$$y = \frac{2}{5}x + (-3)$$

$$y = \frac{2}{5}x - 3$$

b. $(2, 9)$, $m = 5$

Since the line passes through the point $(2, 9)$, we can substitute $x = 2$, $y = 9$, and slope $m = 5$ into $y = mx + b$ and solve for b.

$$y = mx + b$$

$$9 = 5(2) + b$$

$$-1 = b$$

Now substitute the values of m and b into slope-intercept form.

$$y = mx + b$$

$$y = 5x - 1$$

b. $(-1, 4)$, $m = 6$

Objective 3 Practice Exercises

For extra help, see Example 4 on page 261 of your text.

Write an equation for the line passing through the given point and having the given slope. Write the equations in slope-intercept form, if possible.

6. $(-3, 4)$; $m = -\dfrac{3}{5}$ 6. _____

7. $(-4, -3)$; $m = -2$ 7. _____

8. $(0, 2)$; $m = -\dfrac{3}{2}$ 8. _____

Objective 4   Graph and write equations of horizontal and vertical lines.

Video Examples

Review this example for Objective 4:

5a. Graph the line passing through the given point and having the given slope.

    a. $(6, -2)$, $m = 0$

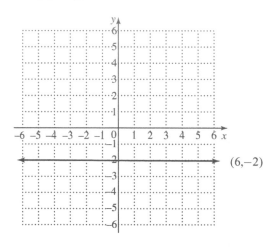

$(6, -2)$

Now Try:

5a. Graph the line passing through the given point and having the given slope.

    a. $(-4, 4)$, $m = 0$

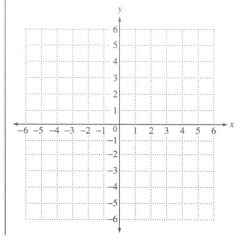

Objective 4 Practice Exercises

For extra help, see Examples 5–6 on pages 261–262 of your text.

Graph each line passing through the given point and having the given slope.

   9. $(1, 4)$, undefined slope   **9.**

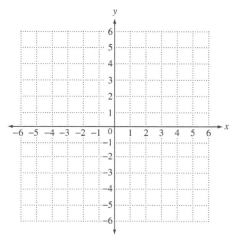

10. $(-2, -2); \ m = 0$ **10.**

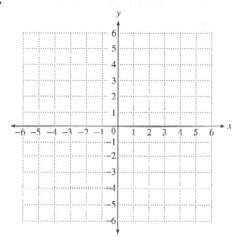

Write an equation of the line passing through the given point and having the given slope.

11. $(3, -5);$ undefined slope   **11.** _____

Objective 5   **Write an equation of a line that models real data.**

Video Examples

Review these examples for Objective 5:

7. A local rental truck costs $35 plus $0.15 per mile the truck is driven. Let x represent the number of miles so that y represents the total cost of the rental truck (in dollars).

   a. Write an equation in the form $y = mx + b$.

   The value for b is 35, and the slope is 0.15.
     $y = 0.15x + 35$

   b. Find and interpret the ordered pair associated with the equation for $x = 20$.

   If $x = 15$, then
     $y = 0.15(20) + 35$

       $= 3 + 35$

       $= 38$

A rental truck that is driven 20 miles has a total cost of $38.

Now Try:

7. A music teacher sets up rows of chair for a concert. There were 8 chairs in each row, plus 15 special reserved seats up from for faculty. Let x represent the number of rows of chairs so that y represents the total number of guests who can be seated.

   a. Write an equation in the form $y = mx + b$.

           _____

   b. Find the ordered pair associated with the equation for $x = 5$.

           _____

Objective 5 Practice Exercises

For extra help, see Example 7 on pages 262–263 of your text.

Solve each problem.

12. x represents the number of funnel cakes sold for $4   12. a. _____
     each at the county faire, and y represents the total
     charge for the funnel cakes.   b._____
     a. Write an equation in the form $y = mx$.
     b. Give three ordered pairs associated with the
     equation for x-values 0, 5, and 10.

13. To run a newspaper ad, there is a $25 set up fee plus   13. a. _____
     a charge of $1.25 per line of type in the ad. Let x
     represent the number of lines in the ad so that y b._____
     represents the total cost of the ad (in dollars).
     a. Write an equation in the form $y = mx + b$.
     b. Give three ordered pairs associated with the
     equation for x-values 0, 5, and 10.

14. Suppose that you are in charge of your office holiday   14. a. _____
     party. You call the local caterer who informs you that
     the standard holiday party package costs $54.95 per   b. _____
     person plus a $150 setup fee.
     a. Write a linear equation in slope-intercept form that
        represents the cost in dollars y, for catering the
        office party.
     b. If 71 people attend, what will be the total cost in
        dollars to cater the party

Chapter 3 GRAPHS OF LINEAR EQUATIONS AND INEQUALITIES IN TWO VARIABLES

3.5   Point-Slope Form of a Linear Equation and Modeling

Learning Objectives
1   Use point-slope form to write an equation of a line.
2   Write an equation of a line using two points on the line.
3   Write an equation of a line parallel or perpendicular to a given line.
4   Write an equation of a line that models real data.

Key Terms

Use the vocabulary terms listed below to complete each statement in exercises 1−3.

slope-intercept form   point-slope form   standard form

1.   A linear equation in the form $y - y_1 = m(x - x_1)$ is written in

_____.

2.   A linear equation in the form $Ax + By = C$ is written in

_____.

3.   A linear equation in the form $y = mx + b$ is written in

_____.

Objective 1   Use point-slope form to write an equation of a line.

Video Examples

Review this example for Objective 1:

1b. Write an equation of each line. Give the final answer in slope-intercept form.

The line passing through $(5, -3)$ with slope $\frac{5}{4}$.

$$y - y_1 = m(x - x_1)$$
$$y - (-3) = \frac{5}{4}(x - 5)$$
$$y + 3 = \frac{5}{4}x - \frac{25}{4}$$
$$y = \frac{5}{4}x - \frac{37}{4}$$

Now Try:

1b. Write an equation of each line. Give the final answer in slope-intercept form.

The line passing through $(6, 11)$ with slope $-\frac{2}{3}$.

_____

Objective 1 Practice Exercises

For extra help, see Example 1 on page 270 of your text.

Write an equation of each line. Give the final answer in slope-intercept form.

1.   The line passing through (8,–7) with slope $\frac{1}{4}$.

                                                                  1. _____

2.   The line passing through (–3, 6) with slope –4.

                                                                    2. _____

3.   The line passing through (5, 2) with slope 2.

                                                                    3. _____

Objective 2   Write an equation of a line by using two points on the line.

Video Examples

Review this example for Objective 2:

2.   Write the equation of the line passing through the point (6, 8) and (–3, 5). Give the final answer in slope-intercept form and then in standard form.

First, find the slope of the line.

$(x_1, y_1) = (6,\ 8)$ and $(x_2, y_2) = (-3,\ 5)$

    slope $m = \dfrac{y_2 - y_1}{x_2 - x_1} = \dfrac{5 - 8}{-3 - 6} = \dfrac{-3}{-9} = \dfrac{1}{3}$

Now use (x_1, y_1), here (6, 8) and point-slope form.

Now Try:

2.   Write the equation of the line passing through the point (7, 15) and (15, 9). Give the final answer in slope-intercept form and then in standard form.

                          _____

$$y - y_1 = m(x - x_1)$$

$$y - 8 = \frac{1}{3}(x - 6)$$

$$y - 8 = \frac{1}{3}x - 2$$

$$y = \frac{1}{3}x + 6 \quad \text{Slope-intercept form}$$

$$3y = x + 18$$

$$-x + 3y = 18$$

$$x \quad 3y = 18 \qquad \text{Standard form}$$

Objective 2 Practice Exercises

For extra help, see Example 2 on page 271 of your text.

Write an equation for the line passing through each pair of points. Write the equations in standard form.

4. $(-2, 1)$ and $(3, 11)$ 4. _____

5. $(2, 3)$ and $(2, 3)$ 5. _____

6. $(3, -4)$ and $(2, 7)$ 6. _____

Objective 3   Write an equation of a line parallel or perpendicular to a given line.

Video Examples

Review these examples for Objective 3:

3. Write an equation in slope-intercept form of the line passing through the point (–4, 5) that satisfies the given condition.

   a. The line is parallel to $5x + 2y = 10$.

First, find the slope of the given line.
$$5x + 2y = 10$$
$$2y = -5x + 10$$
$$y = -\frac{5}{2}x + 5$$

The slope is $-\frac{5}{2}$.

Use point-slope form with $(x_1, y_1) = (-4,\ 5)$ and $m = -\frac{5}{2}$.

$$y - y_1 = m(x - x_1)$$
$$y - 5 = -\frac{5}{2}(x - (-4))$$
$$y - 5 = -\frac{5}{2}(x + 4)$$
$$y - 5 = -\frac{5}{2}x - 10$$
$$y = -\frac{5}{2}x - 5$$

   b. The line is perpendicular to $5x + 2y = 10$.

From part (a), the line in slope-intercept form is $y = -\frac{5}{2}x + 5$.

The line perpendicular to this line must have slope $\frac{2}{5}$, the negative reciprocal of $-\frac{5}{2}$.

Use point-slope form with $(x_1, y_1) = (-4,\ 5)$ and $m = \frac{2}{5}$.

Now Try:

3. Write an equation in slope-intercept form of the line passing through the point (–6, 8) that satisfies the given condition.

   a. The line is parallel to $3x + 4y = 12$.

_____

   b. The line is perpendicular to $3x + 4y = 12$.

_____

$$y - y_1 = m(x - x_1)$$

$$y - 5 = \frac{2}{5}(x - (-4))$$

$$y - 5 = \frac{2}{5}(x + 4)$$

$$y - 5 = \frac{2}{5}x + \frac{8}{5}$$

$$y = \frac{2}{5}x + \frac{33}{5}$$

Objective 3 Practice Exercises

For extra help, see Example 3 on pages 272–273 of your text.

Write the equation in standard form of the line satisfying the given conditions.

7. parallel to $2x + 3y = -12$, through $(9, -3)$　　　　7. _____

8. parallel to $4x - 3y = 8$, through $(-2, 3)$.　　　　8. _____

9. perpendicular to $x - 3y = 0$, through $(-10, 2)$　　　　9. _____

Objective 4   Find an equation of a line that models real data.

Video Examples

Review this example for Objective 4:

4.   The table shows the number of internet users in the world from 1998 to 2005, where year 0 represents 1998.

Year	Number of Internet Users (millions)
0	147
2	361
4	587
6	817
8	1093

Plot the data and find an equation that approximates it.

Letting y represent the number of internet users in year x, we plot the data.

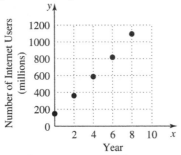

The points appear to lie approximately in a straight line. To find an equation of the line, we choose the ordered pairs (0, 147) and (8, 1093) from the table and find the slope of the line through these points.

$$(x_1, y_1) = (0, 147) \text{ and } (x_2, y_2) = (8, 1093)$$

$$\text{slope } m = \frac{y_2 - y_1}{x_2 - x_1} = \frac{1093 - 147}{8 - 0} = \frac{946}{8}$$

$$= 118.25$$

Use the slope, 118.25, and the point (0, 147) in slope-intercept form.

$$y = mx + b$$

$$147 = 118.25(0) + b$$

$$147 = b$$

Thus, $m = 118.25$ and $b = 147$, so the equation of the line is $y = 118.25x + 147$.

Now Try:

4.   The table shows the average annual telephone expenditures for residential telephones from 2001 to 2006, where year 0 represents 2001.

Year	Annual Telephone Expenditures
0	$686
2	$620
3	$592
4	$570
5	$542

Plot the data and find an equation that approximates it.

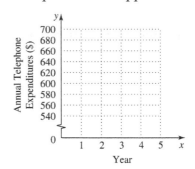

_____

Name: Date:
Instructor: Section:

Objective 4 Practice Exercises

For extra help, see Example 4 on page 274 of your text.

Plot the data and find an equation that approximates it.

10.   The table shows the U.S. municipal solid waste recycling percents since 1985, where year 0 represents 1985.

Year	Recycling Percent
0	10.1
5	16.2
10	26.0
15	29.1
20	32.5

10.

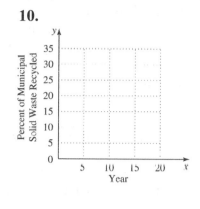

11.   The table shows the approximate consumer expenditures for food in the U.S. in billions of dollars for selected years, where year 0 represents 1985.

Year	Food Expenditures (billions of dollars)
0	233
5	298
10	343
15	417
20	515

11.

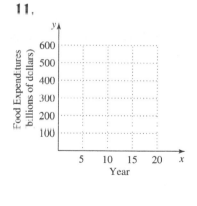

Chapter 3 GRAPHS OF LINEAR EQUATIONS AND INEQUALITIES IN TWO VARIABLES

3.6   Graphing Linear Inequalities in Two Variables

Learning Objectives
1   Graph linear inequalities in two variables.
2   Graph an inequality with a boundary line through the origin.

Key Terms

Use the vocabulary terms listed below to complete each statement in exercises 1–2.

linear inequality in two variables   **boundary line**

1.   In the graph of a linear inequality, the _____ separates the region that satisfies the inequality from the region that does not satisfy the inequality.

2.   An inequality that can be written in the form $Ax + By < C$, $Ax + By > C$, $Ax + By \leq C$, or $Ax + By \geq C$ is called a _____.

Objective 1   Graph linear inequalities in two variables.

Video Examples

Review these examples for Objective 1:

2.   Graph $2x + 5y > -10$.

    This equation does not include equality. Therefore, the points on the line $2x + 5y = -10$ do not belong to the graph. However, the line still serves as a boundary for two regions.

    To graph the inequality, first graph the equation $2x + 5y > -10$. Use a dashed line to show that the points on the line are not solutions of the inequality $2x + 5y > -10$. Then choose a test point to see which region satisfies the inequality.

$$2x + 5y > -10$$
$$2(0) + 5(0) \overset{?}{>} -10$$
$$0 + 0 \overset{?}{>} -10$$
$$0 > -10 \ \text{True}$$

Since $0 > -10$ is true, the graph of the inequality is the region that contains (0, 0). Shade that region.

Now Try:

2.   Graph $5x + 4y > 20$.

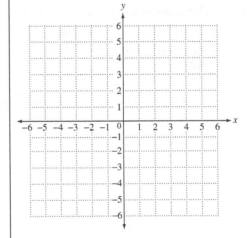

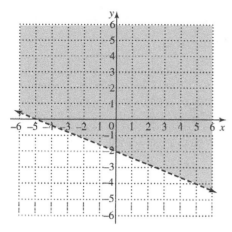

3. Graph $x - 4 < -1$.

First, solve the inequality for x.

$x \leq 3$

Now graph the line $x = 3$, a vertical line through the point $(3, 0)$. Use a solid line, and choose $(0, 0)$ as a test point.

$0 < 3$ True

Because $0 \leq 3$ is true, we shade the region containing $(0, 0)$.

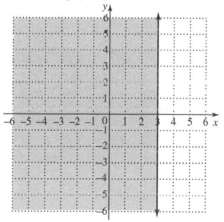

3. Graph $y \geq 1$.

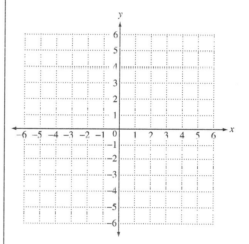

Name: Date:

Instructor: Section:

Objective 1 Practice Exercises

For extra help, see Examples 1–3 on pages 283–285 of your text.

Graph each linear inequality.

1. $y \geq x - 1$

1.

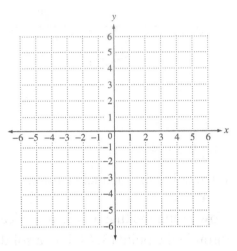

2. $y > -x + 2$

2.

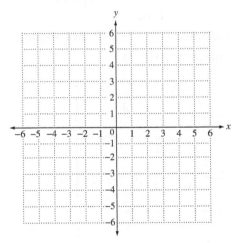

3. $3x - 4y - 12 > 0$

3.

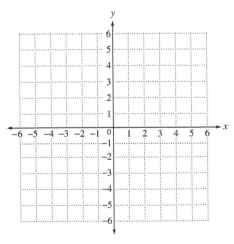

Copyright © 2018 Pearson Education, Inc.

Name: Date:
Instructor: Section:

Objective 2   Graph an inequality with a boundary line through the origin.

Video Examples

Review this example for Objective 2:

4.   Graph $y \geq 3x$.

We graph $y = 3x$ using a solid line through
(0, 0), (1, 3) and (2, 6). Because (0, 0) is on the
line $y \geq 3x$, it cannot be used as a test point.
Instead we choose a test point off the line, say
(3, 0).

$$0 \overset{?}{\geq} 3(3)$$

$$0 \geq 9 \quad \text{False}$$

Because $0 \geq 9$ is false, shade the other region.

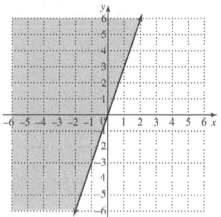

Now Try:

4.   Graph $y \geq x$.

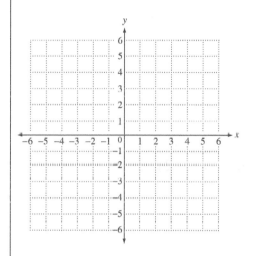

Objective 2 Practice Exercises

For extra help, see Example 4 on page 285 of your text.

Graph each linear inequality.

4. $y \leq \dfrac{2}{5}x$

4.

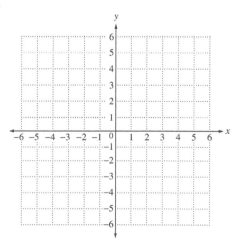

5. $y \geq \dfrac{1}{3}x$

5.

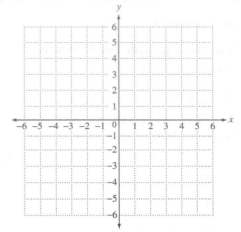

6. $x < -2y$

6.

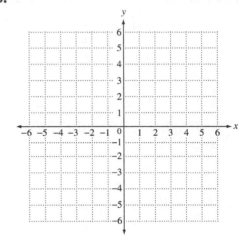

Chapter 4 SYSTEMS OF LINEAR EQUATIONS AND INEQUALITIES

4.1   Solving Systems of Linear Equations by Graphing

Learning Objectives
1   Decide whether a given ordered pair is a solution of a system.
2   Solve linear systems by graphing.
3   Solve special systems by graphing.
4   Identify special systems without graphing.

Key Terms

Use the vocabulary terms listed below to complete each statement in exercises 1–7.

system of linear equations	**solution of a system**
solution set of a system	**consistent system**
inconsistent system	**independent equations**
dependent equations	

1.   Equations of a system that have different graphs are called
      _____.

2.   A system of equations with at least one solution is a
      _____.

3.   The set of all ordered pairs that are solutions of a system is the
      _____.

4.   The _____ of linear equations is an ordered
      pair that makes all the equations of the system true at the same time.

5.   Equations of a system that have the same graph (because they are different forms of
      the same equation) are called _____.

6.   A system with no solution is called a(n) _____.

7.   A(n) _____ consists of two or more linear
      equations with the same variables.

Objective 1   Decide whether a given ordered pair is a solution of a system.

Video Examples

Review this example for Objective 1:

1b. Determine whether the ordered pair (5, –2) is a solution of the system.

$$4x + 5y = 10$$
$$3x + 8y = 6$$

Again, substitute 5 for x and –2 for y in each equation.

$$
\begin{array}{c|c}
4x + 5y = 10 & 3x + 8y = 6 \\
4(5) + 5(-2) \overset{?}{=} 10 & 3(5) + 8(-2) \overset{?}{=} 6 \\
20 - 10 \overset{?}{=} 10 & 15 - 16 \overset{?}{=} 6 \\
10 = 10 \ \text{True} & \text{False} \ -1 = 6
\end{array}
$$

The ordered pair (5, –2) is not a solution of this system because it does not satisfy the second equation.

Now Try:

1b. Determine whether the ordered pair (6, 5) is a solution of the system.

$$5x - 6y = 0$$
$$6x + 5y = 50$$

_____

Objective 1 Practice Exercises

For extra help, see Example 1 on page 302 of your text.

Decide whether the given ordered pair is a solution of the given system.

1. $(2, -4)$

   $2x + 3y = 6$
   $3x - 2y = 14$

1. _____

2. $(-3, -1)$

   $5x - 3y = -12$
   $2x + 3y = -9$

2. _____

3. $(4, 0)$

   $4x + 3y = 16$
   $x - 4y = -4$

3. _____

Name: Date:
Instructor: Section:

Objective 2   Solve linear systems by graphing.

Video Examples

Review this example for Objective 2:

2. Solve the system of equation by graphing both equations on the same axes.

$$6x - 5y = 4$$

$$2x - 5y = 8$$

Graph these equations by plotting several points for each line. To find the x-intercept, let $y = 0$. To find the y-intercept, let $x = 0$.
The tables show the intercepts and a check point for each graph.

$6x - 5y = 4$

x	y
0	$-\dfrac{4}{5}$
$\dfrac{2}{3}$	0
4	4

$2x - 5y = 8$

x	y
0	$-\dfrac{8}{5}$
4	0
-6	-4

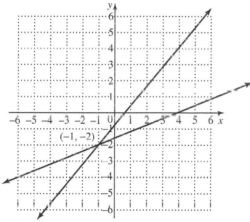

The lines suggest that the graphs intersect at the point $(-1, -2)$. We check by substituting -1 for x and -2 for y in both equations.

$$6x - 5y = 4$$
$$6(-1) - 5(-2) \overset{?}{=} 4$$
$$-6 + 10 \overset{?}{=} 4$$
$$4 = 4 \text{ True}$$

$$2x - 5y = 8$$
$$2(-1) - 5(-2) \overset{?}{=} 8$$
$$-2 + 10 \overset{?}{=} 8$$
$$\text{True } 8 = 8$$

Because $(-1, -2)$ satisfies both equations, the solution set of this system is $\{(-1, -2)\}$.

Now Try:

2. Solve the system of equation by graphing both equations on the same axes.

$$3x - y = -7$$

$$2x + y = -3$$

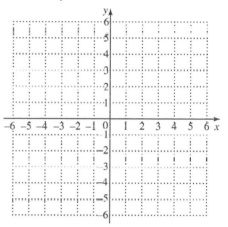

Name: Date:
Instructor: Section:

Objective 2 Practice Exercises

For extra help, see Example 2 on page 303 of your text.

Solve each system by graphing both equations on the same axes.

4. $x - 2y = 6$
      $2x + y = 2$

4.

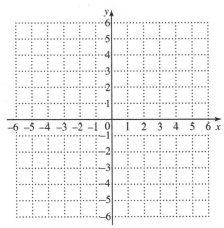

_____

5. $2x = y$
      $5x + 3y = 0$

5.

_____

6. $3x + 2 = y$

$\quad\quad 2x - y = 0$

6.

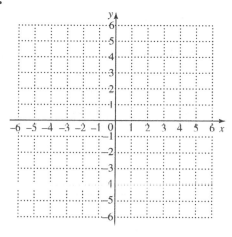

_____

Objective 3   Solve special systems by graphing.

Video Examples

Review these examples for Objective 3:

3. Solve each system by graphing.

   a. $x - y = 1$

$\quad\quad\quad x - y = -1$

The graphs of these two equations are parallel and have no points in common. There is no solution for this system. The solution set is $\varnothing$.

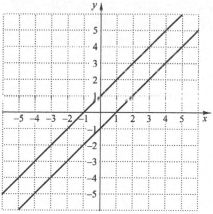

Now Try:

3. Solve each system by graphing.

   a. $4x - 2y = 8$

$\quad\quad\quad 6x - 3y = 12$

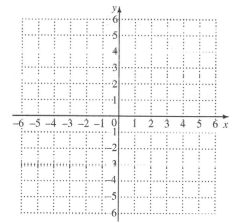

_____

b. $3x - y = 0$

  $2y = 6x$

The graphs of these two equations are the same line.

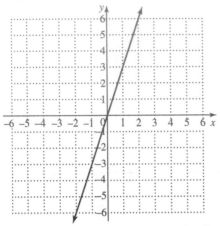

In this case, every point on the line is a solution of the system, and the solution set contains an infinite number of ordered pairs, each of which satisfies both equations of the system. We write the solution set as

  $\{(x, y)|\ 3x - y = 0\}.$

b. $x - 3y = 6$

  $x - 3y = 4$

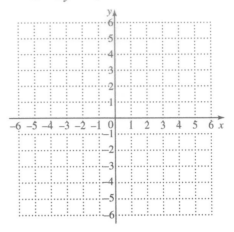

_____

Objective 3 Practice Exercises

For extra help, see Example 3 on pages 301–305 of your text.

Solve each system of equations by graphing both equations on the same axes. If the two equations produce parallel lines, write **no solution**. *If the two equations produce the same line, write* **infinite number of solutions**.

7. $8x + 4y = -1$

  $4x + 2y = 3$

7.

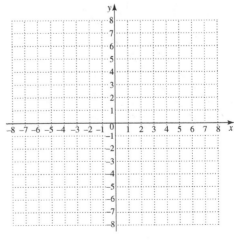

_____

Name: _____ Date: _____
Instructor: _____ Section: _____

8. $-3x + 2y = 6$

  $-6x + 4y = 12$

8.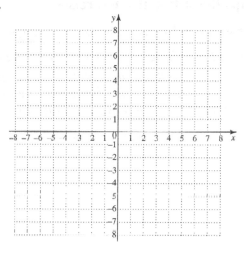

_____

Objective 4   Identify special systems without graphing.

Video Examples

Review this example for Objective 4:

4c. Describe the system without graphing. State the number of solutions.

$3x - 4y = 12$

$2x \ 3y = 6$

Write each equation in slope-intercept form.

$3x - 4y = 12$	$2x - 3y = 6$
$-4y = -3x + 12$	$-3y = -2x + 6$
$y = \dfrac{3}{4}x - 3$	$y = \dfrac{2}{3}x - 2$

The graphs are neither parallel nor the same line, since the slopes are different. This system has exactly one solution.

Now Try:

4c. Describe the system without graphing. State the number of solutions.

$x - 6y = 4$

$6x - y = 9$

_____

Objective 4 Practice Exercises

For extra help, see Example 4 on pages 306–307 of your text.

Without graphing, answer the following equations for each linear system.
(a) Is the system inconsistent, are the equations dependent, or neither?
(b) Is the graph a pair of intersecting lines, a pair of parallel lines, or one line?
(c) Does the system have one solution, no solution, or an infinite number of solutions?

9. $y = 2x + 1$
    $3x - y = 7$

9. (a)_____

   (b)_____

   (c)_____

10. $-2x + y = 4$
    $-4x + 2y = -2$

10. (a)_____

    (b)_____

    (c)_____

11. $4x + 3y = 12$
    $-12x = -36 + 9y$

11. (a)_____

    (b)_____

    (c)_____

Chapter 4 SYSTEMS OF LINEAR EQUATIONS AND INEQUALITIES

4.2   Solving Systems of Linear Equations by Substitution

Learning Objectives
1   Solve linear systems by substitution.
2   Solve special systems by substitution.
3   Solve linear systems with fractions and decimals.

Key Terms

Use the vocabulary terms listed below to complete each statement in exercises 1−4.

substitution   **ordered pair**   **inconsistent system**

dependent system

1.   The solution of a linear system of equations is written as a(n) _____.

2.   When one expression is replaced by another, _____ is being used.

3.   A system of equations in which all solutions of the first equation are also solutions of the second equation is a(n) _____.

4.   A system of equations that has no common solution is called a(n) _____.

Objective 1   Solve linear systems by substitution.

Video Examples

Review these examples for Objective 1:

1.   Solve the system by the substitution method.
$$2x + 5y = 22 \quad (1)$$
$$y = 4x \quad (2)$$

Equation (2) is already solved for y. We substitute $4x$ for y in equation (1).
$$2x + 5y = 22$$
$$2x + 5(4x) = 22$$
$$2x + 20x = 22$$
$$22x = 22$$
$$x = 1$$

Find the value of y by substituting 1 for x in either equation. We use equation (2).
$$y = 4x$$
$$y = 4(1) = 4$$

We check the solution (1, 4) by substituting 1 for

Now Try:

1.   Solve the system by the substitution method.
$$x + y = 7$$
$$y - 6x$$

_____

x and 4 for y in both equations.

$$2x + 5y = 22 \qquad\qquad y = 4x$$
$$2(1) + 5(4) \overset{?}{=} 22 \qquad 4 \overset{?}{=} 4(1)$$
$$2 + 20 \overset{?}{=} 22 \qquad \text{True } 4 = 4$$
$$22 = 22 \text{ True}$$

Since (1, 4) satisfies both equations, the solution set of the system is $\{(1, 4)\}$.

2. Solve the system by the substitution method.

$$4x + 5y = 13 \qquad (1)$$
$$x = -y + 2 \qquad (2)$$

Equation (2) gives x in terms of y. We substitute $-y + 2$ for x in equation (1).

$$4x + 5y = 13$$
$$4(-y + 2) + 5y = 13$$
$$-4y + 8 + 5y = 13$$
$$y + 8 = 13$$
$$y = 5$$

Find the value of x by substituting 5 for y in either equation. We use equation (2).

$$x = -y + 2$$
$$x = -5 + 2 = -3$$

We check the solution (–3, 5) by substituting –3 for x and 5 for y in both equations.

$$4x + 5y = 13 \qquad\qquad x = -y + 2$$
$$4(-3) + 5(5) \overset{?}{=} 13 \qquad -3 \overset{?}{=} -5 + 2$$
$$-12 + 25 \overset{?}{=} 13 \qquad \text{True } -3 = -3$$
$$13 = 13 \text{ True}$$

Both results are true, so the solution set of the system is $\{(-3, 5)\}$.

3. Solve the system by the substitution method.

$$4x = 5 - y \qquad (1)$$
$$7x + 3y = 15 \qquad (2)$$

Step 1 Solve one of the equations for x or y. Solve equation (1) for y to avoid fractions.

$$4x = 5 - y$$
$$y + 4x = 5$$
$$y = -4x + 5$$

2. Solve the system by the substitution method.

$$2x + 3y = 6$$
$$x = 5 - y$$

_____

3. Solve the system by the substitution method.

$$2x + 7y = 2$$
$$3y = 2 - x$$

_____

Step 2 Now substitute $-4x+5$ for y in equation (2).

$$7x+3y=15$$
$$7x+3(-4x+5)=15$$

Step 3 Solve the equation from Step 2.

$$7x-12x+15=15$$
$$-5x+15=15$$
$$-5x=0$$
$$x=0$$

Step 4 Equation (1) solved for y is $y=-4x+5$. Substitute 0 for x.

$$y=-4(0)+5=5$$

Step 5 Check that (0, 5) is the solution.

$$4x=5-y \qquad\qquad 7x+3y=15$$
$$4(0)\overset{?}{=}5-5 \qquad 7(0)+3(5)\overset{?}{=}15$$
$$0=0 \ \text{True} \qquad\quad \text{True} \ 15=15$$

Since both results are true, the solution set of the system is {(0, 5)}.

Objective 1 Practice Exercises

For extra help, see Examples 1–3 on pages 312–314 of your text.

Solve each system by the substitution method. Check each solution.

1. $3x+2y=14$
   $y=x+2$

   1. _____

2. $x+y=9$
   $5x-2y=-4$

   2. _____

3. $\quad 3x - 21 = y$ **3.** _____

$\qquad y + 2x = -1$

Objective 2   Solve special systems by substitution.

Video Examples

Review these examples for Objective 2:

4. Use substitution to solve the system.

$$x = 7 - 3y \quad (1)$$

$$5x + 15y = 1 \qquad (2)$$

Because equation (1) is already solved for x, we substitute $7 - 3y$ for x in equation (2).

$$5x + 15y = 1$$

$$5(7 - 3y) + 15y = 1$$

$$35 - 15y + 15y = 1$$

$$35 = 1 \quad \text{False}$$

A false result, here $35 = 1$, means that the equations in the system have graphs that are parallel lines. The system is inconsistent and has no solution, so the solution set is $\varnothing$.

5. Use substitution to solve the system.

$$14x - 7y = 21 \quad (1)$$

$$-2x + y = -3 \quad (2)$$

Begin by solving equation (2) for y to get $y = 2x - 3$. Substitute $2x - 3$ for y in equation (1).

$$14x - 7y = 21$$

$$14x - 7(2x - 3) = 21$$

$$14x - 14x + 21 = 21$$

$$0 = 0$$

This true result means that every solution of one equation is also a solution of the other, so the system has an infinite number of solutions. The solution set is $\{(x, y) | 14x - 7y = 21\}$.

Now Try:

4. Use substitution to solve the system.

$$5x - 10y = 8$$

$$x = 2y + 5$$

_____

5. Use substitution to solve the system.

$$5x + 4y = 20$$

$$-10x + 40 = 8y$$

_____

Objective 2 Practice Exercises

For extra help, see Examples 4–5 on pages 314–315 of your text.

Solve each system by the substitution method. Use set-builder notation for dependent equations.

4. $y = -\dfrac{1}{3}x + 5$

    $3y + x = -9$

    4. _____

5. $\dfrac{1}{2}x + 3 = y$

    $6 = -x + 2y$

    5. _____

6. $4x + 3y = 2$

    $8x + 6y = 6$

    6. _____

Objective 3   Solve linear systems with fractions and decimals.

Video Examples

Review this example for Objective 3:

6. Solve the system by the substitution method.

$$\frac{1}{2}x - y = 3 \quad (1)$$

$$\frac{1}{5}x + \frac{1}{2}y = \frac{3}{10} \quad (2)$$

Clear equation (1) of fractions by multiplying each side by 2.

$$2\left(\frac{1}{2}x - y\right) = 2(3)$$

$$2\left(\frac{1}{2}x\right) - 2y = 2(3)$$

$$x - 2y = 6$$

Clear equation (2) of fractions by multiplying each side by 10.

$$10\left(\frac{1}{5}x + \frac{1}{2}y\right) = 10\left(\frac{3}{10}\right)$$

$$10\left(\frac{1}{5}x\right) + 10\left(\frac{1}{2}y\right) = 10\left(\frac{3}{10}\right)$$

$$2x + 5y = 3$$

The given system of equations has been simplified to an equivalent system.

$$x - 2y = 6 \quad (3)$$

$$2x + 5y = 3 \quad (4)$$

To solve the system by substitution, solve equation (3) for x.

$$x - 2y = 6$$

$$x = 2y + 6$$

Now substitute the result for x in equation (4).

$$2x + 5y = 3$$

$$2(2y + 6) + 5y = 3$$

$$4y + 12 + 5y = 3$$

$$9y + 12 = 3$$

$$9y = -9$$

$$y = -1$$

Substitute -1 for y in $x = 2y + 6$ (equation (3) solved for x).

$$x = 2(-1) + 6 = 4$$

Check $(4, -1)$ in both of the original equations. The solution set is $\{(4, -1)\}$.

Now Try:

6. Solve the system by the substitution method.

$$x + \frac{1}{2}y = \frac{1}{2}$$

$$\frac{1}{2}x + \frac{1}{5}y = 0$$

_____

Name: _____ Date: _____

Instructor: _____ Section: _____

Objective 3 Practice Exercises

For extra help, see Examples 6–7 on pages 316–317 of your text.

Solve each system by the substitution method. Check each solution.

7. $\frac{5}{4}x - y = -\frac{1}{4}$

   $-\frac{7}{8}x + \frac{5}{8}y = 1$

7. _____

8. $\frac{1}{4}x + \frac{3}{8}y = -3$

   $\frac{5}{6}x - \frac{3}{7}y = -10$

8. _____

9. $0.6x + 0.8y = 1$

   $0.4y = 0.5 - 0.3x$

9. _____

Chapter 4 SYSTEMS OF LINEAR EQUATIONS AND INEQUALITIES

4.3   Solving Systems of Linear Equations by Elimination

Learning Objectives

1	Solve linear systems by elimination.
2	Multiply when using the elimination method.
3	Use an alternative method to find the second value in a solution.
4	Solve special systems by elimination.

Key Terms

Use the vocabulary terms listed below to complete each statement in exercises 1–3.

addition property of equality   **elimination method**   **substitution**

1. Using the addition property to solve a system of equations is called the

   _____.

2. The _____ states that the same added quantity to each side of an equation results in equal sums.

3. _____ is being used when one expression is replaced by another.

Objective 1   Solve linear systems by elimination.

Video Examples

Review this example for Objective 1:

1. Use the elimination method to solve the system.

$$x + y = 6 \quad (1)$$
$$-x + y = 4 \quad (2)$$

Add the equations vertically.

$$x + y = 6 \quad (1)$$
$$\underline{-x + y = 4 \quad (2)}$$
$$2y = 10$$
$$y = 5$$

To find the x-value, substitute 5 for y in either of the two equations of the system. We choose equation (1).

$$x + y = 6$$
$$x + 5 = 6$$
$$x = 1$$

Check the solution (1, 5), by substituting 1 for x and 5 for y in both equations of the given system.

Now Try:

1. Use the elimination method to solve the system.

$$x + y = 11$$
$$x - y = 5$$

_____

$$x + y = 6 \qquad \bigg| \qquad -x + y = 4$$
$$1 + 5 \overset{?}{=} 6 \qquad \bigg| \qquad -1 + 5 \overset{?}{=} 4$$
$$6 = 6 \text{ True} \quad \bigg| \quad \text{True } 4 = 4$$

Since both results are true, the solution set of the system is $\{(1, 5)\}$.

Objective 1 Practice Exercises

For extra help, see Examples 1–2 on pages 320–322 of your text.

Solve each system by the elimination method. Check your answers.

1. $x - 4y = -4$ 1. _____
    $-x + y = -5$

2. $2x - y = 10$ 2. _____
    $3x + y = 10$

3. $x - 3y = 5$ 3. _____
    $-x + 4y = -5$

Objective 2   Multiply when using the elimination method.

Video Examples

Review these examples for Objective 2:	**Now Try:**

4.   Solve the system.

$$3x + 8y = -2 \quad (1)$$
$$2x + 7y = 2 \quad (2)$$

To eliminate x, multiply equation (1) by 2 and multiply equation (2) by -3. Then add.

$$6x + 16y = -4$$
$$\underline{-6x - 21y = -6}$$
$$-5y = -10$$
$$y = 2$$

Substituting 2 for y in either equation (1) or (2) gives $x = -6$. Check that the solution set of the system is $\{(-6, 2)\}$.

Now Try:

4.   Solve the system.

$$3x + 4y = 24$$
$$4x + 3y = 11$$

_____

3.   Solve the system.

$$4x + 2y = 0 \quad (1)$$
$$x + 3y = 5 \quad (2)$$

Step 1   The equations are already written in $Ax + By = C$ form.

Step 2   Adding the two equations gives $5x + 5y = 5$, which does not eliminate either variable. However, multiplying equation (2) by -4 and then adding will eliminate the variable x.

Step 3   Add the two equations.

$$4x + 2y = 0$$
$$\underline{-4x - 12y = -20}$$
$$-10y = -20$$

Step 4   Solve. $y = 2$

Step 5   Find the value of x by substituting 2 for y in either of the original equations. We choose equation (2).

$$x + 3y = 5$$
$$x + 3(2) = 5$$
$$x = -1$$

Step 6   A check of the ordered pair $(-1, 2)$ by substituting $x = -1$ and $y = 2$ in both of the original equations shows that the solution set of the system is $\{(-1, 2)\}$.

3.   Solve the system.

$$2x - 3y = 11$$
$$x + 4y = -11$$

_____

Objective 2 Practice Exercises

For extra help, see Examples 3–4 on pages 322–323 of your text.

Solve each system by the elimination method. Check your answers.

4. $6x + 7y = 10$ 4. _____

      $2x - 3y = 14$

5. $8x + 6y = 10$ 5. _____

      $4x - \ y = 1$

6. $6x + y = 1$ 6. _____

      $3x - 4y = 23$

Objective 3   Use an alternative method to find the second value in a solution.

Video Examples

Review this example for Objective 3:

5.   Solve the system.

$$6x = 7 - 3y \quad (1)$$
$$8x - 5y = 5 \quad (2)$$

Write equation (1) in standard form.

$$6x + 3y = 7 \quad (3)$$
$$8x - 5y = 5 \quad (4)$$

One way to proceed is to eliminate y by multiplying each side of equation (3) by 5 and each side of equation (4) by 3, and then adding.

$$30x + 15y = 35$$
$$\underline{24x - 15y = 15}$$
$$54x \qquad = 50$$

$$x = \frac{50}{54} \text{ or } \frac{25}{27}$$

Substituting $\frac{25}{27}$ for x in one of the given

equations would give y, but the arithmetic would be complicated. Instead, solve for y by starting again with the original equations in standard form, and eliminating x. Multiply equation (3) by 4 and equation (4) by –3.

$$24x + 12y = 28$$
$$\underline{-24x + 15y = -15}$$
$$27y = 13$$

$$y = \frac{13}{27}$$

The solution set is $\left\{ \left(\frac{25}{27}, \frac{13}{27} \right) \right\}$.

Now Try:

5.   Solve the system.

$$8x = 5y + 1$$
$$6x - 8y = -2$$

_____

Objective 3 Practice Exercises

For extra help, see Example 5 on pages 323–324 of your text.

Solve each system by the elimination method. Check your answers.

7. $4x - 3y - 20 = 0$
$6x + 5y + 8 = 0$

7. _____

8. $6x = 16 - 7y$

   $4x = 3y + 26$

8. _____

9. $2x = 14 + 4y$

   $6y = -5x + 3$

9. _____

Objective 4   Solve special systems by elimination.

Video Examples

Review these examples for Objective 4:

6.   Solve each system by the elimination method.

   a. $5x + 10y = 9$ (1)

      $3x + 6y = 8$ (2)

   Multiply each side of equation (1) by 3 and each side of equation (2) by –5.

   $$15x + 30y = 27$$
   $$\underline{-15x - 30y = -40}$$
   $$0 = -13 \quad \text{False}$$

   The false statement $0 = -13$ indicates that the system has solution set ∅.

Now Try:

6.   Solve each system by the elimination method.

   a. $2x + 6y = 5$

      $5x + 15y = 8$

   _____

b. $7x + y = 9$ (1)

$-14x - 2y = -18$ (2)

Multiply each side of equation (1) by 2.

$14x + 2y = 18$

$\underline{-14x - 2y = -18}$

$0 = 0$ True

A true statement occurs when the equations are equivalent. This indicates that every solution of one equation is also a solution of the other. The solution set is $\{(x,\ y)\,|\,7x + y = 9\}$.

b. $9x - 7y = 5$

$18x = 14y + 10$

_____

Objective 4 Practice Exercises

For extra help, see Example 5 on page 324 of your text.

Solve each system by the elimination method. Use set-builder notation for dependent equations. Check your answers.

10. $12x - 8y = 3$

$6x - 4y = 6$

10. _____

11. $2x + 4y = -6$

$-x - 2y = 3$

11. _____

12. $15x + 6y = 9$

$10x + 4y = 18$

12. _____

Chapter 4 SYSTEMS OF LINEAR EQUATIONS AND INEQUALITIES

4.4   Applications of Linear Systems

Learning Objectives
1   Solve problems about unknown numbers.
2   Solve problems about quantities and their costs.
3   Solve problems about mixtures.
4   Solve problems about distance, rate (or speed), and time.

Key Terms

Use the vocabulary terms listed below to complete each statement in exercises 1–2.

     system of linear equations $d = rt$

1.   The formula that relates distance, rate, and time is _____.

2.   A _____ consists of at least two linear equations with different variables.

Objective 1   Solve problems about unknown numbers.

Video Examples

Review this example for Objective 1:	**Now Try:**
1. Two towns have a combined population of 9045. There are 2249 more people living in one than in the other. Find the population in each town.	1. A rope 82 centimeters long is cut into two pieces with one piece four more than twice as long as the other. Find the length of each piece.

Step 1 Read the problem carefully. We are to find the population of each town.

Step 2 Assign variables. Let x = the population of the larger town, and y = the population of smaller town.

Step 3 Write two equations. There are 2249 more people living in one town. The total population is 9045.

$$x = y + 2249 \quad (1)$$
$$x + y = 9045 \quad (2)$$

Step 4 Solve the system. We use substitution. Substitute $y + 2249$ for x in equation (2).

$$x + y = 9045$$
$$(y + 2249) + y = 9045$$
$$2249 + 2y = 9045$$
$$2y = 6796$$
$$y = 3398$$

To find x, substitute 3398 into either original equation. We use equation (1).
$$x = 3398 + 2249 = 5647$$

Step 5 State the answer. The population of the larger town is 5647. The population of the smaller town is 3398.

Step 6 Check the answer in the original problem. $3398 + 2249 = 5647$ and $5647 + 3398 = 9045$ The answers check.

Objective 1 Practice Exercises

For extra help, see Example 1 on pages 330–331 of your text.

Write a system of equations for each problem, then solve the problem.

1.   The difference between two numbers is 14. If two times the smaller is added to one-half the larger, the result is 52. Find the numbers.

1.
larger number _____

smaller number _____

2.   There are a total of 49 students in the two second grade classes at Jefferson School. If Carla has 7 more students in her class than Linda, find the number of students in each class.

2.
Carla's class _____

Linda's class _____

3. The perimeter of a rectangular room is 50 feet. The length is three feet greater than the width. Find the dimensions of the rectangle.

3.

length _____

width _____

Objective 2   Solve problems about quantities and their costs.

Video Examples

Review this example for Objective 2:

2. The total receipts for a basketball game were $4690.50. There were 723 tickets sold, some for children and some for adults. If the adult tickets cost $9.50 and the children's tickets cost $4, how many of each type were there?

Step1 Read the problem.

Step 2 Assign variables. Let x = the number of adult tickets and y = the number of children's tickets.

Step 3 Write two equations. The total number of tickets is 723. The total value of the tickets is $4690.50.

$$x + y = 723 \qquad (1)$$
$$9.50x + 4y = 4690.5 \quad (2)$$

Step 4 Solve the system. Solve using elimination. Multiply equation (1) by –4.
$$-4x - 4y = -2892$$
$$9.50x + 4y = 4690.5$$
$$\overline{5.50x \qquad = 1798.50}$$
$$x = 327$$

Substitute 327 for x in equation (1) to find y.
$$x + y = 723$$
$$327 + y = 723$$
$$y = 396$$

Now Try:

2. Twice as many general admission tickets to a basketball game were sold as reserved seat tickets. General admission tickets cost $10 and reserved seat tickets cost $15. If the total value of both kinds of tickets was $26,250, how many tickets of each kind were sold?

_____

Step 5 State the answer. The number of adult tickets is 327 and the number of children's tickets is 396.

Step 6 Check. The sum of the tickets is 723. The value of the tickets is

$$9.50(327) + 4(396) = 4690.50$$

This checks.

Objective 2 Practice Exercises

For extra help, see Example 2 on pages 331–332 of your text.

Write a system of equations for each problem, then solve the problem.

4. There were 411 tickets sold for a soccer game, some for students and some for nonstudents. Student tickets cost $4.25 and nonstudent tickets cost $8.50 each. The total receipts were $3021.75. How many of each type were sold?

    4.

    student tix _____

    nonstudent tix _____

5. A cashier has some $5 bills and some $10 bills. The total value of the money is $750. If the number of tens is equal to twice the number of fives, how many of each type are there?

    5.

    $5 bills _____

    $10 bills _____

6.   Luke plans to buy 10 ties with exactly $162. If some ties cost $14, and the others cost $25, how many ties of each price should he buy?

6.

$14 ties _____

$25 ties _____

Objective 3   Solve problems about mixtures.

Video Examples

Review this example for Objective 3:

3.   A mixture of 75% solution should be mixed with a 55% solution to get 70 liters of 63% solution. Determine the number of liters required of the 55% and 75% solutions.

Step 1 Read the problem carefully.

Step 2 Assign variables. Let x = the number of liters of the 75% liquid and y = the number of liters of the 55% liquid.

Step 3 Write two equations. The total amount of liquid of the final mixture is 70 liters. The amount of 75% solution mixed with 55% solution will equal the 70 liters of 63% solution.

$$x + y = 70 \qquad\qquad (1)$$
$$0.75x + 0.55y = 0.63(70) \quad (2)$$

Step 4 Solve the system. Solve by substitution. Solving equation (1) for x results in $70 - y$.

$$0.75x + 0.55y = 0.63(70)$$
$$0.75(70 - y) + 0.55y = 44.1$$
$$52.5 - 0.75y + 0.55y = 44.1$$
$$-0.20y + 52.5 = 44.1$$
$$-0.20y = -8.4$$
$$y = 42$$

Now Try:

3.   A mixture of 85% solution should be mixed with a 65% solution to get 80 ounces of 77% solution. Determine the number of ounces required of the 85% and 65% solutions.

Substitute 42 for y in equation (1).
$$x = 70 - 42 = 28$$

Step 5 State the answer. There should be 28 liters of 75% solution and 42 liters of 55% solution.

Step 6 Check the answer in the original problems. $28 + 42 = 70$ and $0.75(28) + 0.55(42) = 44.1$ The answer checks.

Objective 3 Practice Exercises

For extra help, see Example 3 on pages 332–333 of your text.

Write a system of equations for each problem, then solve the problem.

7. Jorge wishes to make 150 pounds of coffee blend that can be sold for $8 per pound. The blend will be a mixture of coffee worth $6 per pound and coffee worth $12 per pound. How many pounds of each kind of coffee should be used in the mixture?

7.

$6 coffee_____

$12 coffee_____

8. A solution of 50% acid is mixed with a 10% acid solution to get a 500 mL solution that is 40% acid. Determine the number of milliliters required of the 50% and 10% solutions.

8.

50% solution _____

10% solution _____

9.   Ben wishes to blend candy selling for $1.60 a pound with candy selling for $2.50 a pound to get a mixture that will be sold for $1.90 a pound. How many pounds of the $1.60 and the $2.50 candy should be used to get 30 pounds of the mixture?

9.

$1.60 candy _____

$2.50 candy _____

Objective 4   Solve problems about distance, rate (or speed), and time.

Video Examples

Review this example for Objective 4:

4.   Bill and Hillary start in Washington and fly in opposite directions. At the end of 4 hours, they are 4896 kilometers apart. If Bill flies 60 kilometers per hour faster than Hillary, what are their speeds?

Step 1 Read the problem carefully.

Step 2 Assign variables. Let x = Bill's rate of speed and y − Hillary's rate of speed.

Step 3 Write two equations.

$$x = 60 + y \qquad (1)$$

$$4x + 4y = 4896 \quad (2)$$

Step 4 Solve the system. . Solve by substitution.

$$4(60 + y) + 4y = 4896$$

$$240 + 4y + 4y = 4896$$

$$240 + 8y = 4896$$

$$8y = 4656$$

$$y = 582$$

Substitute 582 for y in equation (1).

$$x = 60 + 582 = 642$$

Step 5 State the answer. Bill's rate is 642 kmh and Hillary's rate is 582 kmh.

Step 6 Check. Since $4(642) + 4(582) = 4896$ and $642 = 582 + 60$ the answers check.

Now Try:

4.   Enid leaves Cherry Hill, driving by car toward New York, which is 90 miles away. At the same time, Jerry, riding his bicycle, leaves New York cycling toward Cherry Hill. Enid is traveling 28 miles per hour faster than Jerry. They pass each other $1\frac{1}{2}$ hours later. What are their speeds?

_____

Objective 4 Practice Exercises

For extra help, see Examples 4–5 on pages 334–335 of your text.

Write a system of equations for each problem, and then solve the problem.

10.   It takes Carla's boat $\frac{1}{2}$ hour to go 8 miles downstream and 1 hour to make the return trip upstream. Find the speed of the current and the speed of Carla's boat in still water.

    10.

    boat speed_____

    current speed _____

11.   Two planes left Philadelphia traveling in opposite directions. Plane A left 15 minutes before plane B. After plane B had been flying for 1 hour, the planes were 860 miles apart. What were the speeds of the two planes if plane A was flying 40 miles per hour faster than plane B?

    11.

    plane A_____

    plane B_____

12.   At the beginning of a fund-raising walk, Steve and Vic are 30 miles apart. If they leave at the same time and walk in the same direction, Steve would overtake Vic in 15 hours. If they walked toward each other, they would meet in 3 hours. What are their speeds?

    12.

    Steve_____

    Vic _____

Chapter 4 SYSTEMS OF LINEAR EQUATIONS AND INEQUALITIES

4.5   Solving Systems of Linear Inequalities

Learning Objectives
1   Solve systems of linear inequalities by graphing.

Key Terms

Use the vocabulary terms listed below to complete each statement in exercises 1–2.

> **system of linear inequalities**
>
> **solution set of a system of linear inequalities**

1.   All ordered pairs that make all inequalities of the system true at the same time is called the _____.

2.   A _____ contains two or more linear inequalities (and no other kinds of inequalities).

Objective 1   Solve systems of linear inequalities by graphing.

Video Examples

Review these examples for Objective 1:

1. Graph the solution set of the system.
$$x + y \le 3$$
$$5x - y \ge 5$$

To graph $x + y \le 3$, graph the solid boundary line $x + y = 3$ using the intercepts (0, 3) and (3, 0). Determine the region to shade using (0, 0) as a test point.
$$x + y \le 3$$
$$0 + 0 \overset{?}{\le} 3$$
$$0 \le 3 \quad \text{True}$$

Shade the region containing (0, 0).

To graph $5x - y \ge 5$, graph the solid boundary line $5x - y = 5$ using the intercepts (0, –5) and (1, 0). Determine the region to shade using (0, 0) as a test point.
$$5x - y \ge 5$$
$$5(0) - 0 \overset{?}{\ge} 5$$
$$0 \ge 5 \quad \text{False}$$

Shade the region that does not contain (0, 0).

Now Try:

1. Graph the solution set of the system.
$$3x - y \le 3$$
$$x + y \le 0$$

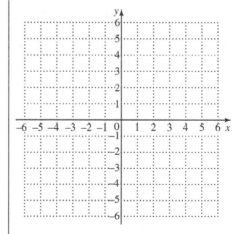

The solution set of this system includes all points in the intersection (overlap) of the graph of the two inequalities. This intersection is the gray shaded region and portions of the two boundary lines that surround it.

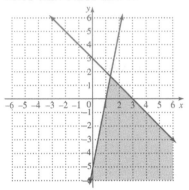

2. Graph the solution set of the system.

$$4x - y > 2$$

$$x + y > -2$$

Graph $4x - y > 2$ using a dashed line for

$4x - y = 2$ and intercepts $(0, -2)$ and $\left(\frac{1}{2}, 0\right)$.

Using the test point $(0, 0)$, we shade the region that does not contain the point $(0, 0)$.

Graph $x + y > -2$ using a dashed line for $x + y = -2$ and intercepts $(-2, 0)$ and $(0, -2)$.

Using the test point $(0, 0)$, we shade the region that does contain the point $(0, 0)$.

The solution set is marked in gray. The solution set does not include either boundary line.

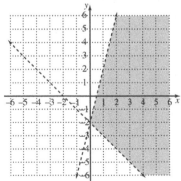

2. Graph the solution set of the system.

$$6x - y > 6$$

$$2x + 5y < 10$$

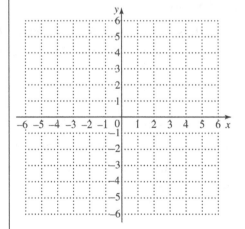

3. Graph the solution set of the system.
$$y \geq -1$$
$$2x - y > -1$$

Recall that $y = -1$ is a horizontal line through the point $(0, -1)$. Use a solid line, and shade the region with the point $(0, 0)$.

The graph of $2x - y > -1$ is created using a dashed line through the intercepts $(0, 1)$ and $\left(-\frac{1}{2}, 0\right)$. Shade the region with the point $(0, 0)$.

The solution set is marked in gray. The solution set includes the boundary line $y \geq -1$.

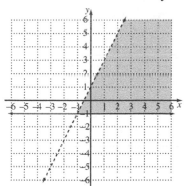

3. Graph the solution set of the system.
$$x - 3y \leq -7$$
$$x < 2$$

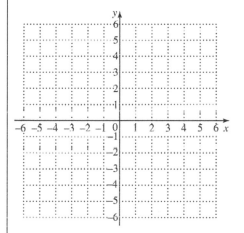

Objective 1 Practice Exercises

For extra help, see Examples 1–3 on pages 342–344 of your text.

Graph the solution of each system of linear inequalities.

1. $4x + 5y \leq 20$
$$y \leq x + 3$$

1.

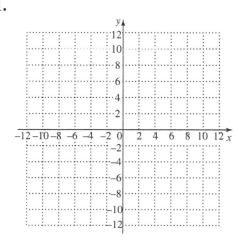

2. $x < 2y + 3$

$0 < x + y$

2.

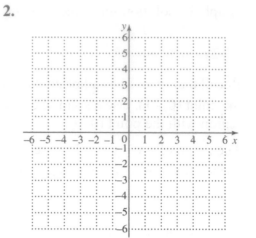

3. $y < 4$

$x \geq -3$

3.

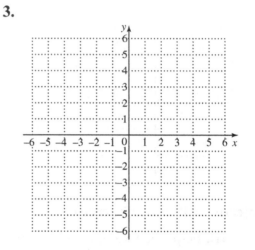

Chapter 5 EXPONENTS AND POLYNOMIALS

5.1   The Product Rule and Power Rules for Exponents

Learning Objectives

1   Use exponents.
2   Use the product rule for exponents.
3   Use the rule $(a^m)^n = a^{mn}$.
4   Use the rule $(ab)^m = a^m b^m$.
5   Use the rule $\left(\dfrac{a}{b}\right)^m = \dfrac{a^m}{b^m}$.
6   Use combinations of the rules for exponents.
7   Use the rules for exponents in a geometry problem.

Key Terms

Use the vocabulary terms listed below to complete each statement in exercises 1–3.

exponential expression   base   power

1. 2^5 is read "2 to the fifth _____".

2.   A number written with an exponent is called a(n) _____.

3.   The _____ is the number being multiplied repeatedly.

Objective 1   Use exponents.

Video Examples

Review these examples for Objective 1:

1.   Write $5 \cdot 5 \cdot 5$ in exponential form and evaluate.

   Since 5 occurs as a factor three times, the base is 5 and the exponent is 3.

   $$5 \cdot 5 \cdot 5 = 5^3 = 125$$

2.   Name the base and exponent of each expression. Then evaluate.

   a. 3^4

   Base: 3
   Exponent: 4
   Value: $3^4 = 3 \cdot 3 \cdot 3 \cdot 3 = 81$

Now Try:

1.   Write $4 \cdot 4 \cdot 4 \cdot 4 \cdot 4$ in exponential form and evaluate.

   _____

2.   Name the base and exponent of each expression. Then evaluate.

   a. 2^6

   _____

b. -3^4

Base: 3
Exponent: 4
Value: $-3^4 = -1 \cdot (3 \cdot 3 \cdot 3 \cdot 3) = -81$

c. $(-3)^4$

Base: -3
Exponent: 4
Value: $(-3)^4 = (-3)(-3)(-3)(-3) = 81$

b. -2^6

_____

c. $(-2)^6$

_____

Objective 1 Practice Exercises

For extra help, see Examples 1–2 on page 358 of your text.

Write the expression in exponential form.

1. $\left(\frac{1}{3}\right)\left(\frac{1}{3}\right)\left(\frac{1}{3}\right)\left(\frac{1}{3}\right)\left(\frac{1}{3}\right)$

1. _____

Evaluate each exponential expression. Name the base and the exponent.

2. $(-4)^4$

2. _____

base_____

exponent_____

3. -3^8

3. _____

base_____

exponent_____

Objective 2   Use the product rule for exponents.

Video Examples

Review these examples for Objective 2:

3. Use the product rule for exponents to simplify, if possible.

a. $8^4 \cdot 8^5$

$8^4 \cdot 8^5 = 8^{4+5}$

$= 8^9$

Now Try:

3. Use the product rule for exponents to simplify, if possible.

a. $9^6 \cdot 9^7$

_____

Copyright © 2018 Pearson Education, Inc.

d. $m^7 m^8 m^9$

$$m^7 m^8 m^9 = m^{7+8+9}$$
$$= m^{24}$$

4. Multiply $5x^4$ and $6x^9$.

$$5x^4 \cdot 6x^9 = (5 \cdot 6) \cdot (x^4 \cdot x^9)$$
$$= 30x^{4+9}$$
$$= 30x^{13}$$

3f. Use the product rule for exponents to simplify, if possible.

$$5^2 + 5^3$$

$$5^2 + 5^3 = 25 + 125$$
$$= 150$$

d. $m^{11} m^9 m^7$

_____

4. Multiply $6x^5$ and $3x^6$.

_____

3f. Use the product rule for exponents to simplify, if possible.

$$3^4 + 3^3$$

_____

Objective 2 Practice Exercises

For extra help, see Examples 3–4 on pages 359–3360 of your text.

Use the product rule to simplify each expression, if possible. Write each answer in exponential form.

4. $7^4 \cdot 7^3$

4. _____

5. $(-2c^7)(-4c^8)$

5. _____

6. $(3k^7)(-8k^2)(-2k^9)$

6. _____

Objective 3   Use the rule $\left(a^m\right)^n = a^{mn}$.

Video Examples

Review these examples for Objective 3:

5.   Use power rule (a) for exponents to simplify.

   a. $\left(3^4\right)^5$

   $$\left(3^4\right)^5 = 3^{4\cdot 5}$$
   $$= 3^{20}$$

   c. $\left(x^3\right)^4$

   $$\left(x^3\right)^4 = x^{3\cdot 4}$$
   $$= x^{12}$$

Now Try:

5.   Use power rule (a) for exponents to simplify.

   a. $\left(4^5\right)^3$

   _____

   c. $\left(x^5\right)^6$

   _____

Objective 3 Practice Exercises

For extra help, see Example 5 on page 360 of your text.

Simplify each expression. Write all answers in exponential form.

7. $\left(7^3\right)^4$

7.   _____

8. $-\left(v^4\right)^9$

8.   _____

9. $\left[(-3)^3\right]^7$

9.   _____

Objective 4   Use the rule $\left(ab\right)^m = a^m b^m$.

Video Examples

Review this example for Objective 4:

6a. Use power rule (b) for exponents to simplify.

   $\left(5xy\right)^3$

   $$\left(5xy\right)^3 = 5^3 x^3 y^3$$
   $$= 125 x^3 y^3$$

Now Try:

6a. Use power rule (b) for exponents to simplify.

   $\left(4ab\right)^3$

   _____

Objective 4 Practice Exercises

For extra help, see Example 6 on page 361 of your text.

Simplify each expression.

10. $\left(5r^3t^2\right)^4$

10. _____

11. $\left(-0.2a^4b\right)^3$

11. _____

12. $\left(-2w^3z^7\right)^4$

12. _____

Objective 5   **Use the rule** $\left(\dfrac{a}{b}\right)^m = \dfrac{a^m}{b^m}$.

Video Examples

Review this example for Objective 5:
7b. Use power rule (c) for exponents to simplify.

$$\left(\frac{1}{8}\right)^3$$

$$\left(\frac{1}{8}\right)^3 = \frac{1^3}{8^3} = \frac{1}{512}$$

Now Try:
7b. Use power rule (c) for exponents to simplify.

$$\left(\frac{1}{4}\right)^5$$

_____

Objective 5 Practice Exercises

For extra help, see Example 7 on page 362 of your text.

Simplify each expression.

13. $\left(-\dfrac{2x}{5}\right)^3$

13. _____

14. $\left(\dfrac{xy}{z^2}\right)^4$

14. _____

15. $\left(\dfrac{-2a}{b^2}\right)^7$

15. _____

Objective 6   Use combinations of the rules for exponents.

Video Examples

Review these examples for Objective 6:

8.   Simplify each expression.

a. $\left(\dfrac{3}{4}\right)^3 \cdot 3^2$

$$\left(\dfrac{3}{4}\right)^3 \cdot 3^2 = \dfrac{3^3}{4^3} \cdot \dfrac{3^2}{1}$$

$$= \dfrac{3^3 \cdot 3^2}{4^3 \cdot 1}$$

$$= \dfrac{3^{3+2}}{4^3}$$

$$= \dfrac{3^5}{4^3}, \quad \text{or} \quad \dfrac{243}{64}$$

d. $\left(-x^5 y\right)^4 \left(-x^6 y^5\right)^3$

$$\left(-x^5 y\right)^4 \left(-x^6 y^5\right)^3$$

$$= \left(-1 x^5 y\right)^4 \left(-1 x^6 y^5\right)^3$$

$$= (-1)^4 \left(x^5\right)^4 \left(y^4\right) \cdot (-1)^3 \left(x^6\right)^3 \left(y^5\right)^3$$

$$= (-1)^4 \left(x^{20}\right)\left(y^4\right) \cdot (-1)^3 \left(x^{18}\right)\left(y^{15}\right)$$

$$= (-1)^7 x^{20+18} y^{4+15}$$

$$= -1 x^{38} y^{19}$$

$$= -x^{38} y^{19}$$

Now Try:

8.   Simplify each expression.

a. $\left(\dfrac{5}{2}\right)^3 \cdot 5^2$

_____

d. $\left(-x^5 y\right)^3 \left(-x^6 y^5\right)^2$

_____

Objective 6 Practice Exercises

For extra help, see Example 8 on pages 362–363 of your text.

Simplify. Write all answers in exponential form.

16. $\left(-x^3\right)^2 \left(-x^5\right)^4$

16. _____

17. $\left(2ab^2c\right)^5 \left(ab\right)^4$

17. _____

18. $\left(5x^2y^3\right)^7 \left(5xy^4\right)^4$

18. _____

Objective 7   Use the rules for exponents in a geometry problem.

Video Examples

Review this example for Objective 7:
9a. Find the area of the figure.

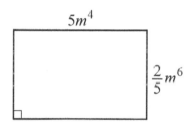
$5m^4$

$\frac{2}{5}m^6$

Use the formula for the area of a rectangle.

$A = LW$

$A = \left(5m^4\right)\left(\frac{2}{5}m^6\right)$

$A = 5 \cdot \frac{2}{5} \cdot m^{4+6}$

$A = 2m^{10}$

Now Try:
9a. Find the area of the figure.

$7x^3$

$4x^2$

_____

Objective 7 Practice Exercises

For extra help, see Example 9 on page 363 of your text.

Find a polynomial that represents the area of each figure.

19.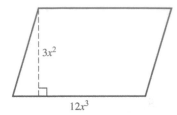

19. _____

20.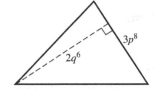

20. _____

21.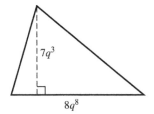

21. _____

Chapter 5 EXPONENTS AND POLYNOMIALS

5.2   **Integer Exponents and the Quotient Rule**

Learning Objectives
1   Use 0 as an exponent.
2   Use negative numbers as exponents.
3   Use the quotient rule for exponents.
4   Use combinations of the rules for exponents.

Key Terms

Use the vocabulary terms listed below to complete each statement in exercises 1–3.

    exponent   **base**   **product rule for exponents**

    power rule for exponents

1.   The statement "If m and n are any integers, then $\left(a^m\right)^n = a^{mn}$" is an example of the

         _____.

2.   In the expression a^m, a is the _____ and m is the _____.

3.   The statement "If m and n are any integers, then $a^m \cdot a^n = a^{m+n}$ is an example of the

         _____.

Objective 1   Use 0 as an exponent.

Video Examples

Review these examples for Objective 1:

1.   Evaluate.

    a. $75^0 - 1$

    b. $(-75)^0 = 1$

    c. $-75^0 = -(1)$ or -1

    d. $x^0 = 1$ $(x \neq 0)$

Now Try:

1.   Evaluate.

    a. 88^0

         _____

    b. $(-88)^0$

         _____

    c. -88^0

         _____

    d. a^0 $(a \neq 0)$

         _____

e. $9x^0 = 9(1)$, or 9 $\quad (x \neq 0)$

e. $88a^0 \quad (a \neq 0)$

_____

f. $(9x)^0 = 1 \quad (x \neq 0)$

f. $(88a)^0 \quad (a \neq 0)$

_____

Objective 1 Practice Exercises

For extra help, see Example 1 on page 366 of your text.

Evaluate each expression.

1. -12^0

1. _____

2. $-15^0 - (-15)^0$

2. _____

3. $\dfrac{0^8}{8^0}$

3. _____

Objective 2   Use negative numbers as exponents.

Video Examples

Review these examples for Objective 2:

2.   Simplify by writing with positive exponents. Assume that all variables represent nonzero real numbers.

a. 5^{-2}

$5^{-2} = \dfrac{1}{5^2}$, or $\dfrac{1}{25}$

d. $\left(\dfrac{3}{5}\right)^{-4}$

$\left(\dfrac{3}{5}\right)^{-4} = \left(\dfrac{5}{3}\right)^4$ The reciprocal of $\dfrac{3}{5}$ is $\dfrac{5}{3}$.

$= \dfrac{5^4}{3^4}$

$= \dfrac{625}{81}$

Now Try:

2.   Simplify by writing with positive exponents. Assume that all variables represent nonzero real numbers.

a. 8^{-2}

_____

d. $\left(\dfrac{6}{7}\right)^{-2}$

_____

f. $5^{-1} - 3^{-1}$

$$5^{-1} - 3^{-1} = \frac{1}{5} - \frac{1}{3}$$

$$= \frac{3}{15} - \frac{5}{15}$$

$$= -\frac{2}{15}$$

h. $\dfrac{1}{x^{-6}}$

$$\frac{1}{x^{-6}} = \frac{1^{6}}{x^{-6}}$$

$$= \left(\frac{1}{x}\right)^{-6}$$

$$= x^{6}$$

3. Simplify. Assume that all variables represent nonzero real numbers.

b. $\dfrac{a^{-6}}{b^{-1}} = \dfrac{b^{1}}{a^{6}}, \quad \text{or} \quad \dfrac{b}{a^{6}}$

c. $\dfrac{x^{-3}y}{4z^{-4}} = \dfrac{yz^{4}}{4x^{3}}$

d. $\left(\dfrac{p}{3q}\right)^{-3}$

$$\left(\frac{p}{3q}\right)^{-3} = \left(\frac{3q}{p}\right)^{3}$$

$$= \frac{3^{3}q^{3}}{p^{3}}$$

$$= \frac{27q^{3}}{p^{3}}$$

f. $4^{-1} - 8^{-1}$

_____

h. $\dfrac{1}{x^{-8}}$

_____

3. Simplify. Assume that all variables represent nonzero real numbers.

b. $\dfrac{x^{-7}}{y^{-1}}$

_____

c. $\dfrac{p^{-3}q}{4r^{-5}}$

_____

d. $\left(\dfrac{a}{5b}\right)^{-4}$

_____

Objective 2 Practice Exercises

For extra help, see Examples 2–3 on pages 367–369 of your text.

Evaluate or simplify each expression, and write it using only positive exponents. Assume that all variables represent nonzero real numbers.

4. $-2k^{-4}$

4. _____

5. $(m^2n)^{-9}$

5. _____

6. $\dfrac{2x^{-4}}{3y^{-7}}$

6. _____

Objective 3   Use the quotient rule for exponents.

Video Examples

Review these examples for Objective 3:

4.   Simplify. Assume that all variables represent nonzero real numbers.

   b. $\dfrac{5^3}{5^7} = 5^{3-7} = 5^{-4} = \dfrac{1}{5^4} = \dfrac{1}{625}$

   d. $\dfrac{p^6}{p^{-4}} = p^{6-(-4)} = p^{10}$

Now Try:

4.   Simplify. Assume that all variables represent nonzero real numbers.

   b. $\dfrac{2^6}{2^{10}}$

   _____

   d. $\dfrac{z^6}{z^{-4}}$

   _____

e. $\dfrac{4^3 x^6}{4^5 x^2}$ **e.** $\dfrac{5^4 a^6}{5^6 a^3}$

$$\dfrac{4^3 x^6}{4^5 x^2} = \dfrac{4^3}{4^5} \cdot \dfrac{x^6}{x^2}$$

$$= 4^{3-5} \cdot x^{6-2}$$

$$= 4^{-2} x^4$$ _____

$$= \dfrac{x^4}{4^2}, \quad \text{or} \quad \dfrac{x^4}{16}$$

Objective 3 Practice Exercises

For extra help, see Example 4 on page 370 of your text.

Use the quotient rule to simplify each expression, and write it using only positive exponents. Assume that all variables represent nonzero real numbers.

7. $\dfrac{4k^7 m^{10}}{8k^3 m^5}$ **7.** _____

8. $\dfrac{a^4 b^3}{a^{-2} b^{-3}}$ **8.** _____

9. $\dfrac{3^{-1} m^{-4} p^6}{3^4 m^{-1} p^{-2}}$ **9.** _____

Objective 4   Use combinations of the rules for exponents.

Video Examples

Review these examples for Objective 4:

5. Simplify each expression. Assume that all variables represent nonzero real numbers.

d. $\left(\dfrac{3x^4}{4}\right)^{-5}$

$$\left(\dfrac{3x^4}{4}\right)^{-5} = \left(\dfrac{4}{3x^4}\right)^{5}$$

$$= \dfrac{4^5}{3^5 x^{20}}$$

$$= \dfrac{1024}{243x^{20}}$$

e. $\left(\dfrac{5x^{-3}}{2^{-1}y^4}\right)^{-2}$

$$\left(\dfrac{5x^{-3}}{2^{-1}y^4}\right)^{-2} = \dfrac{5^{-2}x^6}{2^2 y^{-8}}$$

$$= \dfrac{x^6 y^8}{5^2 \cdot 2^2}$$

$$= \dfrac{x^6 y^8}{100}$$

Now Try:

5. Simplify each expression. Assume that all variables represent nonzero real numbers.

d. $\left(\dfrac{2p^4}{3}\right)^{-5}$

_____

e. $\left(\dfrac{4x^{-4}}{5^{-1}y^5}\right)^{-3}$

_____

Objective 4 Practice Exercises

For extra help, see Example 5 on pages 371–372 of your text.

Simplify each expression, and write it using only positive exponents. Assume that all variables represent nonzero real numbers.

10. $(9xy)^7 (9xy)^{-8}$

10. _____

11. $\dfrac{\left(a^{-1}b^{-2}\right)^{-4}\left(ab^2\right)^6}{\left(a^3b\right)^{-2}}$

11. _____

12. $\left(\dfrac{k^3 t^4}{k^2 t^{-1}}\right)^{-4}$

12. _____

Chapter 5 EXPONENTS AND POLYNOMIALS

5.3   An Application of Exponents: Scientific Notation

Learning Objectives
1   Express numbers in scientific notation.
2   Convert numbers in scientific notation to standard notation.
3   Use scientific notation in calculations.

Key Terms

Use the vocabulary terms listed below to complete each statement in exercises 1–3.

       scientific notation   **quotient rule**   **power rule**

1.   A number written as $a \times 10^n$, where $1 \leq |a| < 10$ and n is an integer, is written in

      _____.

2.   The statement "If m and n are any integers and $b \neq 0$, then $\left(\dfrac{a}{b}\right)^m = \dfrac{a^m}{b^m}$" is an

      example of the _____.

3.   The statement "If m and n are any integers and $b \neq 0$, then $\dfrac{a^m}{a^n} = a^{m-n}$" is an

      example of the _____.

Objective 1   Express numbers in scientific notation.

Video Examples

Review these examples for Objective 1:	Now Try:
1. Write each number in scientific notation.	1. Write each number in scientific notation.
b. 84,300,000,000	**b.** 47,710,000,000
Move the decimal point 10 places to the left. $$84,300,000,000 = 8.43 \times 10^{10}$$	_____
c. 0.00573	**c.** 0.0463
The first nonzero digit is 5. Count the places. Move the decimal point 3 places to the right. $$0.00573 = 5.73 \times 10^{-3}$$	_____

Objective 1 Practice Exercises

For extra help, see Example 1 on page 378 of your text.

Write each number in scientific notation.

1.   23,651   1. _____

2.   −429,600,000,000   2. _____

3.   0.0002208   3. _____

Objective 2   Convert numbers in scientific notation to standard notation.

Video Examples

Review these examples for Objective 2:
2.   Write each number in standard notation.

    b. 3.57×10^6

    Move the decimal point 6 places to the right, and add four zeros.

       $3.57 \times 10^6 = 3,570,000$

    d. 8.98×10^{-3}

    Move the decimal point 3 places to the left.
      $-8.98 \times 10^{-3} = -0.00898$

Now Try:
2.   Write each number in standard notation.

    b. 2.796×10^7

         _____

    d. -1.64×10^{-4}

         _____

Objective 2 Practice Exercises

For extra help, see Example 2 on page 379 of your text.

Write each number in standard notation.

4. -2.45×10^6 4. _____

5. 6.4×10^{-3} 5. _____

6. -4.02×10^0 6. _____

Objective 3 Use scientific notation in calculations.

Video Examples

Review these examples for Objective 3:

3. Perform each calculation. Write answers in scientific notation and standard notation.

a. $(8 \times 10^4)(7 \times 10^3)$

$$(8 \times 10^4)(7 \times 10^3) = (8 \times 7)(10^4 \times 10^3)$$
$$= 56 \times 10^7$$
$$= (5.6 \times 10^1) \times 10^7$$
$$= 5.6 \times 10^8$$
$$= 560,000,000$$

b. $\dfrac{6 \times 10^{-4}}{3 \times 10^2}$

$$\frac{6 \times 10^{-4}}{3 \times 10^2} = \frac{6}{3} \times \frac{10^{-4}}{10^2}$$
$$= 2 \times 10^{-6}$$
$$= 0.000002$$

4. A light-year is the distance that light travels in one year. The speed of light is about 3×10^5 km per second. How many kilometers are in a light-year?

Convert km per second to km per year.

$$\frac{3 \times 10^5 \, \text{km}}{1 \, \text{sec}} \cdot \frac{60 \, \text{sec}}{1 \, \text{min}} \cdot \frac{60 \, \text{min}}{1 \, \text{hr}} \cdot \frac{24 \, \text{hr}}{1 \, \text{day}} \cdot \frac{365 \, \text{day}}{1 \, \text{yr}}$$
$$= (3 \times 60^2 \times 24 \times 365) \times 10^5$$
$$= 94608000 \times 10^5$$
$$= (9.4608 \times 10^7) \times 10^5$$
$$\approx 9.46 \times 10^{12} \, \text{km/yr}$$

Thus, there are 9.46×10^{12} km in a light-year.

Now Try:

3. Perform each calculation. Write answers in scientific notation and standard notation.

a. $(9 \times 10^5)(3 \times 10^2)$

_____

b. $\dfrac{39 \times 10^{-3}}{13 \times 10^5}$

_____

4. Earth has a mass of 6×10^{24} kilograms and a volume of 1.1×10^{21} cubic meters. What is Earth's density in kilograms per cubic meter? Round to the nearest hundredth.

_____

Name: Date:
Instructor: Section:

Objective 3 Practice Exercises

For extra help, see Examples 3–5 on pages 379–380 of your text.

Perform the indicated operations, and write the answers in scientific notation.

7. $(2.3 \times 10^4) \times (1.1 \times 10^{-2})$

7. _____

8. $\dfrac{9.39 \times 10^1}{3 \times 10^3}$

8. _____

Work the problem. Give answer in scientific notation.

9. There are about 6×10^{23} atoms in a mole of atoms. How many atoms are there in 8.1×10^{-5} mole?

9. _____

Chapter 5 EXPONENTS AND POLYNOMIALS

5.4   Adding and Subtracting Polynomials

Learning Objectives
1   Identify terms and coefficients.
2   Combine like terms.
3   Know the vocabulary for polynomials.
4   Evaluate polynomials.
5   Add polynomials.
6   Subtract polynomials.
7   Add and subtract polynomials with more than one variable.

Key Terms

Use the vocabulary terms listed below to complete each statement in exercises 1−9.

term	like terms	polynomial
descending powers		degree of a term
degree of a polynomial		monomial
binomial		trinomial

1.   The _____ is the sum of the exponents on the variables in that term.

2.   A polynomial in x is written in _____ if the exponents on x in its terms are decreasing order.

3.   A _____ is a number, a variable, or a product or quotient of a number and one or more variables raised to powers.

4.   A polynomial with exactly three terms is called a _____.

5.   A _____ is a term, or the sum of a finite number of terms with whole number exponents.

6.   A polynomial with exactly one term is called a _____.

7.   The _____ is the greatest degree of any term of the polynomial.

8.   A _____ is a polynomial with exactly two terms.

9.   Terms with exactly the same variables (including the same exponents) are called _____.

Name: Date:
Instructor: Section:

Objective 1   Identify terms and coefficients.

Video Examples

Review these examples for Objective 1:

1. Identify the coefficient of each term in the expression. Then give the number of terms.

   a. $-4x^3 + x - 7$

   $-4x^3 + x - 7 = -4x^3 + 1x + (-7x^0)$
   The coefficients are –4, 1, and –7.
   There are 3 terms: $-4x^3$, x, and -7.

   b. $6 - y^2$

   $6 - y^2 = 6y^0 + (-1)y^2$
   The coefficients are 6 and –1.
   There are 2 terms.

Now Try:

1. Identify the coefficient of each term in the expression. Then give the number of terms.

   a. $-2w^5 - w + 3$

   _____

   b. $-8 + x^4$

   _____

Objective 1 Practice Exercises

For extra help, see Example 1 on page 385 of your text.

Identify the coefficient of each term in the expression. Then give the number of terms.

1. x^6

   1. _____

2. $\dfrac{m}{7}$

   2. _____

3. $\dfrac{2}{5}y^2 - 3y^3 + y^4$

   3. _____

Objective 2   Combine like terms.

Video Examples

Review these examples for Objective 2:

2.   Simplify each expression by combining like terms.

   e. $19m^3 + 6m + 5m^3$

     $19m^3 + 6m + 5m^3 = (19 + 5)m^3 + 6m$

                    $= 24m^3 + 6m$

   b. $4x^5 - 15x^5 + x^5$

     $4x^5 - 15x^5 + x^5 = (4 - 15 + 1)x^5$

                    $= -10x^5$

Now Try:

2.   Simplify each expression by combining like terms.

   e. $22m^2 + 15m^3 + 7m^2$

   _____

   b. $9x^7 - 18x^7 + x^7$

   _____

Objective 2 Practice Exercises

For extra help, see Example 2 on page 386 of your text.

In each polynomial, combine like terms whenever possible. Write the result with descending powers.

4. $7z^3 - 4z^3 + 5z^3 - 11z^3$

4. _____

5. $-1.3z^7 + 0.4z^7 + 2.6z^8$

5. _____

6. $6c^3 - 9c^2 - 2c^2 + 14 + 3c^2 - 6c - 8 + 2c^3$

6. _____

Name: _____ Date: _____

Instructor: _____ Section: _____

Objective 3 Know the vocabulary for polynomials.

Video Examples

Review these examples for Objective 3:

3. Simplify each polynomial if possible. Then give the degree and tell whether the polynomial is a monomial, a binomial, a trinomial, or none of these.

   a. $5x^4 + 7x$

   We cannot simplify further. This is a binomial of degree 4.

   d. $9x - 7x + 3x$

   $9x - 7x + 3x = 5x$
   The degree is 1. The simplified polynomial is a monomial.

Now Try:

3. Simplify each polynomial if possible. Then give the degree and tell whether the polynomial is a monomial, a binomial, a trinomial, or none of these.

   a. $8x^3 + 4x^2 + 6$

   _____

   d. $x^5 + 3x^5$

   _____

Objective 3 Practice Exercises

For extra help, see Example 3 on page 387 of your text.

For each polynomial, first simplify, if possible, and write the resulting polynomial in descending powers of the variable. Then give the degree of this polynomial, and tell whether it is a monomial, *a* binomial, *a* trinomial, *or* none of these.

7. $3n^8 - n^2 - 2n^8$

7. _____

   degree: _____

   type: _____

8. $-d^2 + 3.2d^3 - 5.7d^8 - 1.1d^5$

8. _____

   degree: _____

   type: _____

9. $-6c^4 - 6c^2 + 9c^4 - 4c^2 + 5c^5$

9. _____

   degree: _____

   type: _____

Objective 4   Evaluate polynomials.

Video Examples

Review these examples for Objective 4:

4. Find the value of $4x^3 + 6x^2 - 5x - 5$ for

    a. $x = -3$

$4x^3 + 6x^2 - 5x - 5$

$= 4(-3)^3 + 6(-3)^2 - 5(-3) - 5$

$= 4(-27) + 6(9) - 5(-3) - 5$

$= -108 + 54 + 15 - 5$

$= -44$

    b. $x = 2$

$4x^3 + 6x^2 - 5x - 5$

$= 4(2)^3 + 6(2)^2 - 5(2) - 5$

$= 4(8) + 6(4) - 5(2) - 5$

$= 32 + 24 - 10 - 5$

$= 41$

Now Try:

4. Find the value of
$5x^4 + 3x^2 - 9x - 7$ for

    a. $x = 4$

    _____

    b. $x = -4$

    _____

Objective 4 Practice Exercises

For extra help, see Example 4 on page 388 of your text.

Find the value of each polynomial (a) *when x = –2 and* (b) *when x = 3.*

10. $3x^3 + 4x - 19$

    10. **a.**_____

        b._____

11. $-4x^3 + 10x^2 - 1$

    11. **a.**_____

        b._____

12. $x^4 - 3x^2 - 8x + 9$

12. a. _____

b. _____

Objective 5   Add polynomials.

Video Examples

Review these examples for Objective 5:

5a. Add vertically.

$5x^4 - 7x^3 + 9$ and $-3x^4 + 8x^3 - 7$

Write like terms in columns.

$5x^4 - 7x^3 + 9$
$\underline{-3x^4 + 8x^3 \quad 7}$

Now add, column by column.

$5x^4 \quad -7x^3 \quad \ 9$
$\underline{-3x^4} \quad \underline{\ 8x^3} \quad \underline{-7}$
$2x^4 \quad \ \ x^3 \quad \ 2$

Add the three sums together to obtain the answer.

$2x^4 + x^3 + 2$

6b. Find the sum.

$(5x^4 - 7x^2 + 6x) + (-3x^3 + 4x^2 - 7)$

$(5x^4 - 7x^2 + 6x) + (-3x^3 + 4x^2 - 7)$

$= 5x^4 - 3x^3 - 7x^2 + 4x^2 + 6x - 7$

$= 5x^4 - 3x^3 - 3x^2 + 6x - 7$

Now Try:

5a. Add vertically.

$8x^3 - 9x^2 + x$ and
$-3x^3 + 4x^2 + 3x$

_____

6b. Find the sum.

$(8x^2 - 6x + 4) + (7x^3 - 8x - 5)$

_____

Objective 5 Practice Exercises

For extra help, see Examples 5–6 on pages 388–389 of your text.

Add.

13. $9m^3 + 4m^2 - 2m + 3$

       $\underline{-4m^3 - 6m^2 - 2m + 1}$

13. _____

14. $(x^2 + 6x - 8) + (3x^2 - 10)$

14. _____

15. $(3r^3 + 5r^2 - 6) + (2r^2 - 5r + 4)$

15. _____

Objective 6   Subtract polynomials.

Video Examples

Review this example for Objective 6:
7b. Perform the subtraction.

    Subtract $8x^3 - 5x^2 + 8$ from $9x^3 + 6x^2 - 7$.

    $(9x^3 + 6x^2 - 7) - (8x^3 - 5x^2 + 8)$

    $= (9x^3 + 6x^2 - 7) + (-8x^3 + 5x^2 - 8)$

    $= x^3 + 11x^2 - 15$

Now Try:
7b. Perform the subtraction.

    $(7x^3 - 3x - 5) - (18x^3 + 4x - 6)$

    _____

Name: _____ Date: _____

Instructor: _____ Section: _____

Objective 6 Practice Exercises

For extra help, see Examples 7–9 on pages 389–390 of your text.

Subtract.

16. $\left(-8w^3 + 11w^2 - 12\right) - \left(-10w^2 + 3\right)$

16. _____

17. $\left(8b^4 - 4b^3 + 7\right) \ \left(2b^2 + b + 9\right)$

17. _____

18. $\left(9x^3 + 7x^2 - 6x + 3\right) - \left(6x^3 - 6x + 1\right)$

18. _____

Objective 7   Add and subtract polynomials with more than one variable.

Video Examples

Review this example for Objective 7:

10b.   Add or subtract as indicated.

$\left(3x^2 y + 5xy + y^2\right) - \left(4x^2 y + xy - 3y^2\right)$

$\left(3x^2 y + 5xy + y^2\right) - \left(4x^2 y + xy - 3y^2\right)$

$= 3x^2 y + 5xy + y^2 - 4x^2 y - xy + 3y^2$

$= -x^2 y + 4xy + 4y^2$

Now Try:

10b.   Add or subtract as indicated.

$\left(7x^2 y + 3xy + 4y^2\right)$

$-\left(6x^2 y - xy + 4y^2\right)$

_____

Name: _____ Date: _____

Instructor: _____ Section: _____

Objective 7 Practice Exercises

For extra help, see Example 10 on page 390 of your text.

Add or subtract as indicated.

19. $\left(-2a^6 + 8a^4b - b^2\right) - \left(a^6 + 7a^4b + 2b^2\right)$ 19. _____

20. $\left(4ab + 2bc - 9ac\right) + \left(3ca - 2cb - 9ba\right)$ 20. _____

21. $\left(2x^2y + 2xy - 4xy^2\right) + \left(6xy + 9xy^2\right) - \left(9x^2y + 5xy\right)$ 21. _____

Chapter 5 EXPONENTS AND POLYNOMIALS

5.5   Multiplying Polynomials

Learning Objectives
1   Multiply monomials.
2   Multiply a monomial and a polynomial.
3   Multiply two polynomials.
4   Multiply binomials using the FOIL method.

Key Terms

Use the vocabulary terms listed below to complete each statement in exercises 1–3.

        FOIL   **outer product**   **inner product**

1.   The _____ of $(2y-5)(y+8)$ is $-5y$.

2.   _____ is a shortcut method for finding the product of two binomials.

3.   The _____ of $(2y-5)(y+8)$ is $16y$.

Objective 1   Multiply monomials.

Video Examples

Review this example for Objective 1:	**Now Try:**
1a. Find the product.	**1a.** Find the product.
$3w^2(-7w)$	$9p^3(-8p)$
Use the commutative and associative properties.	
$3w^2(-7w) = 3(-7) \cdot w^2 \cdot w$	
$\qquad\qquad = -21w^{2+1}$	_____
$\qquad\qquad = -21w^3$	

Objective 1 Practice Exercises

For extra help, see Example 1 on page 395 of your text.

Find each product.

1. $-5m^2(-2m^2)$ 1. _____

2. $4x^4(6x^2)$

2. _____

3. $-3ab^2(9a^3b)$

3. _____

Objective 2   Multiply a monomial and a polynomial.

Video Examples

Review this example for Objective 2:
2a. Find each product.

$5x^2(7x+3)$

Use the distributive property.
$$5x^2(7x+3)=5x^2(7x)+5x^2(3)$$
$$=35x^3+15x^2$$

Now Try:
2a. Find each product.

$8x^3(4x+8)$

_____

Objective 2 Practice Exercises

For extra help, see Example 2 on page 395 of your text.

Find each product.

4. $7z(5z^3+2)$

4. _____

5. $2m(3+7m^2+3m^3)$

5. _____

6. $-3y^2(2y^3+3y^2-4y+11)$

6. _____

Objective 3   Multiply two polynomials.

Video Examples

Review these examples for Objective 3:

3. Multiply $(x^2 + 6)(5x^3 - 4x^2 + 3x)$.

Multiply each term of the second polynomial by each term of the first.

$(x^2 + 6)(5x^3 - 4x^2 + 3x)$

$= x^2(5x^3) + x^2(-4x^2) + x^2(3x)$

$\qquad + 6(5x^3) + 6(-4x^2) + 6(3x)$

$= 5x^5 - 4x^4 + 3x^3 + 30x^3 - 24x^2 + 18x$

$= 5x^5 - 4x^4 + 33x^3 - 24x^2 + 18x$

4. Multiply $(2x^3 + 7x^2 + 5x - 1)(4x + 6)$ vertically.

Write the polynomials vertically.

$$2x^3 + 7x^2 + 5x - 1$$
$$\underline{\qquad 4x + 6}$$

Begin by multiplying each term in the top row by 6.

$$2x^3 + 7x^2 + 5x - 1$$
$$\underline{\qquad 4x + 6}$$
$$12x^3 + 42x^2 + 30x - 6$$

Now multiply each term in the top row by $4x$. Then add like terms.

$$2x^3 + 7x^2 + 5x - 1$$
$$\underline{\qquad 4x + 6}$$
$$12x^3 + 42x^2 + 30x - 6$$
$$\underline{8x^4 + 28x^3 + 20x^2 - 4x}$$
$$8x^4 + 40x^3 + 62x^2 + 26x - 6$$

The product is $8x^4 + 40x^3 + 62x^2 + 26x - 6$.

Now Try:

3. Multiply
$$(x^3 + 9)(4x^4 - 2x^2 + x)$$

_____

4. Multiply
$$(4x^3 - 3x^2 + 6x + 5)(7x - 3)$$
vertically.

_____

Objective 3 Practice Exercises

For extra help, see Examples 3–4 on page 396 of your text.

Find each product.

7. $(x+3)(x^2-3x+9)$

7. _____

8. $(2m^2+1)(3m^3+2m^2-4m)$

8. _____

9. $(3x^2+x)(2x^2+3x-4)$

9. _____

Objective 4   Multiply binomials using the FOIL method.

Video Examples

Review these examples for Objective 4:	Now Try:
5. Use the FOIL method to find the product $(x+7)(x-5)$.	5. Use the FOIL method to find the product $(x+9)(x-6)$.

Step 1 F Multiply the first terms: $x(x)=x^2$.

_____

Step 2 O Find the outer product: $x(-5)=-5x$.

Step 3 I   Find the inner product: $7(x)=7x$.
Add the outer and inner products mentally:
$$-5x+7x=2x$$

Step 4 L Multiply the last terms: $7(-5)=-35$.

The product $(x+7)(x-5)$ is $x^2+2x-35$.

6. Multiply $(7x-3)(4y+5)$.

First $7x(4y)=28xy$

Outer $7x(5)=35x$

Inner $-3(4y)=-12y$

Last $-3(5)=-15$

The product $(7x-3)(4y+5)$ is
$28xy+35x-12y-15$.

7b. Find the product

$(8p-5q)(8p+q)$

$(8p-5q)(8p+q)$

$=64p^2+8pq-40pq-5q^2$

$=64p^2-32pq-5q^2$

6. Multiply $(8y-7)(2x+9)$.

_____

7b. Find the product.

$(9p+4q)(5p-q)$

_____

Objective 3 Practice Exercises

For extra help, see Examples 5–7 on page 398 of your text.

Find each product.

10. $(5a-b)(4a+3b)$

10. _____

11. $(3+4a)(1+2a)$

11. _____

12. $(2m+3n)(-3m+4n)$

12. _____

Chapter 5 EXPONENTS AND POLYNOMIALS

5.6   Special Products

Learning Objectives
1   Square binomials.
2   Find the product of the sum and difference of two terms.
3   Find greater powers of binomials.

Key Terms

Use the vocabulary terms listed below to complete each statement in exercises 1−2.

conjugate   **binomial**

1.   A polynomial with two terms is called a _____.

2.   The _____ of $a + b$ is $a - b$.

Objective 1   Square binomials.

Video Examples

Review these examples for Objective 1:

1.   Find $(m+5)^2$.

$$(m+5)^2 = (m+5)(m+5)$$
$$= m^2 + 5m + 5m + 25$$
$$= m^2 + 10m + 25$$

2.   Square each binomial.

b. $(7z-4)^2$

$$(7z-4)^2 = (7z)^2 - 2(7z)(4) + (-4)^2$$
$$= 7^2 z^2 - 56z + 16$$
$$= 49z^2 - 56z + 16$$

c. $(6x+3y)^2$

$$(6x+3y)^2 = (6x)^2 + 2(6x)(3y) + (3y)^2$$
$$= 36x^2 + 36xy + 9y^2$$

Now Try:

1.   Find $(x+6)^2$.

_____

2.   Square each binomial.

b. $(8a-3b)^2$

_____

c. $(2a+9k)^2$

_____

Objective 1 Practice Exercises

For extra help, see Examples 1–2 on pages 402–403 of your text.

Find each square by using the pattern for the square of a binomial.

1. $(7+x)^2$ 1. _____

2. $(2m-3p)^2$ 2. _____

3. $(4y-0.7)^2$ 3. _____

Objective 2   Find the product of the sum and difference of two terms.

Video Examples

Review these examples for Objective 2:

3a. Find the product.

   $(x+5)(x-5)$

   Use the rule for the product of the sum and difference of two terms.

   $$(x+5)(x-5)=x^2-5^2$$
   $$=x^2-25$$

4. Find each product.

   c. $\left(z-\dfrac{6}{7}\right)\left(z+\dfrac{6}{7}\right)$

   $$\left(z-\frac{6}{7}\right)\left(z+\frac{6}{7}\right)=z^2-\left(\frac{6}{7}\right)^2$$
   $$=z^2-\frac{36}{49}$$

   b. $(6x+w)(6x-w)$

   $$(6x+w)(6x-w)=(6x)^2-w^2$$
   $$=36x^2-w^2$$

Now Try:

3a. Find the product.

   $(x+9)(x-9)$

   _____

4. Find each product.

   c. $\left(x+\dfrac{3}{5}\right)\left(x-\dfrac{3}{5}\right)$

   _____

   b. $(11x-y)(11x+y)$

   _____

d. $3q(q^2+4)(q^2-4)$

First, multiply the conjugates.

$$3q(q^2+4)(q^2-4) = 3q(q^4-16)$$
$$= 3q^5 - 48q$$

d. $4p(p^2+6)(p^2-6)$

_____

Objective 2 Practice Exercises

For extra help, see Examples 3–4 on pages 404–405 of your text.

Find each product by using the pattern for the sum and difference of two terms.

4. $(12+x)(12-x)$

4.   _____

5. $(8k+5p)(8k-5p)$

5.   _____

6. $\left(\frac{4}{7}t+2u\right)\left(\frac{4}{7}t-2u\right)$

6.   _____

Objective 3   Find greater powers of binomials.

Video Examples

Review these examples for Objective 3:
5.   Find each product.

   a. $(x+4)^3$

$(x+4)^3$
$= (x+4)^2(x+4)$
$= (x^2+8x+16)(x+4)$
$= x^3+8x^2+16x+4x^2+32x+64$
$= x^3+12x^2+48x+64$

Now Try:
5.   Find each product.

   a. $(x+6)^3$

_____

b. $(5y-4)^4$

b. $(3x-5)^4$

$(5y-4)^4$

$=(5y-4)^2(5y-4)^2$

$=(25y^2-40y+16)(25y^2-40y+16)$

$=625y^4-1000y^3+400y^2$

$\qquad -1000y^3+1600y^2-640y$

$\qquad +400y^2-640y+256$

$=625y^4-2000y^3+2400y^2-1280y+256$

_____

Objective 3 Practice Exercises

For extra help, see Example 5 on page 405 of your text.

Find each product.

7. $(a-3)^3$

7. _____

8. $(j+3)^4$

8. _____

9. $(4s+3t)^4$

9. _____

Chapter 5 EXPONENTS AND POLYNOMIALS

5.7   Dividing a Polynomial by a Monomial

Learning Objectives
1   Divide a polynomial by a monomial.

Key Terms

Use the vocabulary terms listed below to complete each statement in exercises 1–3.

> **quotient**   **dividend**   **divisor**

1. In the division $\dfrac{5x^5 - 10x^3}{5x^2} = x^3 - 2x$, the expression $5x^5 - 10x^3$ is the

     _____.

2. In the division $\dfrac{5x^5 - 10x^3}{5x^2} = x^3 - 2x$, the expression $x^3 - 2x$ is the _____.

3. In the division $\dfrac{5x^5 - 10x^3}{5x^2} = x^3 - 2x$, the expression $5x^2$ is the _____.

Objective 1   Divide a polynomial by a monomial.

Video Examples

Review these examples for Objective 1:

1. Divide $6x^4 - 18x^3$ by $6x^2$.

$$\frac{6x^4 - 18x^3}{6x^2} = \frac{6x^4}{6x^2} - \frac{18x^3}{6x^2}$$
$$= x^2 - 3x$$

Check Multiply. $6x^2(x^2 - 3x) = 6x^4 - 18x^3$

2. Divide. $\dfrac{25a^6 - 15a^4 + 10a^2}{5a^3}$

Divide each term by $5a^3$.
$$\frac{25a^6 - 15a^4 + 10a^2}{5a^3} = \frac{25a^6}{5a^3} - \frac{15a^4}{5a^3} + \frac{10a^2}{5a^3}$$
$$= 5a^3 - 3a + \frac{2}{a}$$

Now Try:

1. Divide $6x^4 - 18x^3$ by $6x^2$.

     _____

2. Divide. $\dfrac{27n^5 - 36n^4 - 18n^2}{9n^3}$

     _____

Check $5a^3\left(5a^3 - 3a + \dfrac{2}{a}\right)$

$= 5a^3(5a^3) + 5a^3(-3a) + 5a^3\left(\dfrac{2}{a}\right)$

$= 25a^6 - 15a^4 + 10a^2$

3. Divide $-12x^4 + 15x^5 - 5x$ by $-5x$.

Write the polynomial in descending powers before dividing.

$$\dfrac{15x^5 - 12x^4 - 5x}{-5x} = \dfrac{15x^5}{-5x} - \dfrac{12x^4}{-5x} - \dfrac{5x}{5x}$$

$$= -3x^4 + \dfrac{12}{5}x^3 + 1$$

Check $-5x\left(-3x^4 + \dfrac{12}{5}x^3 + 1\right)$

$$= -5x(-3x^4) - 5x\left(\dfrac{12}{5}x^3\right) - 5x(1)$$

$$= 15x^5 - 12x^4 - 5x$$

4. Divide
$225x^5y^9 - 150x^3y^7 + 110x^2y^5 \quad 80xy^3 + 75y^2$
by $-25xy^2$.

$$\dfrac{225x^5y^9 - 150x^3y^7 + 110x^2y^5 - 80xy^3 + 75y^2}{25xy^2}$$

$$= \dfrac{225x^5y^9}{-25xy^2} - \dfrac{150x^3y^7}{-25xy^2} + \dfrac{110x^2y^5}{-25xy^2} - \dfrac{80xy^3}{-25xy^2} + \dfrac{75y^2}{-25xy^2}$$

$$= -9x^4y^7 + 6x^2y^5 - \dfrac{22xy^3}{5} + \dfrac{16y}{5} - \dfrac{3}{x}$$

Check by multiplying the quotient by the divisor.

3. Divide $-8z^5 + 7z^6 - 10z - 6$ by $2z^2$.

_____

4. Divide
$80a^5b^3 + 160a^4b^2 - 120a^2b$ by $-40a^2b$.

_____

Objective 1 Practice Exercises

For extra help, see Examples 1–4 on pages 409–410 of your text.

Perform each division.

1. $\dfrac{16a^5 - 24a^3}{8a^2}$ 1. _____

2. $\dfrac{12z^5 + 28z^4 - 8z^3 + 3z}{4z^3}$ 2. _____

3. $\dfrac{39m^4 - 12m^3 + 15}{-3m^2}$ 3. _____

Chapter 5 EXPONENTS AND POLYNOMIALS

5.8   Dividing a Polynomial by a Polynomial

Learning Objectives
1   Divide a polynomial by a polynomial.
2   Apply polynomial division to a geometry problem.

Key Terms

Use the vocabulary terms listed below to complete each statement in exercises 1–3.

> quotient   dividend   divisor

1. In the division $\dfrac{6x^2 - 9x - 12}{2x - 5} = 3x + 3 + \dfrac{3}{2x - 5}$, the expression $2x - 5$ is the

    _____.

2. In the division $\dfrac{6x^2 - 9x - 12}{2x - 5} = 3x + 3 + \dfrac{3}{2x - 5}$, the expression $3x + 3 + \dfrac{3}{2x - 5}$ is

    the _____.

3. In the division $\dfrac{6x^2 - 9x - 12}{2x - 5} = 3x + 3 + \dfrac{3}{2x - 5}$, the expression $6x^2 - 9x - 12$ is the

    _____.

Objective 1   Divide a polynomial by a polynomial.

Video Examples

Review these examples for Objective 1:

2. Divide $8x + 9x^3 - 7 - 9x^2$ by $3x - 1$.

    Write the dividend in descending powers as
    $9x^3 - 9x^2 + 8x - 7$.

    Step 1 $9x^3$ divided by $3x$ is $3x^2$.
    $3x^2(3x - 1) = 9x^3 - 3x^2$

    Step 2   Subtract. Bring down the next term.

    Step 3 $-6x^2$ divided by $3x$ is $-2x$.
    $-2x(3x - 1) = -6x^2 + 2x$

    Step 4   Subtract. Bring down the next term.

    Step 5 $6x$ divided by $3x$ is 2.
    $2(3x - 1) = 6x - 2$

Now Try:

2. Divide $-12x^2 + 10x^3 - 3 - 8x$
    by $5x - 1$.

    _____

$$\begin{array}{r} 3x^2 - 2x + 2 \\ 3x - 1{\overline{\smash{\big)}\,9x^3 - 9x^2 + 8x - 7}} \\ \underline{9x^3 - 3x^2} \\ -6x^2 + 8x \\ \underline{-6x^2 + 2x} \\ 6x - 7 \\ \underline{6x - 2} \\ -5 \end{array}$$

$$\frac{9x^3 - 9x^2 + 8x - 7}{3x - 1} = 3x^2 - 2x + 2 + \frac{-5}{3x - 1}$$

Step 7 Multiply to check.

Check $(3x - 1)\left(3x^2 - 2x + 2 + \dfrac{-5}{3x - 1}\right)$

$$= (3x - 1)(3x^2) + (3x - 1)(-2x)$$
$$\quad + (3x - 1)(2) + (3x - 1)\left(\frac{-5}{3x - 1}\right)$$
$$= 9x^3 - 3x^2 - 6x^2 + 2x + 6x - 2 - 5$$
$$= 9x^3 - 9x^2 + 8x - 7$$

3. Divide $x^3 - 64$ by $x - 4$.

Here the dividend is missing the x^2-term and the x-term. We use 0 as the coefficient for each missing term.

$$\begin{array}{r} x^2 + 4x + 16 \\ x - 4{\overline{\smash{\big)}\,x^3 + 0x^2 + 0x - 64}} \\ \underline{x^3 - 4x^2} \\ 4x^2 + 0x \\ \underline{4x^2 - 16x} \\ 16x - 64 \\ \underline{16x - 64} \\ 0 \end{array}$$

The remainder is 0. The quotient is $x^2 + 4x + 16$.

Check $(x - 4)(x^2 + 4x + 16)$
$$= x^3 + 4x^2 + 16x - 4x^2 - 16x - 64$$
$$= x^3 - 64$$

3. Divide $x^3 - 1000$ by $x - 10$.

_____

4. Divide $x^4 - 3x^3 + 7x^2 - 8x + 14$ by $x^2 + 2$.

Since $x^2 + 2$ is missing the x-term, we write it as $x^2 + 0x + 2$.

$$
\begin{array}{r}
x^2 - 3x + 5 \\
x^2 + 0x + 2 \overline{\smash{)}\, x^4 - 3x^3 + 7x^2 - 8x + 14} \\
\underline{x^4 + 0x^3 + 2x^2} \\
-3x^3 + 5x^2 - 8x \\
\underline{-3x^3 + 0x^2 - 6x} \\
5x^2 - 2x + 14 \\
\underline{5x^2 \mid 0x \mid 10} \\
-2x + 4
\end{array}
$$

The quotient is $x^2 - 3x + 5 + \dfrac{-2x + 4}{x^2 + 2}$

The check shows that the quotient multiplied by the divisor gives the original dividend.

5. Divide $5x^3 + 8x^2 + 12x - 1$ by $5x + 5$.

$$
\begin{array}{r}
x^2 + \frac{3}{5}x + \frac{9}{5} \\
5x + 5 \overline{\smash{)}\, 5x^3 + 8x^2 + 12x - 1} \\
\underline{5x^3 + 5x^2} \\
3x^2 + 12x \\
\underline{3x^2 \mid 3x} \\
9x - 1 \\
\underline{9x + 9} \\
-10
\end{array}
$$

The answer is $x^2 + \dfrac{3}{5}x + \dfrac{9}{5} - \dfrac{10}{5x + 5}$.

4. Divide $3x^4 + 5x^3 - 7x^2 - 12x + 9$ by $x^2 - 4$.

_____

5. Divide $8x^3 - 7x^2 + 4x + 1$ by $8x - 8$.

_____

Name: _____ Date: _____

Instructor: _____ Section: _____

Objective 1 Practice Exercises

For extra help, see Examples 1–5 on pages 414–417 of your text.

Perform each division.

1. $\dfrac{-6x^2 + 23x - 20}{2x - 5}$

        1. _____

2. $\dfrac{6x^4 - 12x^3 + 13x^2 - 5x - 1}{2x^2 + 3}$

        2. _____

3. $\dfrac{2a^4 + 5a^2 + 3}{2a^2 + 3}$

        3. _____

Objective 2   Apply polynomial division to a geometry problem.

Video Examples

Review this example for Objective 2:

6. The area of a rectangle is given by $12p^3 - 7p^2 + 5p - 1$ square units, and the width is $4p - 1$ units. What is the length of the rectangle?

    For a rectangle, $A = LW$. Solving for L gives $L = \frac{A}{W}$. Divide the area, $12p^3 - 7p^2 + 5p - 1$ by the width $4p - 1$.

$$
\begin{array}{r}
3p^2 - p + 1 \\
4p-1{\overline{\smash{\big)}\,12p^3 - 7p^2 + 5p - 1}} \\
\underline{12p^3 - 3p^2} \\
-4p^2 + 5p \\
\underline{-4p^2 + p} \\
4p - 1 \\
\underline{4p - 1} \\
0
\end{array}
$$

    The length is $3p^2 - p + 1$ units.

Now Try:

6. The area of a rectangle is given by $6r^3 - 5r^2 + 16r - 5$ square units, and the width is $3r - 1$ units. What is the length of the rectangle?

    _____

Objective 2 Practice Exercises

For extra help, see Example 6 on page 417 of your text.

Work each problem.

4. The area of a parallelogram is given by $4y^3 - 44y - 600$ square units, and the height is $y - 6$ units. What is the base of the parallelogram?

4. _____

5. The area of a parallelogram is given by
    $3t^3 + 16t^2 - 32t - 64$ square units, and the base is
    $t^2 + 4t - 16$ units. What is the height of the
    parallelogram?

5.

6. The area of a rectangle is given by
    $20x^3 - 19x^2 + 8x - 1$ square units, and the width is
    $5x - 1$ units. What is the length of the rectangle?

6.

Chapter 6 FACTORING AND APPLICATIONS

6.1   Greatest Common Factors; Factor by Grouping

Learning Objectives
1   Find the greatest common factor of a list of numbers.
2   Find the greatest common factor of a list of variable terms.
3   Factor out the greatest common factor.
4   Factor by grouping.

Key Terms

Use the vocabulary terms listed below to complete each statement in exercises 1–4.

> **factor**   **factored form**   **greatest common factor (GCF)**
>
> **factoring**

1.   The process of writing a polynomial as a product is called _____.

2.   An expression is in _____ when it is written as a product.

3.   The _____ is the largest quantity that is a factor of each of a group of quantities.

4.   An expression A is a _____ of an expression B if B can be divided by A with 0 remainder.

Objective 1   Find the greatest common factor of a list of numbers.

Video Examples

Review these examples for Objective 1:

1.   Find the greatest common factor for each list of numbers.

   b. 90, 36, 108

   Write the prime factored form of each number.
   $$90 = 2 \cdot 3 \cdot 3 \cdot 5$$
   $$36 = 2 \cdot 2 \cdot 3 \cdot 3$$
   $$108 = 2 \cdot 2 \cdot 3 \cdot 3 \cdot 3$$
   There is one factor of 2 and two factors of 3.
   $$\text{GCF} = 2^1 \cdot 3^2 = 18$$

Now Try:

1.   Find the greatest common factor for each list of numbers.

   b. 32, 40, 72

   _____

c. 17, 18, 24   c. 26, 27, 28

Write the prime factored form of each number.

$17 = 17$

$18 = 2 \cdot 3 \cdot 3$

$24 = 2 \cdot 2 \cdot 2 \cdot 3$

There are no primes common to all three numbers, so the GCF is 1.

_____

Objective 1 Practice Exercises

For extra help, see Example 1 on pages 432–433 of your text.

Find the greatest common factor for each group of numbers.

1. 84, 280, 112   1. _____

2. 56, 21, 49   2. _____

3. 42, 48, 72   3. _____

Objective 2   Find the greatest common factor of a list of variable terms.

Video Examples

Review these examples for Objective 2:	**Now Try:**

Review these examples for Objective 2:

2. Find the greatest common factor for each list of terms.

    a. $28m^4$, $35m^6$, $49m^9$, $70m^5$

$$28m^4 = 2 \cdot 2 \cdot 7 \cdot m^4$$
$$35m^6 = 5 \cdot 7 \cdot m^6$$
$$49m^9 = 7 \cdot 7 \cdot m^9$$
$$70m^5 = 2 \cdot 5 \cdot 7 \cdot m^5$$

Then, GCF $= 7m^4$.

    b. a^5b^3, a^7b^4, a^3b^5, b^9

$$a^5b^3 = a^5 \cdot b^3$$
$$a^7b^4 = a^7 \cdot b^4$$
$$a^3b^5 = a^3 \cdot b^5$$
$$b^9 - b^9$$

Then, GCF $- b^3$.

Now Try:

2. Find the greatest common factor for each list of terms.

    a. $54x^5$, $48x^7$, $42x^9$, $30x^4$

         _____

    b. p^5q^7, p^4q^3, p^8q^5, q^9

         _____

Objective 2 Practice Exercises

For extra help, see Example 2 on page 434 of your text.

Find the greatest common factor for each list of terms.

    4. $12ab^3$, $18a^2b^4$, $26ab^2$, $32a^2b^2$ **4.** _____

    5. $6k^2m^4n^5$, $8k^3m^7n^4$, $k^4m^8n^7$ **5.** _____

6. $9xy^4$, $72x^4y^7$, $27xy^2$, $108x^2y^5$ 6. _____

Objective 3   Factor out the greatest common factor.

Video Examples

Review these examples for Objective 3:

3. Write in factored form by factoring out the greatest common factor.

 a. $7a^2 + 14a$

 GCF = $7a$

 $$7a^2 + 14a = 7a(a) + 7a(2)$$
 $$= 7a(a + 2)$$

 Check Multiply the factored form.
 $$7a(a + 2) = 7a(a) + 7a(2)$$
 $$= 7a^2 + 14a$$

 b. $12x^5 + 27x^4 - 15x^3$

 GCF = $3x^3$
 $$12x^5 + 27x^4 - 15x^3$$
 $$= 3x^3(4x^2) + 3x^3(9x) + 3x^3(-5)$$
 $$= 3x^3(4x^2 + 9x - 5)$$
 Check Multiply the factored form.
 $$3x^3(4x^2 + 9x - 5)$$
 $$= 3x^3(4x^2) + 3x^3(9x) + 3x^3(-5)$$
 $$= 12x^5 + 27x^4 - 15x^3$$

5a. Write in factored form by factoring out the greatest common factor.

 $$x(x + 9) + 7(x + 9)$$

 Factor out $x + 9$.
 $$x(x + 9) + 7(x + 9) = (x + 9)(x + 7)$$

Now Try:

3. Write in factored form by factoring out the greatest common factor.

 a. $8x^5 + 24x$

 _____

 b. $20y^4 - 12y^3 + 4y^2$

 _____

5a. Write in factored form by factoring out the greatest common factor.

 $$y(y + 8) + 4(y + 8)$$

 _____

Objective 3 Practice Exercises

For extra help, see Examples 3–5 on pages 434–436 of your text.

Factor out the greatest common factor or a negative common factor if the coefficient of the term of greatest degree is negative.

7. $20x^2 + 40x^2y - 70xy^2$ 7. _____

8. $2a(x-2y)+9b(x-2y)$ 8. _____

9. $26x^8 - 13x^{12} + 52x^{10}$ 9. _____

Objective 4   Factor by grouping.

Video Examples

Review these examples for Objective 4:

6.   Factor by grouping.

   a. $5x+15+bx+3b$

   Group the first two terms and the last two terms.
   $5x+15+bx+3b$

   $= (5x+15)+(bx+3b)$

   $= 5(x+3)+b(x+3)$

   $= (x+3)(5+b)$

   Check   Use the FOIL method.

   $(x+3)(5+b) = 5x+bx+3(5)+3b$

   $= 5x+15+bx+3b$

Now Try:

6.   Factor by grouping.

   a. $2x+xy+14+7y$

   _____

    c. $3x^2 - 18x + 5xy - 30y$ **c.** $4x^2 - 28x + 5xy - 35y$

$3x^2 - 18x + 5xy - 30y$

$= (3x^2 - 18x) + (5xy - 30y)$

$= 3x(x - 6) + 5y(x - 6)$ _____

$= (x - 6)(3x + 5y)$

Check by multiplying using the FOIL method.

Objective 4 Practice Exercises

For extra help, see Examples 6–7 on pages 436–438 of your text.

Factor each polynomial by grouping.

10. $15 - 5x - 3y + xy$ **10.** _____

11. $2x^2 - 14xy + xy - 7y^2$ **11.** _____

12. $3r^3 - 2r^2s + 3s^2r - 2s^3$ **12.** _____

Chapter 6 FACTORING AND APPLICATIONS

6.2   Factoring Trinomials

Learning Objectives
1   Factor trinomials with a coefficient of 1 for the second-degree term.
2   Factor such trinomials after factoring out the greatest common factor.

Key Terms

Use the vocabulary terms listed below to complete each statement in exercises 1–3.

prime polynomial   factoring   greatest common factor

1.   _____ is the process of writing a polynomial as a product.

2.   The _____ of a polynomial is the greatest term that is a factor of all the terms in the polynomial.

3.   A _____ is a polynomial that cannot be factored using only integers.

Objective 1   Factor trinomials with a coefficient of 1 for the second-degree term.

Video Examples

Review these examples for Objective 1:

2.   Factor $x^2 - 11x + 28$.

Look for integers whose product is 28 and whose sum is -11. Since the numbers have a positive product and a negative sum, we consider only pairs of negative integers.

Factors of 28	Sums of Factors
$-28, -1$	$-28 + (-1) = -29$
$-14, -2$	$-14 + (-2) = -16$
$-7, -4$	$-7 + (-4) = -11$

The required integers are -7 and -4.

$x^2 - 11x + 28$ factors as $(x-7)(x-4)$

Check   Use the FOIL method.

$$(x-7)(x-4) = x^2 - 4x - 7x + 28$$
$$= x^2 - 11x + 28$$

Now Try:

2.   Factor $y^2 - 12y + 35$.

_____

4. Factor $x^2 - 10x - 39$.

Look for integers whose product is –39 and whose sum is –10. Because the constant term, –39, is negative, we need pairs of integers with different signs.

Factors of -39	Sums of Factors
39, −1	$39 + (-1) = 38$
−39, 1	$-39 + 1 = -38$
3, −13	$3 + (-13) = -10$

The required integers are –13 and 3.

$x^2 - 10x - 39$ factors as $(x - 13)(x + 3)$

Check   Use the FOIL method.

$$(x - 13)(x + 3) = x^2 + 3x - 13x - 39$$
$$= x^2 - 10x - 39$$

5a. Factor the trinomial.

$x^2 - 7x + 18$

Look for integers whose product is 18 and whose sum is –7. Since the numbers have a positive product and a negative sum, we consider only pairs of negative integers.

Factors of 18	Sums of Factors
−18, −1	$-18 + (-1) = -19$
−9, −2	$-9 + (-2) = -11$
−6, −3	$-6 + (-3) = -9$

None of the pairs of integers has a sum of –7.

$x^2 - 7x + 18$ cannot be factored.

It is a prime polynomial.

6. Factor $x^2 - 6xy - 7y^2$.

Here, the coefficient of x in the middle term is –6y, so we need to find two expressions whose product is $-7y^2$ and whose sum is –6y.

Factors of $-7y^2$	Sums of Factors
$7y, -y$	$7y + (-y) = 6y$
$-7y, y$	$-7y + y = -6y$

$x^2 - 6xy - 7y^2$ factors as $(x - 7y)(x + y)$

4. Factor $a^2 - 15a - 34$.

_____

5a. Factor the trinomial.

$m^2 - 7m + 5$

_____

6. Factor $p^2 - 5pq - 14q^2$.

_____

Check   Use the FOIL method.

$$(x - 7y)(x + y) = x^2 + xy - 7xy + 7y^2$$
$$= x^2 - 6xy - 7y^2$$

Objective 1 Practice Exercises

For extra help, see Examples 1–6 on pages 443–445 of your text.

Factor completely. If a polynomial cannot be factored, write prime.

1. $r^2 + r + 3$ 1. _____

2. $x^2 - 11x + 28$ 2. _____

3. $x^2 - 8x - 33$ 3. _____

Objective 2   **Factor such trinomials after factoring out the greatest common factor.**

Video Examples

Review this example for Objective 2:	Now Try:

Review this example for Objective 2:

7.   Factor $5x^5 - 45x^4 + 90x^3$.

There is no second-degree term. Look for a common factor.

$$5x^5 - 45x^4 + 90x^3 = 5x^3\left(x^2 - 9x + 18\right)$$

Now factor $x^2 - 9x + 18$. The integers –3 and –6 have a product of 18 and a sum of –9.

$5x^5 - 45x^4 + 90x^3$ factors as $5x^3(x-6)(x-3)$

Check   Use the FOIL method.

$$5x^3(x-6)(x-3)$$
$$= 5x^3\left(x^2 - 3x - 6x + 18\right)$$
$$= 5x^3\left(x^2 - 9x + 18\right)$$
$$= 5x^5 - 45x^4 + 90x^3$$

Now Try:

7.   Factor $7x^6 - 49x^5 + 70x^4$.

_____

Objective 2 Practice Exercises

For extra help, see Example 7 on page 446 of your text.

Factor completely. If a polynomial cannot be factored, write **prime**.

4. $2n^4 - 16n^3 + 30n^2$

4. _____

5. $2a^3b - 10a^2b^2 + 12ab^3$

5. _____

6. $10k^6 + 70k^5 + 100k^4$

6. _____

Chapter 6 FACTORING AND APPLICATIONS

6.3   Factoring Trinomials by Grouping

Learning Objectives
1   Factor trinomials by grouping when the coefficient of the second-degree term is not 1.

Key Terms

Use the vocabulary terms listed below to complete each statement in exercises 1–2.

       coefficient   **trinomial**

1.   In the term $6x^2y$, 6 is the _____.

2.   A polynomial with three terms is a _____.

Objective 1   Factor trinomials by grouping when the coefficient of the second-degree term is not 1.

Video Examples

Review these examples for Objective 1:

2.   Factor each trinomial.

     a. $8x^2 - 2x - 1$

We must find two integers with a product of $8(-1) = -8$ and a sum of -2. The integers are -4 and 2. We write the middle term as $-4x + 2x$.

$$8x^2 - 2x - 1 = 8x^2 - 4x + 2x - 1$$
$$= \left(8x^2 - 4x\right) + (2x - 1)$$
$$= 4x(2x - 1) + 1(2x - 1)$$
$$= (2x - 1)(4x + 1)$$

Check   Multiply $(2x - 1)(4x + 1)$ to obtain $8x^2 - 2x - 1$.

     b. $15z^2 + z - 2$

Look for two integers whose product is $15(-2) = -30$ and whose sum is 1.

The integers are 6 and -5.

Now Try:

2.   Factor each trinomial.

     a. $14x^2 - 3x - 5$

     b. $3m^2 - m - 14$

               _____

$$15z^2 + z - 2 = 15z^2 - 5z + 6z - 2$$
$$= (15z^2 - 5z) + (6z - 2)$$
$$= 5z(3z - 1) + 2(3z - 1)$$
$$= (3z - 1)(5z + 2)$$

Check   Multiply $(3z - 1)(5z + 2)$ to obtain $15z^2 + z - 2$.

c. $12r^2 + 5rs - 2s^2$

Two integers whose product is $12(-2) = -24$ and whose sum is 5 are 8 and –3. Rewrite the trinomial with four terms.

$$12r^2 + 5rs - 2s^2 = 12r^2 + 8rs - 3rs - 2s^2$$
$$= (12r^2 + 8rs) + (-3rs - 2s^2)$$
$$= 4r(3r + 2s) - s(3r + 2s)$$
$$= (3r + 2s)(4r - s)$$

Check   Multiply $(3r + 2s)(4r - s)$ to obtain $12r^2 + 5rs - 2s^2$.

3. Factor $100x^5 + 140x^4 - 15x^3$.

Factor out the greatest common factor, $5x^3$.
$$100x^5 + 140x^4 - 15x^3 = 5x^3(20x^2 + 28x - 3)$$

To factor $20x^2 + 28x - 3$, find two integers whose product is $20(-3) = -60$ and whose sum is 28. Factor 60 into prime factors.
$$60 = 2 \cdot 2 \cdot 3 \cdot 5$$
Combine the prime factors in pairs using one positive factor and one negative factor to get –60. The factors of 30 and –2 have the correct sum, 28.

$$100x^5 + 140x^4 - 15x^3$$
$$= 5x^3(20x^2 + 28x - 3)$$
$$= 5x^3(20x^2 + 30x - 2x - 3)$$
$$= 5x^3\left[(20x^2 + 30x) + (-2x - 3)\right]$$
$$= 5x^3\left[10x(2x + 3) - 1(2x + 3)\right]$$
$$= 5x^3(2x + 3)(10x - 1)$$

c. $10x^2 + xy - 3y^2$

_____

3. Factor $30x^5 + 87x^4 - 63x^3$.

_____

Name: Date:
Instructor: Section:

Objective 1 Practice Exercises

For extra help, see Examples 1–3 on pages 449–450 of your text.

Factor each trinomial by grouping.

1. $8b^2 + 18b + 9$ 1. _____

2. $7a^2b + 18ab + 8b$ 2. _____

3. $10c^2 - 29ct + 21t^2$ 3. _____

Chapter 6 FACTORING AND APPLICATIONS

6.4   Factoring Trinomials Using the FOIL Method

Learning Objectives
1   Factor trinomials using the FOIL method.

Key Terms

Use the vocabulary terms listed below to complete each statement in exercises 1–3.

> **FOIL**   **outer product**   **inner product**

1.   The _____ of $(2y-5)(y+8)$ is $-5y$.

2.   _____ is a shortcut method for finding the product of two binomials.

3.   The _____ of $(2y-5)(y+8)$ is $16y$.

Objective 1   Factor trinomials using the FOIL method.

Video Examples

Review these examples for Objective 1:

1.   Factor $5x^2 + 14x + 8$.

The product of the first terms of the binomial is $5x^2$. The possible factors are $5x$ and x, since we consider only positive factors. So we have

$$5x^2 + 14x + 8 = (5x + \underline{\quad})(x + \underline{\quad})$$

The product of the last terms is 8. The positive factors are $1\cdot8$, $8\cdot1$, $4\cdot2$, or $2\cdot4$. We want the middle term of 14.

Try $1\cdot8$ in $(5x + \underline{\quad})(x + \underline{\quad})$.

$$(5x + 1)(x + 8)$$

gives middle term $40x + x = 41x$. Incorrect.

Try $8\cdot1$ in $(5x + \underline{\quad})(x + \underline{\quad})$.

$$(5x + 8)(x + 1)$$

gives middle term $5x + 8x = 13x$. Incorrect.

Try $4\cdot2$ in $(5x + \underline{\quad})(x + \underline{\quad})$.

$$(5x + 4)(x + 2)$$

gives middle term $10x + 4x = 14x$. Correct.

Thus, $5x^2 + 14x + 8 = (5x + 4)(x + 2)$.

Check. Multiply $(5x + 4)(x + 2)$ to obtain $5x^2 + 14x + 8$.

Now Try:

1.   Factor $3x^2 + 17x + 10$.

_____

2. Factor $6x^2 + 13x + 7$.

The number 6 has several possible pairs of factors, but 7 has only 1 and 7, or -1 and -7. We choose positive factors since all coefficients in the trinomial are positive.

$$(\_\_\_+7)(\_\_\_+1)$$

The possible pairs of $6x^2$ are $6x$ and x, or $3x$ and $2x$.

$$(3x+7)(2x+1)$$

gives middle term $3x + 14x = 17x$. Incorrect.

$$(2x+7)(3x+1)$$

gives middle term $2x + 21x = 23x$. Incorrect.

$$(6x+7)(x+1)$$

gives middle term $6x + 7x = 13x$. Correct.

$6x^2 + 13x + 7$ factors as $(6x+7)(x+1)$.

Check. Multiply $(6x+7)(x+1)$ to obtain $6x^2 + 13x + 7$.

3. Factor $10x^2 - 19x + 7$.

Since 7 has only 1 and 7 or -1 and -7 as factors, it is better to begin by factoring 7. We need two negative factors because the product of two negative factors is positive and their sum is negative, as required.
We try -1 and -7.

$$(\_\_\_-1)(\_\_\_-7)$$

The factors of $10x^2$ are $10x$ and x, or $5x$ and $2x$.

$$(10x-1)(x-7)$$

has middle term $-70x - x = -71x$. Incorrect.

$$(5x-1)(2x-7)$$

has middle term $-35x - 2x = -37x$. Incorrect.

$$(2x-1)(5x-7)$$

has middle term $-14x - 5x = -19x$. Correct.

Thus $10x^2 - 19x + 7$ factors as $(2x-1)(5x-7)$.

2. Factor $15x^2 + 26x + 7$.

----

3. Factor $20x^2 - 13x + 2$.

----

4. Factor $6x^2 - x - 15$.

The integer 6 has several possible pairs of factors, as does -15. Since the constant term is negative, one positive factor and one negative factor of -15 are needed. Since the coefficient of the middle term is relatively small, it is wise to avoid large factors. We try $3x$ and $2x$ as factors of $6x^2$ and 5 and -3 as factors of -15.

$(3x+5)(2x-3)$
has middle term $-9x + 10x = x$. Incorrect.
$(3x-5)(2x+3)$
has middle term $9x - 10x = -x$. Correct.

$6x^2 - x - 15$ factors as $(3x-5)(2x+3)$.

5. Factor $18x^2 - 3xy - 28y^2$.

There are several factors of $18x^2$, including
   $18x$ and x, $9x$ and $2x$, and $6x$ and $3x$.
There are many possible pairs of factors of $-28y^2$, including
   $28y$ and $-y$, $-28y$ and y, $14y$ and $-2y$,
   $-14y$ and $2y$, $7y$ and $-4y$, $-7y$ and $4y$.

Once again, since the coefficient of the middle term is relatively small, avoid the larger factors. Try the factors of $6x$ and $3x$, and $4y$ and $-7y$.
   $(6x+4y)(3x-7y)$ Incorrect
The first binomial has a common factor of 2.
   $(6x-7y)(3x+4y)$
has middle term $24xy - 21xy = 3xy$. Incorrect.
Interchange the signs of the last two terms.
   $(6x+7y)(3x-4y)$
has middle term $-24xy + 21xy = -3xy$. Correct.

Thus, $18x^2 - 3xy - 28y^2$ factors as
$(6x+7y)(3x-4y)$

4. Factor $8x^2 + 2x - 21$.

_____

5. Factor $24x^2 - 2xy - 15y^2$.

_____

Objective 1 Practice Exercises

For extra help, see Examples 1–6 on pages 453–456 of your text.

Factor each trinomial completely.

1. $8q^2 + 10q + 3$

1. _____

2. $3a^2 + 8ab + 4b^2$

2. _____

3. $4c^2 + 14cd - 8d^2$

3. _____

Chapter 6 FACTORING AND APPLICATIONS

6.5   Special Factoring Techniques

Learning Objectives	
1	Factor a difference of squares.
2	Factor a perfect square trinomial.
3	Factor a difference of cubes.
4	Factor a sum of cubes.

Key Terms

Use the vocabulary terms listed below to complete each statement in exercises 1–2.

perfect square trinomial   difference

1.   A _____ is the result of a subtraction.

2.   A _____ is a trinomial that can be factored as the square of a binomial.

Objective 1   Factor a difference of squares.

Video Examples

Review these examples for Objective 1:

1.   Factor each binomial, if possible.

   a. $x^2 - 64$

   $x^2 - 64 = x^2 - 8^2 = (x+8)(x-8)$

   b. $y^2 - 27$

   Because 27 is not a square of an integer, this binomial is not a difference in squares. It is a prime polynomial.

   c. $c^2 + 25$

   Since $c^2 + 25$ is a sum of squares, it is a prime polynomial.

Now Try:

1.   Factor each binomial, if possible.

   a. $z^2 - 36$

   _____

   b. $y^2 - 110$

   _____

   c. $x^2 + 100$

   _____

3. Factor completely.

   a. $28y^2 - 175$

   $$28y^2 - 175 = 7(4y^2 - 25)$$
   $$= 7\left[(2y)^2 - 5^2\right]$$
   $$= 7(2y + 5)(2y - 5)$$

   c. $p^4 - 81$

   $$p^4 - 81 = (p^2)^2 - 9^2$$
   $$= (p^2 + 9)(p^2 - 9)$$
   $$= (p^2 + 9)(p + 3)(p - 3)$$

3. Factor completely.

   a. $90x^2 - 490$

   _____

   c. $p^4 - 256$

   _____

Objective 1 Practice Exercises

For extra help, see Examples 1–3 on pages 459–460 of your text.

Factor each binomial completely. If a binomial cannot be factored, write **prime**.

1. $x^2 - 49$

1. _____

2. $81x^4 - 16$

2. _____

3. $9x^2 + 16$

3. _____

Objective 2   Factor a perfect square trinomial.

Video Examples

Review these examples for Objective 2:	Now Try:

Review these examples for Objective 2:

4. Factor $x^2 + 20x + 100$.

The terms x^2 and 100 are perfect squares.
$$x^2 + 20x + 100 = (x+10)^2$$
Check the middle term. $2(x)(10) = 20x$
The trinomial is a perfect square.

5. Factor each trinomial.

a. $x^2 - 26x + 169$

The first and last terms are perfect squares.
Check to see if the middle term is twice the
product of the first and last terms of the binomial
$x - 13$.
$$2 \cdot x \cdot (-13) = -26x$$
Thus, $x^2 - 26x + 169$ is a perfect square
trinomial.
$$x^2 - 26x + 169 \text{ factors as } (x-13)^2.$$

b. $49m^2 - 70m + 25$

$49m^2 - 70m + 25$
$$= (7m)^2 + 2(7m)(-5) + (-5)^2$$
$$= (7m-5)^2$$

d. $128z^3 + 192z^2 + 72z$

$128z^3 + 192z^2 + 72z$
$$= 8z(16z^2 + 24z + 9)$$
$$= 8z\left[(4z)^2 + 2(4z)(3) + 3^2\right]$$
$$= 8z(4z+3)^2$$

Now Try:

4. Factor $p^2 + 16p + 64$.

_____

5. Factor each trinomial.

a. $x^2 - 24x + 144$

_____

b. $64m^2 + 48m + 9$

_____

d. $20x^3 + 100x^2 + 125x$

_____

Name: Date:
Instructor: Section:

Objective 2 Practice Exercises

For extra help, see Examples 4–5 on pages 461–462 of your text.

Factor each trinomial completely. It may be necessary to factor out the greatest common factor first.

4. $z^2 - \dfrac{4}{3}z + \dfrac{4}{9}$ 4. _____

5. $9j^2 + 12j + 4$ 5. _____

6. $-12a^2 + 60ab - 75b^2$ 6. _____

Objective 3   Factor a difference of cubes.

Video Examples

Review these examples for Objective 3:

6.   Factor each difference of cubes.

   a. $m^3 - 1000$

   Use the pattern for a difference of cubes.
$$m^3 - 1000 = m^3 - 10^3$$
$$= (m-10)(m^2 + 10m + 100)$$

   b. $64p^3 - 125$

$$64p^3 - 125 = (4p^3) - 5^3$$
$$= (4p-5)\left[(4p)^2 + (4p)(5) + 5^2\right]$$
$$= (4p-5)(16p^2 + 20p + 25)$$

Now Try:

6.   Factor each difference of cubes.

   a. $t^3 - 216$

           _____

   b. $27k^3 - y^3$

           _____

Objective 3 Practice Exercises

For extra help, see Example 6 on pages 463–464 of your text.

Factor.

7. $8a^3 - 125b^3$

7. _____

8. $216x^3 - 8y^3$

8. _____

9. $3x^3 - 192$

9. _____

Objective 4   Factor a sum of cubes.

Video Examples

Review these examples for Objective 4:
7. Factor each sum of cubes.

   a. $k^3 + 1000$

   $k^3 + 1000 = k^3 + 10^3$
   $\qquad = (k+10)(k^2 - 10k + 100)$

   b. $2m^3 + 250n^3$

   $2m^3 + 250n^3 = 2(m^3 + 125n^3)$
   $\qquad = 2\left[m^3 + (5n)^3\right]$
   $\qquad = 2(m+5n)\left[m^2 - m(5n) + (5n)^2\right]$
   $\qquad = 2(m+5n)(m^2 - 5mn + 25n^2)$

Now Try:
7. Factor each sum of cubes.

   a. $216x^3 + 1$

   _____

   b. $6x^3 + 48y^3$

   _____

Objective 4 Practice Exercises

For extra help, see Example 7 on page 464 of your text.

Factor.

10. $27r^3 + 8s^3$ **10.** _____

11. $8a^3 + 64b^3$ **11.** _____

12. $64x^3 + 343y^3$ **12.** _____

Name: _____ Date: _____

Instructor: _____ Section: _____

Chapter 6 FACTORING AND APPLICATIONS

6.6   A General Approach to Factoring

Learning Objectives
1   Factor any polynomial.

Key Terms

Use the vocabulary terms listed below to complete each statement in exercises 1–2.

FOIL   factoring by grouping

1.   When there are more than three terms in a polynomial, use a process called
      _____ to factor the polynomial.

2.   _____ is a shortcut method for finding the product of two binomials.

Objective 1   Factor any polynomial.

Video Examples

Review these examples for Objective 1:

1c. Factor the polynomial.

$$9x(a+c)-z(a+c)$$

Factor out $(a + c)$
$$9x(a+c)-z(a+c)=(a+c)(9x-z)$$

2. Factor each binomial if possible.

d. $16y^2 + 49$

$16y^2 + 49$ is prime. It is the sum of squares. There is no common factor.

b. $216p^3 - 125z^3$

Difference of cubes
$$216p^3 - 125z^3$$
$$= (6p)^3 - (5z)^3$$
$$= (6p-5z)\left[(6p)^2 + (6p)(5z) + (5z)^2\right]$$
$$= (6p-5z)(36p^2 + 30pz + 25z^2)$$

Now Try:

1c. Factor the polynomial.

$$7x(y+z)-5(y+z)$$

_____

2. Factor each binomial if possible.

d. $36p^2 + 169$

_____

b. $27t^3 - 64w^3$

_____

3d. Factor the trinomial.

$y^2 - 3y - 4$

$y^2 - 3y - 4 = (y-4)(y+1)$

4a. Factor the polynomial.

$ad - 8d + af - 8f$

Group the terms and factor each group.
$ad - 8d + af - 8f$

$= (ad - 8d) + (af - 8f)$

$= d(a-8) + f(a-8)$

$= (a-8)(d+f)$

3d. Factor the trinomial.

$4k^2 - 7k - 2$

_____

4a. Factor the polynomial.

$cm - 3m + cx - 3x$

_____

Objective 1 Practice Exercises

For extra help, see Examples 1 4 on pages 468–470 of your text.

Factor completely.

1. $-12x^2 - 6x$

1. _____

2. $12a^2b^2 + 3a^2b - 9ab^2$

2. _____

3. $(x+1)(2x+3) - (x+1)$

3. _____

4. $2x^3y^4 - 72xy^2$

4. _____

5. $128x^3 - 2y^3$

5. _____

6. $y^6 + 1$

6. _____

7. $4x^2 - 12xy + 9y^2$

7. _____

8. $2a^2 - 17a + 30$

8. _____

9. $14w^2 + 6wx - 35wx - 15x^2$

9. _____

10. $x^3 - 3x^2 + 7x - 21$

10. _____

11. $a^2 - 6ab + 9b^2 - 25$

11. _____

Chapter 6 FACTORING AND APPLICATIONS

6.7 Solving Quadratic Equations using the Zero-Factor Property

Learning Objectives
1 Solve quadratic equations using the zero-factor property.
2 Solve other equations using the zero-factor property.

Key Terms

Use the vocabulary terms listed below to complete each statement in exercises 1–2.

quadratic equation   **standard form**

1. An equation written in the form $ax^2 + bx + c = 0$ is written in the
    _____ of a quadratic equation.

2. An equation that can written in the form $ax^2 + bx + c = 0$, with $a \neq 0$, is a
    _____ .

Objective 1   Solve quadratic equations using the zero-factor property.

Video Examples

Review these examples for Objective 1:	Now Try:
1a. Solve the equation	1a. Solve the equation.

1a. Solve the equation

$$(x+9)(5x-6)=0$$

By the zero-factor property, either $x+9=0$ or $5x-6=0$, or both.

$$x+9=0 \quad \text{or} \quad 5x-6=0$$
$$x=-9 \quad \text{or} \qquad 5x=6$$
$$x=\frac{6}{5}$$

Check:
Let $x = -9$.
$$(x+9)(5x-6)=0$$
$$(-9+9)[5(-9)-6]\overset{?}{=}0$$
$$0(-51)\overset{?}{=}0$$
$$0=0 \quad \text{True}$$

Now Try:

1a. Solve the equation.

$$(x+12)(4x-7)=0$$

_____

Let $x = \dfrac{6}{5}$.

$$(x+9)(5x-6) = 0$$

$$\left(\frac{6}{5}+9\right)\left[5\left(\frac{6}{5}\right)-6\right] \overset{?}{=} 0$$

$$\left(\frac{51}{5}\right)(6-6) \overset{?}{=} 0$$

$$0 = 0 \quad \text{True}$$

Both values check, so the solution set is

$$\left\{-9, \ \frac{6}{5}\right\}.$$

2a. Solve the equation.

$$x^2 - 6x = -5$$

First, write the equation in standard form.

$$x^2 - 6x = -5$$

$$x^2 - 6x + 5 = 0$$

Factor and use the zero-factor property.

$$(x-1)(x-5) = 0$$

$$x - 1 = 0 \quad \text{or} \quad x - 5 = 0$$

$$x = 1 \quad \text{or} \quad x = 5$$

Check these solutions by substituting each in the original equation. The solution set is $\{1, 5\}$.

3. Solve $8p^2 + 30 = 46p$.

$$8p^2 + 30 = 46p$$

$$8p^2 - 46p + 30 = 0 \quad \text{Standard form}$$

$$2\left(4p^2 - 23p + 15\right) = 0 \quad \text{Factor out 2.}$$

$$4p^2 - 23p + 15 = 0 \quad \text{Divide each side by 2}$$

$$(4p-3)(p-5) = 0 \quad \text{Factor.}$$

$$4p - 3 = 0 \quad \text{or} \quad p - 5 = 0 \quad \text{Zero-factor}$$

$$p = \frac{3}{4} \quad \text{or} \quad p = 5 \quad \text{property}$$

The solution set is $\left\{\frac{3}{4}, \ 5\right\}$.

2a. Solve the equation.

$$x^2 - 5x = 24$$

_____

3. Solve $15p^2 + 36 = 57p$.

_____

4. Solve each equation.

    c. $k(3k+5)=2$

$$k(3k+5)=2$$
$$3k^2+5k=2$$
$$3k^2+5k-2=0$$
$$(3k-1)(k+2)=0$$
$$3k-1=0 \quad \text{or} \quad k+2=0$$
$$3k-1 \quad \text{oi} \quad k--2$$
$$k=\frac{1}{3}$$

The solution set is $\left\{-2, \frac{1}{3}\right\}$.

    b. $y^2=9y$

$$y^2=9y$$
$$y^2-9y=0$$
$$y(y-9)=0$$
$$y=0 \quad \text{or} \quad y-9=0$$
$$y=9$$

The solution set is $\{0, 9\}$.

4. Solve each equation.

    c. $k(4k-23)=6$

            _____

    b. $y^2=11y$

            _____

Objective 1 Practice Exercises

For extra help, see Examples 1–5 on pages 474–477 of your text.

Solve each equation and check your solutions.

1. $2x^2-3x-20-0$

1. _____

2. $25x^2=20x$

2. _____

3. $c(5c+17)=12$

3. _____

Objective 2   **Solve other equations using the zero-factor property.**

Video Examples

Review this example for Objective 2:

6. Solve $12z^3 - 3z = 0$.

$$12z^3 - 3z = 0$$
$$3z(4z^2 - 1) = 0$$
$$3z(2z+1)(2z-1) = 0$$

By an extension of the zero-factor property, we have

$$3z = 0 \quad \text{or} \quad 2z+1=0 \quad \text{or} \quad 2z-1=0$$
$$z = 0 \quad \text{or} \quad z = -\frac{1}{2} \quad \text{or} \quad z = \frac{1}{2}$$

Check by substituting each value in the original equation. The solution set is $\left\{-\frac{1}{2},\ 0,\ \frac{1}{2}\right\}$.

Now Try:

6. Solve $3r^3 = 75r$.

_____

Objective 2 Practice Exercises

For extra help, see Examples 6–7 on page 478 of your text.

Solve each equation and check your solutions.

4. $x^3 + 2x^2 - 8x = 0$

4. _____

5. $z^4 + 8z^3 - 9z^2 = 0$ **5.** _____

6. $\left(y^2 - 5y + 6\right)\left(y^2 - 36\right) = 0$ **6.** _____

Chapter 6 FACTORING AND APPLICATIONS

6.8   Applications of Quadratic Equations

Learning Objectives	
1	Solve problems involving geometric figures.
2	Solve problems involving consecutive integers.
3	Solve problems by applying the Pythagorean theorem.
4	Solve problems using given quadratic models.

Key Terms

Use the vocabulary terms listed below to complete each statement in exercises 1–2.

hypotenuse   legs

1. In a right triangle, the sides that form the right angle are the _____.

2. The longest side of a right triangle is the _____.

Objective 1   Solve problems involving geometric figures.

Video Examples

Review this example for Objective 1:

1. The length of a rectangle is three times its width. If the width was increased by 4 and the length remained the same, the resulting rectangle would have an area of 231 square inches. Find the dimensions of the original rectangle.

    Step 1 Read the problem carefully. Find the dimensions of the original rectangle.

    Step 2 Assign a variable.
    Let x = width.
        $3x$ = length
      $x + 4$ = new width
        $3x$ = length

    Step 3 Write an equation. The area of the rectangle is given by $\text{Area} = \text{Length} \times \text{Width}$. Substitute 231 for area, $3x$ for length, and $x + 4$ for width.
        $231 = 3x(x+4)$

Now Try:

1. Mr. Fixxall is building a box which will have a volume of 60 cubic meters. The height of the box will be 4 meters, and the length will be 2 meters more than the width. Find the width and length of the box.

    _____

Step 4 Solve.

$$231 = 3x(x+4)$$

$$231 = 3x^2 + 12x$$

$$3x^2 + 12x - 231 = 0$$

$$3(x^2 + 4x - 77) = 0$$

$$x^2 + 4x - 77 = 0$$

$$(x+11)(x-7) = 0$$

$$x+11 = 0 \quad \text{or} \quad x-7 = 0$$

$$x = -11 \quad \text{or} \quad x = 7$$

Step 5 State the answer. The solutions are −11 and 7. A rectangle cannot have a side of negative length, so we discard −11. The width is 7 inches. The length is $3(7) = 21$ inches.

Step 6 Check. The new width is $7 + 4 = 11$. The new area of the rectangle is $11(21) = 231$ square inches.

Objective 1 Practice Exercises

For extra help, see Example 1 on pages 482–483 of your text.

Solve each problem. Check your answers to be sure they are reasonable.

1. A book is three times as long as it is wide. Find the length and width of the book in inches if its area is numerically 128 more than its perimeter.

    1. width_____

       length _____

2.   The area of a triangle is 42 square centimeters. The base is 2 centimeters less than twice the height. Find the base and height of the triangle.

                                               2. base_____

                                              height _____

3.   The volume of a box is 192 cubic feet. If the length of the box is 8 feet and the width is 2 feet more than the height, find the height and width of the box.

                                               3. height _____

                                              width_____

Objective 2   Solve problems involving consecutive integers.

Video Examples

Review these examples for Objective 2:

3. Find two consecutive positive even integers whose product is 168.

    Step 1 Read carefully. Note that the integers are positive consecutive even integers.

    Step 2 Assign a variable.
        Let x = the first integer.
        Then $x + 2$ = the next even integer.

    Step 3 Write an equation. The product is 168.
        $x(x+2) - 168$

    Step 4 Solve.
        $$x^2 + 2x - 168$$
        $$x^2 + 2x - 168 = 0$$
        $$(x+14)(x-12) = 0$$
        $$x+14 = 0 \quad \text{or} \quad x-12 = 0$$
        $$x = -14 \quad \text{or} \qquad x = 12$$

    Step 5 State the answer. The solutions are −14 and 12. We discard −14 since the integers are required to be positive. If $x - 12$, then $x + 2 - 14$. So, the numbers are 12 and 14.

    Step 6 Check. The numbers 12 and 14 are consecutive positive even integers, and their product is 168.

2. Find two consecutive integers such that the sum of the squares of the two integers is 3 more than the opposite (additive inverse) of the smaller integer.

    Step 1 Read the problem. Note that the numbers are consecutive.

    Step 2 Assign a variable.
        Let x = the first number.
        Then $x + 1$ = the second number.

    Step 3 Write an equation. The sum of squares is three more than the opposite of the smaller integer.
        $$x^2 + (x+1)^2 = 3 + (-x)$$

Now Try:

3. The product of two consecutive even positive integers is 10 more than seven times the larger. Find the integers.

    _____

2. If the square of the sum of two consecutive integers is reduced by twice their product, the result is 25. Find the integers.

    _____

Step 4 Solve.

$$x^2 + x^2 + 2x + 1 = 3 - x$$
$$2x^2 + 3x - 2 = 0$$
$$(2x - 1)(x + 2) = 0$$
$$2x + 1 = 0 \quad \text{or} \quad x + 2 = 0$$
$$x = -\frac{1}{2} \quad \text{or} \quad x = -2$$

Step 5 State the answer. The solutions are $-\frac{1}{2}$

and –2. We discard $-\frac{1}{2}$ since it is not an integer.
If $x = -2$, then $x + 1 = -2 + 1 = -1$. So, the
numbers are –2 and –1.

Step 6 Check. $(-2)^2 + (-1)^2 = 3 + [-(-2)]$
$$4 + 1 = 3 + 2$$
$$5 = 5$$

Objective 2 Practice Exercises

For extra help, see Examples 2–3 on pages 484–485 of your text.

Solve each problem.

4. Find all possible pairs of consecutive odd integers   4. _____
    whose sum is equal to their product decreased by 47.

5. Find two consecutive positive even integers whose product is six more than three times its sum.

5. _____

6. Find three consecutive positive odd integers such that four times the sum of all three equals 13 more than the product of the smaller two.

6. _____

Objective 3   Solve problems by applying the Pythagorean theorem.

Video Examples

Review this example for Objective 3:

4. Penny and Carla started biking from the same corner. Penny biked east and Carla biked south. When they were 26 miles apart, Carla had biked 14 miles further than Penny. Find the distance each biked.

Step 1 Read carefully. Find the two distances.

Step 2 Assign a variable.
    Let $x =$ Penny's distance.

Now Try:

4. A ladder is leaning against a building. The distance from the bottom of the ladder to the building is 8 feet less than the length of the ladder. How high up the side of the building is the top of the ladder if that distance is 4 feet less than the length of

Then $x + 14 =$ Carla's distance.   the ladder?

Step 3 Write an equation. Substitute into the Pythagorean theorem.

$$a^2 + b^2 = c^2$$

$$x^2 + (x+14)^2 = 26^2$$

Step 4 Solve.

$$x^2 + x^2 + 28x + 196 = 676$$

$$2x^2 + 28x - 480 = 0$$

$$2(x^2 + 14x - 240) = 0$$

$$x^2 + 14x - 240 = 0$$

$$(x+24)(x-10) = 0$$

$$x + 24 = 0 \quad \text{or} \quad x - 10 = 0$$

$$x = -24 \quad \text{or} \quad x = 10$$

Step 5 State the answer. Since –24 cannot be a distance, 10 is the distance for Penny, and 10 + 14 = 24 is the distance for Carla.

Step 6 Check. Since $10^2 + 24^2 = 26^2$ is true, the answer is correct.

Objective 3 Practice Exercises

For extra help, see Example 4 on page 486 of your text.

Solve each problem.

7. A field is in the shape of a right triangle. The shorter leg measures 45 meters. The hypotenuse measures 45 meters less than twice the longer the leg. Find the dimensions of the lot.

7. _____

8. A train and a car leave a station at the same time, the train traveling due north and the car traveling west. When they are 100 miles apart, the train has traveled 20 miles farther than the car. Find the distance each has traveled.

8. car_____

train _____

9. Two ships left a dock at the same time. When they were 25 miles apart, the ship that sailed due south had gone 10 miles less than twice the distance traveled by the ship that sailed due west. Find the distance traveled by the ship that sailed due south.

9. _____

Objective 4   Solve problems using given quadratic models.

Video Examples

Review this example for Objective 4:

5.   Jeff threw a stone straight upward at 46 feet per second from a dock 6 feet above a lake. The height of the stone above the lake t seconds after it is thrown is given by $h = -16t^2 + 46t + 6$. How long will it take for the stone to reach a height of 39 feet?

Substitute 39 for h.
$$39 = -16t^2 + 46t + 6$$
Solve for t.
$$16t^2 - 46t + 33 = 0$$
$$(8t - 11)(2t - 3) = 0$$
$$8t - 11 = 0 \quad \text{or} \quad 2t - 3 = 0$$
$$t = \frac{11}{8} \quad \text{or} \quad t = \frac{3}{2}$$

Since we have found two acceptable answers, the stone will be at height of 39 feet twice (once on its way up and once on its way down) —at $\frac{11}{8}$ sec or $\frac{3}{2}$ sec.

Now Try:

5.   A ball is dropped from the roof of a 19.6 meter high building. Its height h (in meters) t seconds later is given by the equation $h = -4.9t^2 + 19.6$. After how many second is the height 14.7 meters?

_____

Objective 4 Practice Exercises

For extra help, see Examples 5–6 on pages 487–488 of your text.

Solve each problem.

10.   If an object is propelled upward from a height of 16 feet with an initial velocity of 48 feet per second, its height h (in feet) t seconds later is given by the equation $h = -16t^2 + 48t + 16$.

10.   a._____

   b._____

   (a)   After how many seconds is the height 52 feet?

   (b)   After how many seconds is the height 48 feet?

11. A company determines that its daily revenue R (in dollars) for selling x items is modeled by the equation $R = x(150 - x)$. How many items must be sold for its revenue to be \$4400?

11. _____

12. If a ball is batted at an angle of $35°$, the distance that the ball travels is given approximately by

$D = 0.029v^2 + 0.021v - 1$, where v is the bat speed in miles per hour and D is the distance traveled in feet. Find the distance a batted ball will travel if the ball is batted with a velocity of 90 miles per hour. Round your answer to the nearest whole number.

12. _____

Chapter 7 RATIONAL EXPRESSIONS AND APPLICATIONS

7.1 The Fundamental Property of Rational Expressions

Learning Objectives
1   Find the numerical value of a rational expression.
2   Find the values of the variable for which a rational expression is undefined.
3   Write rational expressions in lowest terms.
4   Recognize equivalent forms of rational expressions.

Key Terms

Use the vocabulary terms listed below to complete each statement in exercises 1–2.

> **rational expression**   **lowest terms**

1. The quotient of two polynomials with denominator not 0 is called a

    _____ .

2. A rational expression is written in _____ if the greatest
    common factor of its numerator and denominator is 1.

Objective 1   Find the numerical value of a rational expression.

Video Examples

Review these examples for Objective 1:

1. Find the numerical value of $\dfrac{4x+8}{3x-6}$ for each value of x.

    a. $x = 1$

    $$\frac{4x+8}{3x-6} = \frac{4(1)+8}{3(1)-6} = \frac{12}{-3} = -4$$

    b. $x = 0$

    $$\frac{4x+8}{3x-6} = \frac{4(0)+8}{3(0)-6} = \frac{8}{-6}, \text{ or } -\frac{4}{3}$$

    c. $x = 2$

    $$\frac{4x+8}{3x-6} = \frac{4(2)+8}{3(2)-6} = \frac{16}{0} \quad \text{undefined}$$

    d. $x = -2$

    $$\frac{4x+8}{3x-6} = \frac{4(-2)+8}{3(-2)-6} = \frac{0}{-12} = 0$$

Now Try:

1. Find the numerical value of $\dfrac{x-7}{x-5}$ for each value of x.

    a. $x = 1$

    _____

    b. $x = 0$

    _____

    c. $x = 5$

    _____

    d. $x = 7$

    _____

Objective 1 Practice Exercises

For extra help, see Example 1 on page 508 of your text.

Find the numerical value of each expression when (a) $x = 4$ *and (b)* $x = -1$.

1. $\dfrac{-3x+1}{2x+1}$

    1. (a) _____

       (b) _____

2. $\dfrac{2x^2-4}{x^2-2}$

    2. (a) _____

       (b) _____

3. $\dfrac{2x-5}{2-x-x^2}$

    3. (a) _____

       (b) _____

Objective 2   **Find the values of the variable for which a rational expression is undefined.**

Video Examples

Review these examples for Objective 2:

2. Find any values of the variable for which each rational expression is undefined.

  a. $\dfrac{3x+8}{5x+4}$

  Step 1 Set the denominator equal to 0.
$$5x + 4 = 0$$
  Step 2 Solve.
$$5x = -4$$
$$x = -\frac{4}{5}$$

Now Try:

2. Find any values of the variable for which each rational expression is undefined.

  a. $\dfrac{y+6}{7y-1}$

        _____

Step 3 The given expression is undefined for $-\dfrac{4}{5}$, so $x \neq -\dfrac{4}{5}$.

b. $\dfrac{3m^2}{m^2 - 4m - 5}$

Set the denominator equal to 0.

$$m^2 - 4m - 5 = 0$$

$$(m + 1)(m - 5) = 0$$

$$m + 1 = 0 \quad \text{or} \quad m - 5 = 0$$

$$m = -1 \quad \text{or} \quad m = 5$$

The given expression is undefined for -1 and 5, so $m \neq -1$, $m \neq 5$.

c. $\dfrac{6r}{r^2 + 49}$

This denominator will not equal 0 for any value of r. There are no values for which this expression is undefined.

b. $\dfrac{15m^2}{m^2 - m - 20}$

_____

c. $\dfrac{12t^2}{t^2 + 100}$

_____

Objective 2 Practice Exercises

For extra help, see Example 2 on page 509 of your text.

Find any value(s) of the variable for which each rational expression is undefined. Write answers with $\neq$.

4. $\dfrac{4x^2}{x + 7}$

4. _____

5. $\dfrac{2x^2}{x^2 + 4}$

5. _____

6. $\dfrac{2y - 5}{2y^2 + 4y - 16}$

6. _____

Name: Date:
Instructor: Section:

Objective 3   Write rational expressions in lowest terms.

Video Examples

Review these examples for Objective 3:	Now Try:

3b. Write the expression $\dfrac{15k^3}{3k^4}$ in lowest terms.

Write k^3 as $k \cdot k \cdot k$ and k^4 as $k \cdot k \cdot k \cdot k$.

$$\dfrac{15k^3}{3k^4} = \dfrac{3 \cdot 5 \cdot k \cdot k \cdot k}{3 \cdot k \cdot k \cdot k \cdot k}$$

$$= \dfrac{5 \cdot (3 \cdot k \cdot k \cdot k)}{k \cdot (3 \cdot k \cdot k \cdot k)}$$

$$= \dfrac{5}{k}$$

3b. Write the expression $\dfrac{12k^5}{4k^8}$ in lowest terms.

_____

4. Write each rational expression in lowest terms.

  a. $\dfrac{6x-18}{5x-15}$

$$\dfrac{6x-18}{5x-15} = \dfrac{6(x-3)}{5(x-3)} = \dfrac{6}{5}$$

  c. $\dfrac{m^2+5m-24}{3m^2-5m-12}$

$$\dfrac{m^2+5m-24}{3m^2-5m-12} = \dfrac{(m+8)(m-3)}{(3m+4)(m-3)}$$

$$= \dfrac{m+8}{3m+4}$$

4. Write each rational expression in lowest terms.

  a. $\dfrac{7x-35}{9x-45}$

_____

  c. $\dfrac{m^2-3m-54}{2m^2-15m-27}$

_____

5. Write $\dfrac{3x-2y}{2y-3x}$ in lowest terms.

Factor -1 from the denominator.

$$\dfrac{3x-2y}{2y-3x} = \dfrac{3x-2y}{-1(-2y+3x)}$$

$$= \dfrac{3x-2y}{-1(3x-2y)}$$

$$= -1$$

Alternatively, we could factor -1 from the numerator.

$$\dfrac{3x-2y}{2y-3x} = \dfrac{-1(-3x+2y)}{2y-3x}$$

$$= \dfrac{-1(2y-3x)}{2y-3x}$$

$$= -1$$

5. Write $\dfrac{4y-5x}{5x-4y}$ in lowest terms.

_____

Name: _____   Date: _____
Instructor: _____   Section: _____

Objective 3 Practice Exercises

For extra help, see Examples 3–6 on pages 510–513 of your text.

Write each rational expression in lowest terms. Assume that no values of any variable make any denominator zero.

7. $\dfrac{15ab^3c^9}{-24ab^2c^{10}}$

7. _____

8. $\dfrac{16-x^2}{2x-8}$

8. _____

9. $\dfrac{9x^2-9x\ \ 108}{2x-8}$

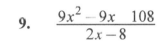

9. _____

Objective 4   Recognize equivalent forms of rational expressions.

Video Examples

Review this example for Objective 4:

7.   Write four equal forms of the following rational expression.

$$-\frac{4x+3}{x-8}$$

If we apply the negative sign to the numerator, we obtain the first two equivalent forms.

$$\frac{-(4x+3)}{x-8} \quad \text{and} \quad \frac{-4x-3}{x-8}$$

If we apply the negative sign to the denominator, we obtain the last two equivalent forms.

$$\frac{4x+3}{-(x-8)} \quad \text{and} \quad \frac{4x+3}{-x+8}$$

Now Try:

7.   Write four equal forms of the following rational expression.

$$-\frac{10x-7}{4x-3}$$

_____

Objective 4 Practice Exercises

For extra help, see Example 7 on page 514 of your text.

Write four equivalent forms of the following rational expressions. Assume that no values of any variable make any denominator zero.

10. $-\dfrac{4x+5}{3-6x}$

10. _____

11. $\dfrac{2p-1}{1-4p}$

11. _____

12. $-\dfrac{2x-3}{x+2}$

12. _____

Chapter 7 RATIONAL EXPRESSIONS AND APPLICATIONS

7.2   Multiplying and Dividing Rational Expressions

Learning Objectives
1   Multiply rational expressions.
2   Find reciprocals.
3   Divide rational expressions.

Key Terms

Use the vocabulary terms listed below to complete each statement in exercises 1–3.

      rational expression   **reciprocal**   **lowest terms**

1. The _____ of the expression $\frac{4x-5}{x+2}$ is $\frac{x+2}{4x-5}$.

2. A _____ is the quotient of two polynomials with denominator not 0.

3. A rational expression is written in _____ when the numerator and denominator have no common terms.

Objective 1   Multiply rational expressions.

Video Examples

Review these examples for Objective 1:

1b. Multiply. Write each answer in lowest terms.

$$\frac{9}{x^2}\cdot\frac{x^3}{15}$$

$$\frac{9}{x^2}\cdot\frac{x^3}{15}=\frac{9\cdot x^3}{x^2\cdot 15}$$

$$=\frac{3\cdot 3\cdot x\cdot x\cdot x}{3\cdot 5\cdot x\cdot x}$$

$$=\frac{3x}{5}$$

2. Multiply. Write the answer in lowest terms.
$$\frac{3x+2y}{5x}\cdot\frac{x^2}{(3x+2y)^2}$$

Multiply numerators, multiply denominators, factor, and identify the common factors.

Now Try:

1b. Multiply. Write each answer in lowest terms.

$$\frac{8}{x^3}\cdot\frac{x^2}{6}$$

_____

2. Multiply. Write the answer in lowest terms.
$$\frac{r-s}{6s}\cdot\frac{s^3}{(r-s)^2}$$

_____

$$\frac{3x+2y}{5x} \cdot \frac{x^2}{(3x+2y)^2} = \frac{(3x+2y)x^2}{5x(3x+2y)^2}$$

$$= \frac{(3x+2y)\cdot x \cdot x}{5x(3x+2y)(3x+2y)}$$

$$= \frac{x}{5(3x+2y)}$$

Objective 1 Practice Exercises

For extra help, see Examples 1–3 on pages 518–519 of your text.

Multiply. Write each answer in lowest terms.

1. $\dfrac{8m^4n^3}{3} \cdot \dfrac{5}{4mn^2}$

1. _____

2. $\dfrac{m^2-16}{m-3} \cdot \dfrac{9-m^2}{4-m}$

2. _____

3. $\dfrac{3x+12}{6x-30} \cdot \dfrac{x^2-x-20}{x^2-16}$

3. _____

Objective 2   Find reciprocals.

Video Examples

Review these examples for Objective 2:

4.   Find the reciprocal of each rational expression.

    a. $\dfrac{7x^3}{8y}$

    $\dfrac{7x^3}{8y}$ has reciprocal $\dfrac{8y}{7x^3}$.

    b. $\dfrac{m^2-16}{m^2-2m-15}$

    $\dfrac{m^2-16}{m^2-2m-15}$ has reciprocal $\dfrac{m^2-2m-15}{m^2-16}$.

Now Try:

4.   Find the reciprocal of each rational expression.

    a. $\dfrac{11a^3}{10b}$

    _____

    b. $\dfrac{w^2-6w}{w^2+6w-7}$

    _____

Objective 2 Practice Exercises

For extra help, see Example 4 on page 519 of your text.

Write the reciprocal of each rational expression.

4. $\dfrac{7}{2y}$

                                                4. _____

5. $4r^2+2r+3$

                                                 5. _____

6. $\dfrac{3x+4y}{5x-2y}$

                                                 6. _____

Name: _____ Date: _____

Instructor: _____ Section: _____

Objective 3 Divide rational expressions.

Video Examples

Review these examples for Objective 3:

6. Divide. Write the answer in lowest terms.

$$\frac{(5m)^2}{(3p)^3} \div \frac{10m^4}{24p^2}$$

Multiply by the reciprocal.

$$\frac{(5m)^2}{(3p)^3} \div \frac{10m^4}{24p^2} = \frac{(5m)^2}{(3p)^3} \cdot \frac{24p^2}{10m^4}$$

$$= \frac{25m^2}{27p^3} \cdot \frac{24p^2}{10m^4}$$

$$= \frac{25 \cdot 24m^2 p^2}{27 \cdot 10m^4 p^3}$$

$$= \frac{5 \cdot 5 \cdot 3 \cdot 2 \cdot 4 \cdot m^2 \cdot p^2}{3 \cdot 9 \cdot 2 \cdot 5 \cdot m^2 \cdot m^2 \cdot p^2 \cdot p}$$

$$= \frac{20}{9m^2 p}$$

7. Divide. Write the answer in lowest terms.

$$\frac{m^2 - 16}{(m-5)(m-4)} \div \frac{(m+4)(m-5)}{-6m}$$

Multiply by the reciprocal.

$$\frac{m^2 - 16}{(m-5)(m-4)} \div \frac{(m+4)(m-5)}{-6m}$$

$$= \frac{m^2 - 16}{(m-5)(m-4)} \cdot \frac{-6m}{(m+4)(m-5)}$$

$$= \frac{-6m(m^2 - 16)}{(m-5)(m-4)(m+4)(m-5)}$$

$$= \frac{-6m(m+4)(m-4)}{(m-5)(m-4)(m+4)(m-5)}$$

$$= \frac{-6m}{(m-5)^2}, \text{ or } -\frac{6m}{(m-5)^2}$$

8. Divide. Write the answer in lowest terms.

$$\frac{x^2 - 25}{x^2 - 9} \div \frac{3x^2 - 15x}{3 - x}$$

Multiply by the reciprocal.

Now Try:

6. Divide. Write the answer in lowest terms.

$$\frac{(7c)^2}{(5d)^3} \div \frac{49c^2}{10d^3}$$

_____

7. Divide. Write the answer in lowest terms.

$$\frac{x^2 - 49}{(x-7)(x-3)} \div \frac{(x+7)(x-3)}{8x}$$

_____

8. Divide. Write the answer in lowest terms.

$$\frac{m^2 - 64}{m^2 - 81} \div \frac{5m^2 + 40m}{9 - m}$$

_____

$$\frac{x^2-25}{x^2-9} \div \frac{3x^2-15x}{3-x}$$

$$=\frac{x^2-25}{x^2-9} \cdot \frac{3-x}{3x^2-15x}$$

$$=\frac{(x^2-25)(3-x)}{(x^2-9)(3x^2-15x)}$$

$$=\frac{(x+5)(x-5)(3-x)}{(x+3)(x-3)(3x)(x-5)}$$

$$=\frac{-1(x+5)}{3x(x+3)}, \quad \text{or} \quad \frac{x \quad 5}{3x(x+3)}$$

Objective 3 Practice Exercises

For extra help, see Examples 5–8 on pages 520–521 of your text.

Divide. Write each answer in lowest terms.

7. $\dfrac{b-7}{16} \div \dfrac{7-b}{8}$ 7. _____

8. $\dfrac{m^2+2mn+n^2}{m^2 \mid m} \div \dfrac{m^2-n^2}{m^2-1}$ 8. _____

9. $\dfrac{27-3k^2}{3k^2+8k-3} \div \dfrac{k^2-6k+9}{6k^2-19k+3}$ 9. _____

Chapter 7 RATIONAL EXPRESSIONS AND APPLICATIONS

7.3   Least Common Denominators

Learning Objectives
1   Find the least common denominator for a list of fractions.
2   Write equivalent rational expressions.

Key Terms

Use the vocabulary terms listed below to complete each statement in exercises 1–2.

> **least common denominator**   **equivalent expressions**

1. $\dfrac{24x-8}{9x^2-1}$ and $\dfrac{8}{3x+1}$ are _____.

2.   The simplest expression that is divisible by all denominators is called the

   _____.

Objective 1   Find the least common denominator for a list of fractions.

Video Examples

Review these examples for Objective 1:	**Now Try:**
2. Find the LCD for $\dfrac{9}{28s^3}$ and $\dfrac{5}{42s^2}$.	**2.** Find the LCD for $\dfrac{17}{40a^2}$ and $\dfrac{13}{24a^4}$.

2.   *Step 1*

$$28s^3 = 2 \cdot 2 \cdot 7 \cdot s^3$$

$$42s^2 = 2 \cdot 3 \cdot 7 \cdot s^2$$

Step 2

Here s appears three times, 2 appears twice, and 3 and 7 each appear once.

Step 3

$$\text{LCD} = 2^2 \cdot 3 \cdot 7 \cdot s^3 = 84s^3$$

3.   Find the LCD for the fractions in each list.

    a. $\dfrac{5}{7b}$, $\dfrac{8}{b^2-5b}$

$$7b = 7 \cdot b$$

$$b^2 - 5b = b(b-5)$$

$$\text{LCD} = 7 \cdot b(b-5) = 7b(b-5)$$

3.   Find the LCD for the fractions in each list.

    a. $\dfrac{7}{9w}$, $\dfrac{13}{w^2-2w}$

_____

b. $\dfrac{6}{c^2-5c-6}$, $\dfrac{10}{c^2+3c-54}$, $\dfrac{4}{c^2-12c+36}$

$$c^2-5c-6=(c-6)(c+1)$$

$$c^2+3c-54=(c-6)(c+9)$$

$$c^2-12c+36=(c-6)^2$$

Use each factor the greatest number of times it appears as a factor.

$$\text{LCD}=(c+1)(c+9)(c-6)^2$$

b. $\dfrac{12}{b^2-16}$, $\dfrac{6}{b^2-3b-4}$,

$\dfrac{9}{b^2-8b+16}$

_____

Objective 1 Practice Exercises

For extra help, see Examples 1–3 on pages 524–525 of your text.

Find the least common denominator for each list of rational expressions.

1. $\dfrac{13}{36b^4}$, $\dfrac{17}{27b^2}$

1. _____

2. $\dfrac{-7}{a^2-2a}$, $\dfrac{3a}{2a^2+a-10}$

2. _____

3. $\dfrac{8}{w^3-9w}$, $\dfrac{4w}{w^2+w-6}$

3. _____

Objective 2   Write equivalent rational expressions.

Video Examples

Review these examples for Objective 2:

4a. Write the rational expression as an equivalent expression with the indicated denominator.

$$\frac{7}{9} = \frac{?}{45}$$

Step 1 Factor both denominators.

$$\frac{7}{9} = \frac{?}{5 \cdot 9}$$

Step 2 A factor of 5 is missing.

Step 3 Multiply $\frac{7}{9}$ by $\frac{5}{5}$.

$$\frac{7}{9} = \frac{7}{9} \cdot \frac{5}{5} = \frac{35}{45}$$

5. Write each rational expression as an equivalent expression with the indicated denominator.

a. $\dfrac{9}{5x+2} = \dfrac{?}{15x+6}$

Factor the denominator on the right.

$$\frac{9}{5x+2} = \frac{?}{3(5x+2)}$$

The missing factor is 3, so multiply by $\frac{3}{3}$.

$$\frac{9}{5x+2} \cdot \frac{3}{3} = \frac{27}{15x+6}$$

b. $\dfrac{7}{q^2+6q} = \dfrac{?}{q^3+q^2-30q}$

Factor the denominator in each rational expression.

$$\frac{7}{q(q+6)} = \frac{?}{q(q+6)(q-5)}$$

The missing factor is $(q-5)$, so multiply by $\frac{q-5}{q-5}$.

$$\frac{7}{q(q+6)} = \frac{7}{q(q+6)} \cdot \frac{q-5}{q-5}$$

$$= \frac{7(q-5)}{q^3-q^2-30q}$$

$$= \frac{7q-35}{q^3-q^2-30q}$$

Now Try:

4a. Write the rational expression as an equivalent expression with the indicated denominator.

$$\frac{13}{6} = \frac{?}{30}$$

_____

5. Write each rational expression as an equivalent expression with the indicated denominator.

a. $\dfrac{19}{6c-5} = \dfrac{?}{24c-20}$

_____

b. $\dfrac{3}{z^2-7z} = \dfrac{?}{z^3-5z^2-14z}$

_____

Name: Date:

Instructor: Section:

Objective 2 Practice Exercises

For extra help, see Examples 4–5 on pages 526–527 of your text.

Rewrite each rational expression with the indicated denominator. Give the numerator of the new fraction.

4. $\dfrac{5a}{8a-3} = \dfrac{?}{6-16a}$

4. _____

5. $\dfrac{3}{5r-10} = \dfrac{?}{50r^2-100r}$

5. _____

6. $\dfrac{3}{k^2+3k} = \dfrac{?}{k^3+10k^2+21k}$

6. _____

Chapter 7 RATIONAL EXPRESSIONS AND APPLICATIONS

7.4   Adding and Subtracting Rational Expressions

Learning Objectives	
1	Add rational expressions having the same denominator.
2	Add rational expressions having different denominators.
3	Subtract rational expressions.

Key Terms

Use the vocabulary terms listed below to complete each statement in exercises 1–2.

least common multiple   **greatest common factor**

1. The _____ of $2m^2 - 5m - 3$ and $2m - 6$ is $m - 3$.

2. The _____ of $2m^2 - 5m - 3$ and $2m - 6$ is
    $2(m-3)(2m+1)$.

Objective 1   Add rational expressions having the same denominator.

Video Examples

Review this example for Objective 1:
1b. Add. Write each answer in lowest terms.

$$\frac{4x}{x+2} + \frac{8}{x+2}$$

$$\frac{4x}{x+2} + \frac{8}{x+2} = \frac{4x+8}{x+2} = \frac{4(x+2)}{x+2} = 4$$

Now Try:
1b. Add. Write each answer in lowest terms.

$$\frac{2x^2}{x+4} + \frac{8x}{x+4}$$

_____

Objective 1 Practice Exercises

For extra help, see Example 1 on page 530 of your text.

Add. Write each answer in lowest terms.

1. $\dfrac{5}{3w^2} + \dfrac{7}{3w^2}$

1. _____

2. $\dfrac{b}{b^2 - 4} + \dfrac{2}{b^2 - 4}$

2. _____

3. $\dfrac{2x+3}{x^2+3x-10}+\dfrac{2-x}{x^2+3x-10}$

3. _____

Objective 2   Add rational expressions having different denominators.

Video Examples

Review these examples for Objective 2:

2a. Add. Write each answer in lowest terms.

$$\frac{5}{18}+\frac{7}{24}$$

Step 1 Find the LCD.

$$18 = 2\cdot3\cdot3 = 2\cdot3^2$$

$$24 = 2\cdot2\cdot2\cdot3 = 2^3\cdot3$$

$$LCD = 2^3\cdot3^2 = 72$$

Step 2 Now write each expression as an equivalent expression with the LCD as the denominator.

$$\frac{5}{18}+\frac{7}{24}=\frac{5(4)}{18(4)}+\frac{7(3)}{24(3)}$$

$$=\frac{20}{72}+\frac{21}{72}$$

Step 3 Add the numerators.

Step 4 Write in lowest terms, if necessary.

$$=\frac{20+21}{72}$$

$$=\frac{41}{72}$$

3. Add. Write the answer in lowest terms.

$$\frac{10x}{x^2-9}+\frac{-5}{x+3}$$

Step 1 Find the LCD.

$$x^2-9=(x+3)(x-3)$$

$x+3$ is prime

$$LCD=(x+3)(x-3)$$

Step 2 Now write each expression as an equivalent expression with the LCD as the

Now Try:

2a. Add. Write each answer in lowest terms.

$$\frac{9}{35}+\frac{8}{45}$$

_____

3. Add. Write the answer in lowest terms.

$$\frac{8x}{x^2-100}+\frac{-4}{x+10}$$

_____

denominator.

$$\frac{10x}{x^2-9}+\frac{-5}{x+3}$$

$$=\frac{10x}{(x+3)(x-3)}+\frac{-5(x-3)}{(x+3)(x-3)}$$

$$=\frac{10x}{(x+3)(x-3)}+\frac{-5x+15}{(x+3)(x-3)}$$

Step 3 Add the numerators.

$$=\frac{10x-5x+15}{(x+3)(x-3)}$$

$$=\frac{5x+15}{(x+3)(x-3)}$$

Step 4 Write in lowest terms, if necessary.

$$=\frac{5(x+3)}{(x+3)(x-3)}$$

$$=\frac{5}{x-3}$$

4. Add. Write the answer in lowest terms.

$$\frac{3x}{x^2-x-20}+\frac{5}{x^2-2x-15}$$

The LCD is $(x+3)(x+4)(x-5)$.

$$=\frac{3x}{(x+4)(x-5)}+\frac{5}{(x+3)(x-5)}$$

$$=\frac{3x(x+3)}{(x+3)(x+4)(x-5)}+\frac{5(x+4)}{(x+3)(x+4)(x-5)}$$

$$=\frac{3x(x+3)+5(x+4)}{(x+3)(x+4)(x-5)}$$

$$=\frac{3x^2+9x+5x+20}{(x+3)(x+4)(x-5)}$$

$$=\frac{3x^2+14x+20}{(x+3)(x+4)(x-5)}$$

4. Add. Write the answer in lowest terms.

$$\frac{7x}{x^2-2x-8}+\frac{4}{x^2-3x-4}$$

4. _____

Objective 2 Practice Exercises

For extra help, see Examples 2–5 on pages 531–533 of your text.

Add. Write each answer in lowest terms.

4. $\dfrac{7}{x-5}+\dfrac{4}{x+5}$

4. _____

5. $\dfrac{3z}{z^2-4}+\dfrac{4z-3}{z^2-4z+4}$

5. _____

6. $\dfrac{4z}{z^2+6z+8}+\dfrac{2z-1}{z^2+5z+6}$

6. _____

Objective 3   **Subtract rational expressions.**

Video Examples

Review these examples for Objective 3:

8. Subtract. Write the answer in lowest terms.

$$\dfrac{5x}{x-7}-\dfrac{x-42}{7-x}$$

The denominators are opposites.

$$\dfrac{5x}{x-7}-\dfrac{x-42}{7-x}=\dfrac{5x}{x-7}-\dfrac{(x-42)(-1)}{(7-x)(-1)}$$

$$=\dfrac{5x}{x-7}-\dfrac{-x+42}{x-7}$$

$$=\dfrac{5x+x-42}{x-7}$$

$$=\dfrac{6x-42}{x-7}$$

$$=\dfrac{6(x-7)}{x-7}$$

$$=6$$

Now Try:

8. Subtract. Write the answer in lowest terms.

$$\dfrac{4x}{x-9}-\dfrac{3x-63}{9-x}$$

_____

Name: Date:

Instructor: Section:

9. Subtract. Write the answer in lowest terms.

$$\frac{7x}{x^2-4x+4}-\frac{2}{x^2-4}$$

$$=\frac{7x}{(x-2)^2}-\frac{2}{(x+2)(x-2)}$$

The LCD is $(x+2)(x-2)^2$.

$$=\frac{7x}{(x-2)^2}-\frac{2}{(x+2)(x-2)}$$

$$=\frac{7x(x+2)}{(x+2)(x-2)^2}-\frac{2(x-2)}{(x+2)(x-2)^2}$$

$$=\frac{7x(x+2)-2(x-2)}{(x+2)(x-2)^2}$$

$$=\frac{7x^2+14x-2x+4}{(x+2)(x-2)^2}$$

$$=\frac{7x^2+12x+4}{(x+2)(x-2)^2}$$

10. Subtract. Write the answer in lowest terms.

$$\frac{2q}{q^2-2q-8}-\frac{7}{3q^2-14q+8}$$

The LCD is $(q+2)(3q-2)(q-4)$.

$$=\frac{2q}{(q+2)(q-4)}-\frac{7}{(3q-2)(q-4)}$$

$$=\frac{2q(3q-2)}{(q+2)(3q-2)(q-4)}-\frac{7(q+2)}{(q+2)(3q-2)(q-4)}$$

$$=\frac{6q^2-4q-7q-14}{(q+2)(3q-2)(q-4)}$$

$$=\frac{6q^2-11q-14}{(q+2)(3q-2)(q-4)}$$

9. Subtract. Write the answer in lowest terms.

$$\frac{8x}{x^2-10x+25}-\frac{3}{x^2-25}$$

_____

10. Subtract. Write the answer in lowest terms.

$$\frac{5}{p^2-6p+5}-\frac{4}{p^2-1}$$

_____

Name: _____ Date: _____

Instructor: _____ Section: _____

Objective 3 Practice Exercises

For extra help, see Examples 6–10 on pages 533–535 of your text.

Subtract. Write each answer in lowest terms.

7. $\dfrac{z+2}{z-2} - \dfrac{z-2}{z+2}$

7. _____

8. $\dfrac{-4}{x^2-4} - \dfrac{3}{4-2x}$

8. _____

9. $\dfrac{m}{m^2-1} - \dfrac{1-m}{m^2+4m+4}$

9. _____

Chapter 7 RATIONAL EXPRESSIONS AND APPLICATIONS

7.5   Complex Fractions

Learning Objectives
1   Define and recognize a complex fraction.
2   Simplify a complex fraction by writing it as a division problem (Method 1).
3   Simplify a complex fraction by multiplying numerator and denominator by the least common denominator (Method 2).
4   Simplify rational expressions with negative exponents.

Key Terms

Use the vocabulary terms listed below to complete each statement in exercises 1−2.

complex fraction   **LCD**

1.   A _____ is a rational expression with one or more fractions in the numerator, denominator, or both.

2.   To simplify a complex fraction, multiply the numerator and denominator by the _____ of all the fractions within the complex fraction.

Objective 1   Define and recognize a complex fraction.
For extra help, see page 540 of your text.

Objective 2   Simplify a complex fraction by writing it as a division problem (Method 1).

Video Examples

Review these examples for Objective 2:
1b. Simplify the complex fraction.

$$\dfrac{8+\dfrac{4}{x}}{\dfrac{x}{6}+\dfrac{1}{12}}$$

Step 1 Write the numerator as a single fraction.
$$8+\frac{4}{x}=\frac{8}{1}+\frac{4}{x}=\frac{8x}{x}+\frac{4}{x}=\frac{8x+4}{x}$$
Do the same with each denominator.
$$\frac{x}{6}+\frac{1}{12}=\frac{x(2)}{6(2)}+\frac{1}{12}=\frac{2x}{12}+\frac{1}{12}=\frac{2x+1}{12}$$
Step 2 Write the equivalent complex fraction as a division problem.
$$\dfrac{\dfrac{8x+4}{x}}{\dfrac{2x+1}{12}}=\frac{8x+4}{x}\cdot\frac{2x+1}{12}$$

Now Try:
1b. Simplify the complex fraction.

$$\dfrac{9+\dfrac{3}{x}}{\dfrac{x}{10}+\dfrac{1}{30}}$$

_____

Step 3 Divide by multiplying by the reciprocal.

$$\frac{8x+4}{x} \div \frac{2x+1}{12} = \frac{8x+4}{x} \cdot \frac{12}{2x+1}$$

$$= \frac{4(2x+1)}{x} \cdot \frac{12}{2x+1}$$

$$= \frac{48}{x}$$

2. Simplify the complex fraction.

$$\frac{\dfrac{rs^2}{t^3}}{\dfrac{s^3}{r^3t}}$$

Use the definition of division and then the fundamental property.

$$= \frac{rs^2}{t^3} \div \frac{s^3}{r^3t}$$

$$= \frac{rs^2}{t^3} \cdot \frac{r^3t}{s^3}$$

$$= \frac{r^4}{st^2}$$

2. Simplify the complex fraction.

$$\frac{\dfrac{a^3b^2}{c}}{\dfrac{a^3b}{c^3}}$$

_____

3. Simplify the complex fraction.

$$\frac{\dfrac{30}{x+4} - 6}{\dfrac{4}{x+4} + 1} = \frac{\dfrac{30}{x+4} - \dfrac{6(x+4)}{x+4}}{\dfrac{4}{x+4} + \dfrac{1(x+4)}{x+4}}$$

$$= \frac{\dfrac{30 - 6(x+4)}{x+4}}{\dfrac{4 + 1(x+4)}{x+4}}$$

$$= \frac{\dfrac{30 - 6x - 24}{x+4}}{\dfrac{4 + x + 4}{x+4}}$$

$$= \frac{\dfrac{6 - 6x}{x+4}}{\dfrac{x+8}{x+4}}$$

$$= \frac{6 - 6x}{x+4} \cdot \frac{x+4}{x+8}$$

$$= \frac{6 - 6x}{x+8}$$

3. Simplify the complex fraction.

$$\frac{\dfrac{20}{x-5} - 9}{\dfrac{5}{x-5} + 2}$$

_____

Objective 2 Practice Exercises

For extra help, see Examples 1–3 on pages 541–542 of your text.

Simplify each complex fraction by writing it as a division problem.

1. $\dfrac{\dfrac{49m^3}{18n^5}}{\dfrac{21m}{27n^2}}$

1. _____

2. $\dfrac{\dfrac{p}{2} - \dfrac{1}{3}}{\dfrac{p}{3} + \dfrac{1}{6}}$

2. _____

3. $\dfrac{3 + \dfrac{4}{s}}{2s + \dfrac{2}{3}}$

3. _____

Name:

Date:

Instructor:

Section:

Objective 3   Simplify a complex fraction by multiplying numerator and denominator by the least common denominator (Method 2).

Video Examples

Review these examples for Objective 3:

4b. Simplify the complex fraction.

$$\frac{12+\dfrac{4}{x}}{\dfrac{x}{5}+\dfrac{1}{15}}$$

Step 1 Find the LCD for all the denominators.

The LCD for x, 5, and 15 is $15x$.

Step 2 Multiply the numerator and denominator of the complex fraction by the LCD.

$$\frac{12+\dfrac{4}{x}}{\dfrac{x}{5}+\dfrac{1}{15}}=\frac{15x\left(12+\dfrac{4}{x}\right)}{15x\left(\dfrac{x}{5}+\dfrac{1}{15}\right)}$$

$$=\frac{180x+60}{3x^2+x}$$

$$=\frac{60(3x+1)}{x(3x+1)}$$

$$=\frac{60}{x}$$

5. Simplify the complex fraction.

$$\frac{\dfrac{9}{7n}-\dfrac{3}{n^2}}{\dfrac{8}{3n}+\dfrac{5}{6n^2}}$$

The LCD is $42n^2$.

$$=\frac{42n^2\left(\dfrac{9}{7n}-\dfrac{3}{n^2}\right)}{42n^2\left(\dfrac{8}{3n}+\dfrac{5}{6n^2}\right)}$$

$$=\frac{42n^2\left(\dfrac{9}{7n}\right)-42n^2\left(\dfrac{3}{n^2}\right)}{42n^2\left(\dfrac{8}{3n}\right)+42n^2\left(\dfrac{5}{6n^2}\right)}$$

$$=\frac{54n-126}{112n+35},\ \text{ or }\ \frac{18(3n-7)}{7(16n+5)}$$

Now Try:

4b. Simplify the complex fraction.

$$\frac{4+\dfrac{2}{x}}{\dfrac{x}{3}+\dfrac{1}{6}}$$

————————   ————————

5. Simplify the complex fraction.

$$\frac{\dfrac{2}{9n}-\dfrac{2}{5n^2}}{\dfrac{4}{5n}+\dfrac{2}{3n^2}}$$

————————   ————————

Objective 3 Practice Exercises

For extra help, see Examples 4–6 on pages 543–545 of your text.

Simplify each complex fraction by multiplying numerator and denominator by the least common denominator.

4. $\dfrac{\dfrac{9}{x^2}-1}{\dfrac{3}{x}-1}$

4. _____

5. $\dfrac{\dfrac{x-2}{x+2}}{\dfrac{x}{x-2}}$

5. _____

6. $\dfrac{\dfrac{6}{k+1}-\dfrac{5}{k-3}}{\dfrac{3}{k-3}+\dfrac{2}{k+2}}$

6. _____

Objective 4   Simplify rational expressions with negative exponents.

Video Examples

Review this example for Objective 4:

7. Simplify, using only positive exponents in the answer.

$$\frac{2x^{-1}+y^2}{z^{-3}}$$

$$= \frac{\dfrac{2}{x}+y^2}{\dfrac{1}{z^3}} = \frac{xz^3\left(\dfrac{2}{x}+y^2\right)}{xz^3\left(\dfrac{1}{z^3}\right)} = \frac{2z^3+xy^2z^3}{x}$$

Now Try:

7. Simplify, using only positive exponents in the answer.

$$\frac{x^{-1}}{y-x^{-1}}$$

_____

Objective 4 Practice Exercises

For extra help, see Example 7 on page 546 of your text.

Simplify each expression, using only positive exponents in the answer.

10. $\dfrac{4x^{-2}}{2+6y^{-3}}$

10. _____

11. $\dfrac{s^{-1}+r}{r^{-1}+s}$

11. _____

12. $\dfrac{(m+n)^{-2}}{m^{-2}-n^{-2}}$

12. _____

Chapter 7 RATIONAL EXPRESSIONS AND APPLICATIONS

7.6   Solving Equations with Rational Expressions

Learning Objectives
1   Distinguish between operations with rational expressions and equations with terms that are rational expressions.
2   Solve equations with rational expressions.
3   Solve a formula for a specified variable.

Key Terms

Use the vocabulary terms listed below to complete each statement in exercises 1–2.

proposed solution   extraneous solution

1.   A solution that is not an actual solution of a given equation is called a(n)

   _____.

2.   A value of the variable that appears to be a solution after both sides of an equation with rational expressions are multiplied by a variable expression is called a(n)

   _____ .

Objective 1   Distinguish between operations with rational expressions and equations with terms that are rational expressions.

Video Examples

Review these examples for Objective 1:

1.   Identify each of the following as an expression or an equation. Then simplify the expression or solve the equation.

   a. $\dfrac{8}{9}x - \dfrac{5}{6}x$

   This is a difference of two terms. It represents an expression since there is no equality symbol.
   The LCD is 18.

   $$\dfrac{8}{9}x - \dfrac{5}{6}x = \dfrac{2 \cdot 8}{2 \cdot 9}x - \dfrac{3 \cdot 5}{3 \cdot 6}x$$

   $$= \dfrac{16}{18}x - \dfrac{15}{18}x$$

   $$= \dfrac{1}{18}x$$

Now Try:

1.   Identify each of the following as an expression or an equation. Then simplify the expression or solve the equation.

   a. $\dfrac{4}{5}x - \dfrac{3}{10}x$

   _____

b. $\frac{8}{9}x - \frac{5}{6}x = \frac{2}{3}$

Because there is an equality symbol, this is an equation to be solved. The LCD is 18.

$$18\left(\frac{8}{9}x - \frac{5}{6}x\right) = 18\left(\frac{2}{3}\right)$$

$$18\left(\frac{8}{9}x\right) - 18\left(\frac{5}{6}x\right) = 18\left(\frac{2}{3}\right)$$

$$16x - 15x = 12$$

$$x = 12$$

Check $\frac{8}{9}x - \frac{5}{6}x = \frac{2}{3}$

$$\frac{8}{9}(12) - \frac{5}{6}(12) \overset{?}{=} \frac{2}{3}$$

$$\frac{32}{3} - 10 \overset{?}{=} \frac{2}{3}$$

$$\frac{2}{3} = \frac{2}{3} \quad \text{True}$$

The solution set is {12}.

b. $\frac{4}{5}x - \frac{3}{10}x = 7$

_____

Objective 1 Practice Exercises

For extra help, see Example 1 on page 550 of your text.

Identify each of the following as an expression or an equation. Then simplify the expression or solve the equation.

1. $\frac{3x}{5} - \frac{4x}{3} = \frac{22}{15}$

1. _____

2. $\frac{4x}{5} - \frac{5x}{10}$

2. _____

3. $\dfrac{2x}{5} + \dfrac{7x}{3}$

Objective 2 Solve equations with rational expressions.

Video Examples

Review these examples for Objective 2:

7. Solve, and check the proposed solution(s).

$$\dfrac{6}{x^2 - 1} = 1 - \dfrac{3}{x+1}$$

$x \neq 1, -1$ or a denominator is 0.
Factor the denominator.

$$\dfrac{6}{(x+1)(x-1)} = 1 - \dfrac{3}{x+1}$$

The LCD is $(x+1)(x-1)$.

$$(x+1)(x-1)\dfrac{6}{(x+1)(x-1)}$$

$$= (x+1)(x-1)\left(1 - \dfrac{3}{x+1}\right)$$

$$(x+1)(x-1)\dfrac{6}{(x+1)(x-1)}$$

$$= (x+1)(x-1) - (x+1)(x-1)\dfrac{3}{x+1}$$

$$6 = (x+1)(x-1) - 3(x-1)$$

$$6 = x^2 - 1 - 3x + 3$$

$$0 = x^2 - 3x - 4$$

$$0 = (x+1)(x-4)$$

$$x + 1 = 0 \quad \text{or} \quad x - 4 = 0$$

$$x = -1 \quad \text{or} \quad x = 4$$

Since -1 makes the original denominator equal 0, the proposed solution -1 is an extraneous value.
A check shows that $\{4\}$ is the solution set.

Now Try:

7. Solve, and check the proposed solution(s).

$$\dfrac{x}{x+1} + \dfrac{4}{x} = \dfrac{4}{x^2 + x}$$

_____

6. Solve, and check the proposed solution.

$$\frac{2x}{x^2-25}+\frac{5}{x-5}=\frac{3}{x+5}$$

$x \neq 5, -5$ or a denominator is 0.

Factor the denominator.

$$\frac{2x}{(x+5)(x-5)}+\frac{5}{x-5}=\frac{3}{x+5}$$

The LCD is $(x+5)(x-5)$.

$$(x+5)(x-5)\left(\frac{2x}{(x+5)(x-5)}+\frac{5}{x-5}\right)$$

$$=(x+5)(x-5)\frac{3}{x+5}$$

$$(x+5)(x-5)\frac{2x}{(x+5)(x-5)}+(x+5)(x-5)\frac{5}{x-5}$$

$$=(x+5)(x-5)\frac{3}{x+5}$$

$$2x+5(x+5)=3(x-5)$$

$$2x+5x+25=3x-15$$

$$7x+25=3x-15$$

$$4x=-40$$

$$x=-10$$

A check will verify that $\{-10\}$ is the solution set.

5. Solve, and check the proposed solution.

$$\frac{3}{x^2-4}=\frac{5}{x^2-2x}$$

Step 1 Factor the denominators to find the LCD, $x(x+2)(x-2)$. Multiply by the LCD.

$$\frac{3}{(x+2)(x-2)}=\frac{5}{x(x-2)}$$

$$x(x+2)(x-2)\frac{3}{(x+2)(x-2)}$$

$$=x(x+2)(x-2)\frac{5}{x(x-2)}$$

Step 2 Solve for x.

$$3x=5(x+2)$$

$$3x=5x+10$$

$$-2x=10$$

$$x=-5$$

6. Solve, and check the proposed solution.

$$\frac{8r}{9r^2-1}=\frac{4}{3r+1}+\frac{4}{3r-1}$$

———

5. Solve, and check the proposed solution.

$$\frac{4}{x^2-3x}=\frac{6}{x^2-9}$$

———

Step 3 Check the proposed solution.

$$\frac{3}{x^2 - 4} = \frac{5}{x^2 - 2x}$$

$$\frac{3}{(-5)^2 - 4} \stackrel{?}{=} \frac{5}{(-5)^2 - 2(-5)}$$

$$\frac{3}{25 - 4} \stackrel{?}{=} \frac{5}{25 + 10}$$

$$\frac{1}{7} = \frac{1}{7} \quad \text{True}$$

The solution set is $\{-5\}$.

Objective 2 Practice Exercises

For extra help, see Examples 2–8 on pages 551–556 of your text.

Solve each equation and check your solutions.

4. $\dfrac{4}{n+2} - \dfrac{2}{n} = \dfrac{1}{6}$

4. _____

5. $\dfrac{x}{3x+16} = \dfrac{4}{x}$

5. _____

6. $\dfrac{-16}{n^2 - 8n + 12} = \dfrac{3}{n-2} + \dfrac{n}{n-6}$

6. _____

Objective 3   Solve a formula for a specified variable.

Video Examples

Review this example for Objective 3:

9. Solve the formula for the specified variable.

$r = \dfrac{s+w}{v}$ for w

Isolate w. Multiply by v.

$r = \dfrac{s+w}{v}$

$rv = s + w$

$rv - s = w$

Check $r = \dfrac{s+w}{v}$

$r \stackrel{?}{=} \dfrac{s + rv - s}{v}$

$r \stackrel{?}{=} \dfrac{rv}{v}$

$r = r$ True

Now Try:

9. Solve the formula for the specified variable.

$a = \dfrac{b-c}{q}$ for b

_____

Name: Date:
Instructor: Section:

Objective 3 Practice Exercises

For extra help, see Examples 9–11 on pages 556–557 of your text.

Solve each formula for the specified variable.

7. $\dfrac{1}{f} = \dfrac{1}{d_0} + \dfrac{1}{d_1}$ for f

7. _____

8. $m = \dfrac{y_2 - y_1}{x_2 - x_1}$ for y_1

8. _____

9. $A = \dfrac{2pf}{b(q+1)}$ for q

9. _____

Chapter 7 RATIONAL EXPRESSIONS AND APPLICATIONS

7.7   Applications of Rational Expressions

Learning Objectives
1   Solve problems about numbers.
2   Solve problems about distance, rate, and time.
3   Solve problems about work.

Key Terms

Use the vocabulary terms listed below to complete each statement in exercises 1–3.

   reciprocal   numerator   denominator

1.   In the fraction $\frac{x+5}{x-2}$, $x + 5$ is the _____.

2.   In the fraction $\frac{x+5}{x-2}$, $x - 2$ is the _____.

3.   The fraction $\frac{x+5}{x-2}$ is the _____ of the fraction $\frac{x-2}{x+5}$.

Objective 1   Solve problems about numbers.

Video Examples

Review this example for Objective 1:

1.   If a certain number is added to the numerator and twice that number is subtracted from the denominator of the fraction $\frac{3}{5}$, the result is equal to 5. Find the number.

Step 1 Read the problem carefully. We are trying to find a number.

Step 2 Assign a variable.
    Let x = the number.

Step 3 Write an equation. The fraction $\frac{3+x}{5-2x}$ represents adding the number to the numerator and twice that number is subtracted from the denominator of the fraction $\frac{3}{5}$. The result is equal to 5.

$$\frac{3+x}{5-2x} = 5$$

Now Try:

1.   If the same number is added to the numerator and denominator of the fraction $\frac{5}{9}$, the value of the resulting fraction is $\frac{2}{3}$. Find the number.

_____

Step 4 Solve. Multiply by the LCD, $5 - 2x$.

$$(5 - 2x)\frac{3 + x}{5 - 2x} = (5 - 2x)5$$

$$3 + x = 25 - 10x$$

$$11x = 22$$

$$x = 2$$

Step 5 State the answer. The number is 2.

Step 6 Check the solution in the original problem. If 2 is added to the numerator, and twice 2 is subtracted from the denominator of $\frac{3}{5}$,

the result is $\frac{3 + 2}{5 - 2(2)} = \frac{5}{1}$, or 5, as required.

Objective 1 Practice Exercises

For extra help, see Example 1 on page 564 of your text.

Solve each problem. Check your answers to be sure they are reasonable.

1. If three times a number is subtracted from twice its reciprocal, the result is –1. Find the number.

    1. _____

2. If two times a number is added to one-half of its reciprocal, the result is $\frac{13}{6}$. Find the number.

    2. _____

3. The denominator of a fraction is 1 less than twice the numerator. If the numerator and the denominator are each increased by 3, the resulting fraction simplifies to $\frac{3}{4}$. Find the original fraction.

3. _____

Objective 2 Solve problems about distance, rate, and time.

Video Examples

Review this example for Objective 2:

4. A boat goes 6 miles per hour in still water. It takes as long to go 40 miles upstream as 80 miles downstream. Find the speed of the current.

Step 1 Read the problem carefully. Find the speed of the current.

Step 2 Assign a variable.

Let x = the speed of the current.

The rate of traveling upstream is $6 - x$.

The rate of traveling downstream is $6 + x$.

	d	r	t
Upstream	40	$6 - x$	$\dfrac{40}{6-x}$
Downstream	80	$6 + x$	$\dfrac{80}{6+x}$

The times are equal.

Step 3 Write an equation.

$$\frac{40}{6-x} = \frac{80}{6+x}$$

Now Try:

4. The Cuyahoga River has a current of 2 miles per hour. Ali can paddle 10 miles downstream in the time it takes her to paddle 2 miles upstream. How fast can Ali paddle?

_____

Step 4 Solve. The LCD is $(6+x)(6-x)$.

$$(6+x)(6-x)\frac{40}{6-x} = (6+x)(6-x)\frac{80}{6+x}$$

$$40(6+x) = 80(6-x)$$

$$240+40x = 480-80x$$

$$240+120x = 480$$

$$120x = 240$$

$$x = 2$$

Step 5 State the answer. The speed of the current is 2 miles per hour.

Step 6 Check.

Upstream: $\dfrac{40}{6-2} = 10$ hr

Downstream: $\dfrac{80}{6+2} = 10$ hr

The time upstream is the same as the time downstream, as required.

Objective 2 Practice Exercises

For extra help, see Examples 2–4 on pages 565–568 of your text.

Solve each problem.

4. A boat travels 15 miles per hour in still water. The boat travels 20 miles downstream in the same time it takes the boat to travel 10 miles upstream. How fast is the current?

4. _____

5.   A ship goes 120 miles downriver in $2\frac{2}{3}$ hours less than it takes to go the same distance upriver. If the speed of the current is 6 miles per hour, find the speed of the ship.

5. _____

6.   On Saturday, Pablo jogged 6 miles. On Monday, jogging at the same speed, it took him 30 minutes longer to cover 10 miles. How fast did Pablo jog?

6. _____

Objective 3   Solve problems about work.

Video Examples

Review this example for Objective 3.	Now Try:
5. One pipe can fill a swimming pool in 8 hours and another pipe can fill the pool in 12 hours. How long will it take to fill the pool if both pipes are open? *Step 1* Read the problem carefully. Find the time working together. *Step 2* Assign a variable. Let x = the number of hours working together. The rate for the first pipe is $\frac{1}{8}$. The rate for the second pipe is $\frac{1}{12}$.	5. Chuck can weed the garden in $\frac{1}{2}$ hour, but David takes 2 hours. How long does it take them to weed the garden if they work together? _____

Step 3 Write an equation. The sum of the fractional part for each pipe multiplied by the time working together is the whole job.

$$\frac{1}{8}x + \frac{1}{12}x = 1$$

Step 4 Solve. The LCD is 48.

$$48\left(\frac{1}{8}x + \frac{1}{12}x\right) = 48(1)$$

$$48\left(\frac{1}{8}x\right) + 48\left(\frac{1}{12}x\right) = 48$$

$$6x + 4x = 48$$

$$10x = 48$$

$$x = \frac{48}{10} = \frac{24}{5} = 4\frac{4}{5}$$

Step 5 State the answer. Working together, it takes $4\frac{4}{5}$ hours to fill the pool.

Step 6 Check. Substitute $\frac{24}{5}$ for x in the equation from Step 3.

$$\frac{1}{8}x + \frac{1}{12}x = 1$$

$$\frac{1}{8}\left(\frac{24}{5}\right) + \frac{1}{12}\left(\frac{24}{5}\right) = 1$$

$$\frac{3}{5} + \frac{2}{5} = 1 \quad \text{True}$$

Objective 3 Practice Exercises

For extra help, see Example 5 on page 569 of your text.

Solve each problem.

7. Kelly can clean the house in 6 hours, but it takes Linda 4 hours. How long would it take them to clean the house if they worked together?

7. _____

8. Michael can type twice as fast as Sharon. Together they can type a certain job in 2 hours. How long would it take Michael to type the entire job by himself?

8. _____

9. A swimming pool can be filled by an inlet pipe in 18 hours and emptied by an outlet pipe in 24 hours. How long will it take to fill the empty pool if the outlet pipe is accidentally left open at the same time as the inlet pipe is opened?

9. _____

Chapter 8 EQUATIONS, INEQUALITIES, GRAPHS, AND SYSTEMS REVISITED

8.1   Review of Solving Linear Equations and Inequalities in One Variable

Learning Objectives
1   Distinguish between expressions and equations.
2   Solve linear equations.
3   Review interval notation.
4   Solve linear inequalities.
5   Solve three-part linear inequalities

Key Terms

Use the vocabulary terms listed below to complete each statement in exercises 1–8.

> **linear (first-degree) equation in one variable**   **solution**
>
> **solution set**   **equivalent equations**
>
> **conditional equation**   **identity**   **contradiction**
>
> **linear inequality in one variable**

1.   Equations that have exactly the same solution sets are called

      _____.

2.   An equation that can be written in the form $Ax + B = C$, where A, B, and C are real numbers and $A \neq 0$, is called a _____.

3.   The set of all numbers that satisfy an equation is called its

      _____.

4.   An equation with no solution is called a(n) _____.

5.   A(n) _____ is an equation that is true for some values of the variable and false for other values.

6.   An equation that is true for all values of the variable is called a(n)

      _____.

7.   A _____ of an equation is a number that makes the equation true when substituted for the variable.

8.   A(n) _____ can be written in the form $Ax + B > C$, $Ax + B \geq C$, $Ax + B < C$, or $Ax + B \leq C$, where A, B, and C are real numbers with $A \neq 0$.

Objective 1 Distinguish between expressions and equations.

Video Examples

Review these examples for Objective 1:

1. Decide whether each of the following is an *expression* or an *equation*.

    a. $5x + 6$

    $5x + 6$ is an *expression* because there is no equality symbol.

    b. $5x + 6 = 10$

    $5x + 6 = 10$ is an *equation* because there is an equality symbol.

Now Try:

1. Decide whether each of the following is an *expression* or an *equation*.
    a. $9b - 20 = 98$

    _____

    b. $9b - 20$

    _____

Objective 1 Practice Exercises

Decide whether each of the following is an expression or an equation.

1. $2(x + 5) = 7$

2. $2(3t - 8) - 7t$

1. _____

2. _____

Objective 2 Solve linear equations.

Video Examples

Review these examples for Objective 2:

3. Solve $5(k - 4) + 2k = 3k + 8$.

$$5(k - 4) + 2k = 3k + 8$$

Step 1 $5(k) + 5(-4) + 2k = 3k + 8$
$$5k - 20 + 2k = 3k + 8$$
$$7k - 20 = 3k + 8$$

Step 2 $7k - 20 - 3k = 3k + 8 - 3k$
$$4k - 20 = 8$$
$$4k - 20 + 20 = 8 + 20$$
$$4k = 28$$

Step 3 $\dfrac{4k}{4} = \dfrac{28}{4}$
$$k = 7$$

Now Try:

3. Solve $7(k + 3) - 5k = 6 - 3k$.

_____

Step 4 $5(k-4)+2k=3k+8$

$$5(7-4)+2(7)\overset{?}{=}3(7)+8$$

$$15+14\overset{?}{=}21+8$$

$$29=29$$

The proposed solution checks, so the solution set is $\{7\}$.

5.   Solve $\dfrac{x-6}{3}+\dfrac{3x+9}{5}=-3.$

$$\frac{x-6}{3}+\frac{3x+9}{5}=-3$$

Step 1 $15\left(\dfrac{x-6}{3}+\dfrac{3x+9}{5}\right)=15(-3)$

$$15\left(\frac{x-6}{3}\right)+15\left(\frac{3x+9}{5}\right)=-45$$

$$5(x-6)+3(3x+9)=-45$$

$$5x-30+9x+27=-45$$

$$14x-3=-45$$

Step 2 $14x-3+3=-45+3$

$$14x=-42$$

Step 3 $\dfrac{14x}{14}=\dfrac{-42}{14}$

$$x=-3$$

Step 4   A check verifies that the solution set is $\{-3\}$.

7.   Solve each equation. Decide whether it is a *conditional equation*, an *identity*, or a *contradiction*.

    a. $9(x+3)-7=5(x+4)$

$$9(x+3)-7=5(x+4)$$

$$9x+27-7=5x+20$$

$$9x+20=5x+20$$

$$9x+20-5x-20=5x+20-5x-20$$

$$4x=0$$

$$\frac{4x}{4}=\frac{0}{4}$$

$$x=0$$

5.   Solve $\dfrac{x-5}{2}-\dfrac{x+6}{3}=-4.$

————————————

7.   Solve each equation. Decide whether it is a *conditional equation*, an *identity*, or a *contradiction*.

    a. $10(x-4)=3(x+8)-1$

————————————

Check $9(x+3)-7=5(x+4)$

$$9(0+3)-7\overset{?}{=}5(0+4)$$

$$27-7\overset{?}{=}5(4)$$

$$20=20$$

The solution 0 checks, so the solution set is $\{0\}$. Since the solution set has only one element, the equation is conditional.

b. $8x-9=8(x-2)+6$

$$8x-9=8(x-2)+6$$

$$8x-9=8x-16+6$$

$$8x-9=8x-10$$

$$8x-9-8x=8x-10-8x$$

$$-9=-10 \quad \text{False}$$

Since the result is false, the equation has no solution. The solution set is $\varnothing$, so the equation is a contradiction.

c. $6x-17=6(x-4)+7$

$$6x-17=6(x-4)+7$$

$$6x-17=6x-24+7$$

$$6x-17=6x-17$$

$$6x-17-6x+17=6x-17-6x+17$$

$$0=0 \quad \text{True}$$

The true statement $0=0$, indicates that the solution set is {all real numbers}. The equation is an identity.

b. $4x-13=x+3(x+2)$

_____

c. $7(2x-3)=5(3x-4)-(x+1)$

_____

Objective 2 Practice Exercises

For extra help, see Examples 2–7 on pages 591–595 of your text.

Solve and check each equation.

3. $18-3y+11=7y+6y-27-9y$ 3. _____

4. $\dfrac{2x+5}{5} - \dfrac{3x+1}{2} = \dfrac{7-x}{2}$ 4. _____

*Decide whether each equation is a **conditional equation**, an **identity**, or a **contradiction**.
Give the solution set.*

5. $7(2-5b) - 32 = 10b - 3(6+15b)$ 5. _____

6. $7(3-4q) - 10(q-2) = 19(5-2q)$ 6. _____

Objective 3   Review interval notation.

For extra help, see page 596 of your text.

Objective 4   Solve linear inequalities.

Video Examples

Review this example for Objective 4:
10. Solve $-4(x+3)+5 \geq 3-2x$, and graph the
    solution set.

   Step 1 $-4(x+3)+5 \geq 3-2x$
             $-4x-12+5 \geq 3-2x$
             $-4x-7 \geq 3-2x$

   Step 2 $-4x-7+2x \geq 3-2x+2x$
             $-2x-7 \geq 3$
             $-2x-7+7 \geq 3+7$
             $-2x \geq 10$

Now Try:
10. Solve $x-5(x+2) \geq 3x+4$, and
    graph the solution set.

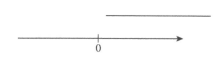

Step 3 $\dfrac{-2x}{-2} \le \dfrac{10}{-2}$

$x \le -5$

The solution set is $(-\infty, -5]$.

Objective 4 Practice Exercises

For extra help, see Examples 8–10 on pages 597–599 of your text.

Solve each inequality, giving its solution set in both interval and graph forms. Check your answers.

7. $-2 + 8b \ge 7b - 1$

7. _____

$\begin{array}{c} \longleftarrow\!+\!+\!+\!+\!+\!+\!+\!+\!+\!+\!+\!+\!+\!+\!\longrightarrow \end{array}$

8. $3 + 5p \le 4p + 3$

8. _____

$\begin{array}{c} \longleftarrow\!+\!+\!+\!+\!+\!+\!+\!+\!+\!+\!+\!+\!+\!+\!\longrightarrow \end{array}$

9. $-9m > -36$

9. _____

$\begin{array}{c} \longleftarrow\!+\!+\!+\!+\!+\!+\!+\!+\!+\!+\!+\!+\!+\!+\!\longrightarrow \end{array}$

Objective 5   Solve three-part inequalities.

Video Examples

Review this example for Objective 5:

11. Solve $-7 \le -4k - 3 \le 9$, and graph the solution set.

$$-7 \le \quad -4k - 3 \quad \le 9$$
$$-7 + 3 \le -4k - 3 + 3 \le 9 + 3$$
$$-4 \le \quad -4k \quad \le 12$$
$$\frac{-4}{-4} \ge \quad \frac{-4k}{-4} \quad \ge \frac{12}{-4}$$
$$1 \ge \quad k \quad \ge -3$$

The solution set is $[-3, 1]$.

Now Try:

11. Solve $-8 \le -5k + 2 \le 22$, and graph the solution set.

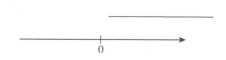

Objective 5 Practice Exercises

For extra help, see Example 11 on page 600 of your text.

Solve each inequality, giving its solution set in both interval and graph forms. Check your answers.

10. $9 \le 2x + 1 \le 15$

10. _____

11. $-17 \le 3x - 2 < -11$

11. _____

12. $1 < 3z + 4 < 19$

12. _____

Chapter 8 EQUATIONS, INEQUALITIES, GRAPHS, AND SYSTEMS REVISITED

8.2   Set Operations and Compound Inequalities

Learning Objectives	
1	Recognize set intersection and union.
2	Find the intersection of two sets.
3	Solve compound inequalities with the word *and*.
4	Find the union of two sets.
5	Solve compound inequalities with the word *or*.

Key Terms

Use the vocabulary terms listed below to complete each statement in exercises 1–3.

      intersection   **compound inequality**   **union**

1. The _____ of two sets, A and B, is the set of elements that belong to either A or B or both.

2. A _____ is formed by joining two inequalities with a connective word such as *and* or *or*.

3. The _____ of two sets, A and B, is the set of elements that belong to both A and B.

Objective 1   Recognize set intersection and union.

For extra help, see page 605 of your text.

Objective 2   Find the intersection of two sets.

Video Examples

Review this example for Objective 2:

1. Let $A = \{6, 7, 8, 9\}$ and $B = \{6, 8, 10\}$. Find $A \cap B$.

   The set $A \cap B$ contains those elements that belong to both A and B.
   $$A \cap B = \{6, 7, 8, 9\} \cap \{6, 8, 10\}$$
   $$= \{6, 8\}$$

Now Try:

1. Let $A = \{20, 30, 40, 50\}$ and $B = \{30, 50, 70\}$. Find $A \cap B$.

   _____

Name:
Instructor:

Date:
Section:

Objective 2 Practice Exercises

For extra help, see Example 1 on page 605 of your text.

Let A = {0, 1, 2, 3, 4, 5}, *B* = {2, 4, 6, 8, 10}, *C* = {1, 3, 5, 7, 9}, *and D* = {0, 2, 4}. *Specify each set.*

1. $A \cap D$

1. _____

2. $B \cap C$

2. _____

3. $A \cap C$

3. _____

Objective 3   Solve compound inequalities with the word *and*.

Video Examples

Review these examples for Objective 3:

3. Solve $-5x - 6 > 8$ and $6x - 3 \le -21$, and graph the solution set.

Now Try:

3. Solve $-8x + 6 > 30$ and $4x - 7 \le 11$, and graph the solution set.

_____

Step 1 Solve each inequality individually.

$$-5x - 6 > 8 \qquad \text{and} \quad 6x - 3 \le -21$$
$$-5x > 14 \qquad \text{and} \qquad 6x \le -18$$
$$x < -\frac{14}{5} \quad \text{and} \qquad x \le -3$$

The graphs of $x < -\frac{14}{5}$ and $x \le -3$ are shown below. $-\frac{14}{5}$

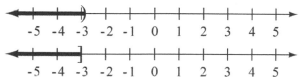

Step 2 Now find all the values of *x* that are less than $-\frac{14}{5}$ and also less than or equal to –3. The solution set is $(-\infty, -3]$.

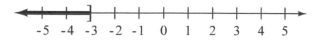

2. Solve $x+6\leq15$ and $x+3\geq7$, and graph the solution set.

Step 1 Solve each inequality individually.
$$x+6\leq15 \quad \text{and} \quad x+3\geq7$$
$$x+6-6\leq15-6 \quad \text{and} \quad x+3-3\geq7-3$$
$$x\leq9 \quad \text{and} \quad x\geq4$$

Step 2 The solution set of the compound inequality includes all numbers that satisfy both inequalities from Step 1.
The graphs of $x\leq9$ and $x\geq4$ are shown below.

The intersection of these two graphs is the solution set [4, 9].

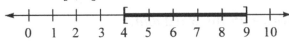

4. Solve the compound inequality, and graph the solution set.
$$x+7<10 \quad \text{and} \quad x-13>4$$

Step 1 Solve each inequality individually.
$$x+7<10 \quad \text{and} \quad x-13>4$$
$$x<3 \quad \text{and} \quad x>17$$
The graphs of $x<3$ and $x>17$ are shown below.

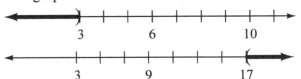

Step 2 No number is less than 4 and greater than 17, so the compound inequality has no solution. The solution set is $\varnothing$. The graph is below.

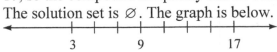

2. Solve $x+4\leq20$ and $x-3\geq9$ and graph the solution set.

_____

4. Solve the compound inequality, and graph the solution set.
$$2x+6<8 \quad \text{and} \quad 5x-11>9$$

_____

Name: Date:

Instructor: Section:

Objective 3 Practice Exercises

For extra help, see Examples 2–4 on pages 606–607 of your text.

For each compound inequality, give the solution set in both interval and graph forms.

4. $1 - 2s \le 7$ and $2s + 7 \ge 11$ **4.** _____

5. $3x + 2 < 11$ and $2 - 3x \le 14$ **5.** _____

6. $5t > 0$ and $5t + 4 \le 9$ **6.** _____

Objective 4   Find the union of two sets.

Video Examples

Review this example for Objective 4:

5.   Let $A = \{6, 7, 8, 9\}$ and $B = \{6, 8, 10\}$. Find $A \cup B$.

The set $A \cup B$ contains those elements that belong to either A or B (or both).

$$A \cup B = \{6, 7, 8, 9\} \cup \{6, 8, 10\}$$
$$= \{6, 7, 8, 9, 10\}$$

Now Try:

5.   Let $A = \{20, 30, 40, 50\}$ and $B = \{30, 50, 70\}$. Find $A \cup B$.

_____

Objective 4 Practice Exercises

For extra help, see Example 5 on page 607 of your text.

Let $A = \{0, 1, 2, 3, 4, 5\}$, $B = \{2, 4, 6, 8, 10\}$, $C = \{1, 3, 5, 7, 9\}$, $D = \{0, 2, 4\}$, and $E = \{0\}$. Specify each set.

7. $A \cup D$ 7. _____

8. $B \cup C$ 8. _____

9. $A \cup E$ 9. _____

Objective 5   Solve compound inequalities with the word *or*.

Video Examples

Review these examples for Objective 5:	**Now Try:**
6. Solve the compound inequality, and graph the solution set.	6. Solve the compound inequality, and graph the solution set.

Review these examples for Objective 5:

6. Solve the compound inequality, and graph the solution set.

$$7x - 6 < 4x \quad \text{or} \quad -4x \leq -16$$

Step 1 Solve each inequality individually.

$$7x - 6 < 4x \quad \text{or} \quad -4x \leq -16$$
$$3x < 6$$
$$x < 2 \quad \text{or} \quad x \geq 4$$

The graphs of $x < 2$ and $x \geq 4$ are shown below.

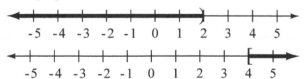

Step 2 Since the inequalities are joined with *or*, find the union of the two solution sets. The union is shown below. $(-\infty, 2) \cup [4, \infty)$

Now Try:

6. Solve the compound inequality, and graph the solution set.

$$5x - 6 < 2x \quad \text{or} \quad -7x \leq -35$$

_____

8. Solve the compound inequality, and graph the solution set.

$$-5x+4\geq -16 \quad \text{or} \quad 6x-11\geq -29$$

Solve each inequality.

$$-5x+4\geq -16 \quad \text{or} \quad 6x-11\geq -29$$
$$-5x\geq -20 \quad \text{or} \quad 6x\geq -18$$
$$x\leq 4 \quad \text{or} \quad x\geq -3$$

Graph each inequality.

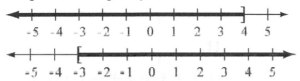

The solution is the union of the two sets, $(-\infty,\ \infty)$.

7. Solve the compound inequality, and graph the solution set.

$$-7x+4\geq 32 \quad \text{or} \quad 8x+6\leq -26$$

Solve each inequality.

$$-7x+4\geq 32 \quad \text{or} \quad 8x+6\leq -18$$
$$-7x\geq 28 \quad \text{or} \quad 8x\leq -24$$
$$x\leq -4 \quad \text{or} \quad x\leq -3$$

Graph each inequality.

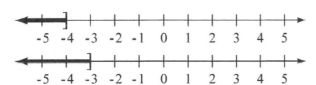

The solution set is the union of the two sets, $(-\infty,-3]$.

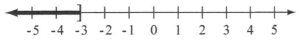

8. Solve the compound inequality, and graph the solution set.

$$-9x+3\geq -15 \quad \text{or} \quad 8x-7\geq -15$$

_____

7. Solve the compound inequality, and graph the solution set.

$$-6x+3\geq 15 \quad \text{or} \quad 9x+7\leq -20$$

_____

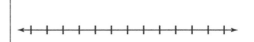

Objective 5 Practice Exercises

For extra help, see Examples 6–9 on pages 608–609 of your text.

For each compound inequality, give the solution set in both interval and graph forms.

10. $q + 3 > 7$ or $q + 1 \leq -3$

10. _____

11. $3 > 4m + 2$ or $4m - 3 \geq -2$

11. _____

12. $2r + 4 \geq 8$ or $4r - 3 < 1$

12. _____

Chapter 8   EQUATIONS, INEQUALITIES, GRAPHS, AND SYSTEMS REVISITED

8.3   Absolute Value Equations and Inequalities

Learning Objectives	
1	Use the distance definition of absolute value.
2	Solve equations of the form $|ax + b| = k$, for $k > 0$.
3	Solve inequalities of the form $|ax + b| < k$ and of the form $|ax + b| > k$, for $k > 0$.
4	Solve absolute value equations that involve rewriting.
5	Solve equations of the form $|ax + b| = |cx + d|$.
6	Solve special cases of absolute value equations and inequalities.
7	Solve an application involving relative error.

Key Terms

Use the vocabulary terms listed below to complete each statement in exercises 1–2.

       absolute value equation   **absolute value inequality**

1.   An _____ is an equation that involves the absolute value of a variable expression.

2.   An _____ is an inequality that involves the absolute value of a variable expression.

Objective 1   Use the distance definition of absolute value.

Objective 1 Practice Exercises

For extra help, see pages 614–615 of your text.

Graph the solution set of each equation or inequality.

1. $|m| = 7$

     1.
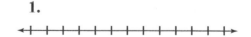

2. $|p| < 3$

     2.

3. $|x| \geq 6$

     3.

Objective 2   **Solve equations of the form $|ax + b| = k$, for $k > 0$.**

Video Examples

Review this example for Objective 2:

1.   Solve $|3x + 2| = 14$. Graph the solution set.

This is Case 1.

$$3x + 2 = 14 \quad \text{or} \quad 3x + 2 = -14$$
$$3x = 12 \quad \text{or} \quad 3x = -16$$
$$x = 4 \quad \text{or} \quad x = -\frac{16}{3}$$

Check

Let $x = 4$ Let $x = -\dfrac{16}{3}$

$$3(4) + 2 \overset{?}{=} 14 \quad \Big| \quad 3\left(-\frac{16}{3}\right) + 2 \overset{?}{=} -14$$
$$12 + 2 \overset{?}{=} 14 \quad \Big| \quad -16 + 2 \overset{?}{=} -14$$
$$14 = 14 \quad \Big| \quad -14 = -14$$

The check confirms that the solution set is

$\left\{-\dfrac{16}{3},\ 4\right\}$.

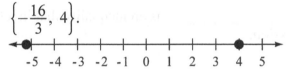

Now Try:

1.   Solve $|5x + 4| = 11$. Graph the solution set.

_____

Objective 2 Practice Exercises

For extra help, see Example 1 on pages 615–617 of your text.

Solve each equation.

4. $|2x + 3| = 10$ 4. _____

5. $|5r - 15| = 0$ 5. _____

6. $\left|\dfrac{1}{2}x - 3\right| = 4$

6. _____

Objective 3   **Solve inequalities of the form $|ax + b| < k$ and of the form $|ax + b| > k$, for $k > 0$.**

Video Examples

Review these examples for Objective 3:

2.   Solve $|4x + 2| > 10$. Graph the solution set.

This is Case 2.

$$4x + 2 > 10 \quad \text{or} \quad 4x + 2 < -10$$
$$4x > 8 \quad \text{or} \quad 4x < -12$$
$$x > 2 \quad \text{or} \quad x < -3$$

A check confirms that the solution set is $(-\infty, -3) \cup (2, \infty)$.

3.   Solve $|10x + 5| < 25$. Graph the solution set.

This is Case 3.

$$-25 < 10x + 5 < 25$$
$$-30 < \quad 10x \quad < 20$$
$$-3 < \quad x \quad < 2$$

A check confirms that the solution set is $(-3, 2)$.

Now Try:

2.   Solve $|8x + 4| > 12$. Graph the solution set.

_____

◄ | | | | | | | | | ├ ├ ├ ├ ├►

3.   Solve $|6x + 3| < 15$. Graph the solution set.

_____

◄├ ├ ├ ├ ├ ├ ├ ├ ├ ├ ├ ├ ├ ├►

Objective 3 Practice Exercises

For extra help, see Examples 2–4 on pages 616–617 of your text.

Solve each inequality and graph the solution set.

7. $|x-2|>8$

7. _____

8. $|2r-9|\geq 23$

8. _____

9. $|5r+2|<18$

9. _____

Objective 4   Solve absolute value equations that involve rewriting.

Video Examples

Review this example for Objective 4:

5. Solve $|x-7|+3=15$.

Isolate the absolute value.
$$|x-7|+3=15$$
$$|x-7|+3-3=15-3$$
$$|x-7|=12$$
Now solve, using Case 1.
$$x-7=12 \quad \text{or} \quad x-7=-12$$
$$x=19 \quad \text{or} \qquad x=-5$$

Now Try:

5. Solve $|x-3|+4=11$.

_____

Check

Let $x = 19$ Let $x = -5$

$$|19-7|+3 \overset{?}{=} 15 \quad | \quad |-5-7|+3 \overset{?}{=} 15$$

$$|12|+3 \overset{?}{=} 15 \quad | \quad |-12|+3 \overset{?}{=} 15$$

$$15 = 15 \quad | \quad 15 = 15$$

The check confirms that the solution set is $\{-5, 19\}$.

Objective 4 Practice Exercises

For extra help, see Examples 5–6 on page 618 of your text.

Solve each equation.

10. $|2w-1|+7 = 12$

10. _____

11. $\left|2 - \frac{1}{2}x\right| - 5 = 18$

11. _____

12. $|4t+3|+8 = 10$

12. _____

Objective 5   Solve equations of the form $|ax + b| = |cx + d|$.

Video Examples

Review this example for Objective 5:

7. Solve $|z+5| = |3z-9|$.

$$z+5 = 3z-9 \quad \text{or} \quad z+5 = -(3z-9)$$
$$z+14 = 3z \quad\;\; \text{or} \quad z+5 = -3z+9$$
$$14 = 2z \quad\;\;\;\; \text{or} \quad\;\;\; 4z = 4$$
$$z = 7 \quad\;\;\;\;\; \text{or} \quad\;\;\;\; z = 1$$

Check

    Let $z = 7$ Let $z = 1$

$$|7+5| \overset{?}{=} |3(7)-9| \quad\Big|\quad |1+5| \overset{?}{=} |3(1)-9|$$
$$|12| \overset{?}{=} |21-9| \quad\;\;\Big|\quad\;\; |6| \overset{?}{=} |-6|$$
$$12 = 12 \quad\quad\;\;\;\Big|\quad\quad 6 = 6$$

The check confirms that the solution set is $\{1, 7\}$.

Now Try:

7. Solve $|z+7| = |2z+8|$.

_____

Objective 5 Practice Exercises

For extra help, see Example 7 on page 619 of your text.

Solve each problem.

13. $|y+5| = |3y+1|$

13. _____

14. $|2p-4| = |7-p|$

14. _____

15. $|3x-2|=|5x+8|$ 15. _____

Objective 6   Solve special cases of absolute value equations and inequalities.

Video Examples

Review these examples for Objective 6:

8. Solve each equation.

a. $|9x-1|=-13$

The absolute value of an expression can never be negative, so there are no solutions for this equation. The solution set is $\varnothing$.

b. $|8x-12|=0$

$$8x-12=0$$
$$8x=12$$
$$x=\frac{3}{2}$$

The solution set is $\left\{\frac{3}{2}\right\}$.

9. Solve each inequality.

a. $|x|\geq-12$

The absolute value of a number is always greater than or equal to 0. Thus, $|x|\geq-12$ is always true, and the solution set is $(-\infty, \infty)$.

b. $|x-11|+7<2$

$$|x-11|+7<2$$
$$|x-11|<-5$$

There is no number whose absolute value is less than −5, so this inequality has no solution. The solution set is $\varnothing$.

Now Try:

8. Solve each equation.

a. $|6x+13|=-5$

_____

b. $|7x+35|=0$

_____

9. Solve each inequality.

a. $|x|\geq-2$

_____

b. $|x+12|+6<3$

_____

   c. $|2x-8|-5 \le -5$

       $|2x-8|-5 \le -5$

          $|2x-8| \le 0$

The value of $|2x-8|$ will never be less than zero. However, $|2x-8|$ will equal 0.

       $2x-8=0$

         $2x=8$

          $x=4$

The solution set is $\{4\}$.

   c. $|5x-20|+7 \le 7$

       _____

Objective 6 Practice Exercises

For extra help, see Examples 8–9 on pages 619–620 of your text.

Solve each problem.

16. $\left|7+\dfrac{1}{2}x\right| = 0$

16. _____

17. $\left|m-2\right| \ge -1$

17. _____

18. $\left|k+5\right| \le -2$

18. _____

Objective 7   Solve an application involving relative error.

Video Examples

Review this example for Objective 7:	**Now Try:**

Review this example for Objective 7:

10. Suppose a machine filling 16.9 oz water bottles is set for a relative error that is no greater than 0.025 oz. How many ounces may a filled water bottle contain?

$$\left|\frac{16.9-x}{16.9}\right| \le 0.025$$

$$-0.025 \le \frac{16.9-x}{16.9} \le 0.025$$

$$0.4225 \le 16.9 - x \le 0.4225$$

$$-17.3225 \le \quad -x \quad \le -16.4775$$

$$17.3225 \ge \quad x \quad \ge 16.4775$$

$$16.4775 \le \quad x \quad \le 17.3225$$

The bottle may contain between 16.4775 and 17.3225 oz, inclusive.

Now Try:

10. Suppose a machine filling 16.9 oz water bottles is set for a relative error that is no greater than 0.05 oz. How many ounces may a filled water bottle contain?

_____

Objective 7 Practice Exercises

For extra help, see Example 10 on page 620 of your text.

Determine the number of ounces a filled 16.9 oz water bottle may contain for the given relative error.

19.   no greater than 0.04 oz

19.   _____

20.   no greater than 0.015 oz

20.   _____

21.   no greater than 0.03 oz   **21.** _____

Chapter 8   EQUATIONS, INEQUALITIES, GRAPHS, AND SYSTEMS REVISITED

8.4   Review of Graphing Linear Equations in Two Variables; Slope

Learning Objectives
1   Plot ordered pairs.
2   Find ordered pairs that satisfy a given equation.
3   Graph lines and find intercepts.
4   Recognize equations of horizontal and vertical lines.
5   Find the slope of a line.
6   Graph a line given its slope and a point on the line.

Key Terms

Use the vocabulary terms listed below to complete each statement in exercises 1–17.

ordered pair	**origin**	**x-axis**	**y-axis**
rectangular (Cartesian) coordinate system		**plot**	
components	**coordinate**	**quadrant**	**graph of an equation**
first-degree equation		**linear equation in two variables**	
y-intercept	**x-intercept**	**rise**	**run** **slope**

1.   The _____ of a line is the ratio of the change in y compared to the change in x when moving along the line from one point to another.

2.   The vertical change between two different points on a line is called the _____.

3.   The horizontal change between two different points on a line is called the _____.

4.   If a graph intersects the y-axis at k, then the _____ is $(0, k)$.

5.   An equation that can be written in the form $Ax + By = C$, where A, B, and C are real numbers and $A, B \neq 0$, is called a _____.

6.   Each number in an ordered pair represents a _____ of the corresponding point.

7.   The axis lines in a coordinate system intersect at the _____.

8.   If a graph intersects the x-axis at k, then the _____ is $(k, 0)$.

9.   In a rectangular coordinate system, the horizontal number line is called the _____.

10.   A _____ is one of the four regions in the plane
      determined by a rectangular coordinate system.

11.   A pair of numbers written between parentheses in which order is important is called
      a(n) _____.

12.   In a rectangular coordinate system, the vertical number line is called the

      _____.

13.   The two numbers in an ordered pair are the _____ of the
      ordered pair.

14.   To _____ an ordered pair is to locate the corresponding point on a
      coordinate system.

15.   Together, the x-axis and the y-axis form a _____.

16.   The _____ is the set of points corresponding to all
      ordered pairs that satisfy the equation.

17.   A _____ has no term with a variable to a power
      greater than one.

Objective 1   Plot ordered pairs.

Objective 1 Practice Exercises

For extra help, see page 628 of your text.

Plot each ordered pair on a coordinate system.

1. $(0,-2)$ **1.–3.**

2. $(-3, 4)$

3. $(2,-5)$

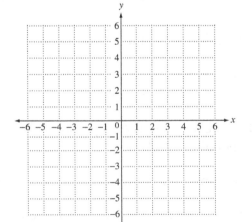

Objective 2   Find ordered pairs that satisfy a given equation.

Video Examples

Review these examples for Objective 2:	**Now Try:**
1. Complete each ordered pair for $5x + 4y = 20$.	1. Complete each ordered pair for $2x - 5y = 10$.

a. $(0, \underline{\quad})$

$5x + 4y = 20$

$5(0) + 4y = 20$

$4y = 20$

$y = 5$

The ordered pair is $(0, 5)$.

a. $(0, \underline{\quad})$

_____

b. $(\underline{\quad}, 0)$

$5x + 4y = 20$

$5x + 4(0) = 20$

$5x = 20$

$x = 4$

The ordered pair is $(4, 0)$.

b. $(\underline{\quad}, 0)$

_____

c. $(-4, \underline{\quad})$

$5x + 4y = 20$

$5(-4) + 4y = 20$

$-20 + 4y = 20$

$4y = 40$

$y = 10$

The ordered pair is $(-4, 10)$.

c. $(-5, \underline{\quad})$

_____

d. $(\underline{\quad}, -5)$

$5x + 4y = 20$

$5x + 4(-5) = 20$

$5x - 20 = 20$

$5x = 40$

$x = 8$

The ordered pair is $(8, -5)$.

d. $(\underline{\quad}, 2)$

_____

Objective 2 Practice Exercises

For extra help, see Example 1 on page 629 of your text.

For each of the given equations, complete the ordered pairs beneath it.

4. $5x + 4y = 10$ **4.**

      (a) $(2, \ \ )$ (a) _____

      (b) $(4, \ \ )$ (b) _____

      (c) $(\ ,3)$ (c) _____

      (d) $(0, \ \ )$ (d) _____

      (e) $(\ ,2)$ (e) _____

Complete each table of values. Write the results as ordered pairs.

5. $-7x + 2y = -14$ **5.** _____

x	y
	0
0	
3	
	7

6. $y - 4 = 0$ **6.** _____

x	y
-4	
0	
6	
-12	

Name: _____   Date: _____

Instructor: _____   Section: _____

Objective 3   Graph lines and find intercepts.

Video Examples

Review this example for Objective 3:

2. Find the *x*- and *y*-intercepts of $3x + y = 6$ and graph the equation.

To find the *y*-intercept, let $x = 0$.
To find the *x*-intercept, let $y = 0$.

$$
\begin{array}{c|c}
3(0) + y = 6 & 3x + 0 = 6 \\
0 + y = 6 & 3x = 6 \\
y = 6 & x = 2
\end{array}
$$

The intercepts are (0, 6) and (2, 0). To find a third point, as a check, we let $x = 1$.

$$3(1) + y = 6$$
$$3 + y = 6$$
$$y - 3$$

This gives the ordered pair (1, 3).

x	y
0	6
2	0
1	3

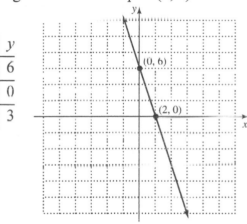

Now Try:

2. Find the *x*- and *y*-intercepts of $5x - 2y = -10$ and graph the equation.

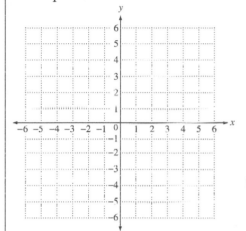

Name: Date:

Instructor: Section:

Objective 3 Practice Exercises

For extra help, see Examples 2–3 on pages 631–632 of your text.

Find the intercepts, then graph the equation.

7. $4x - y = 4$ 7. _____

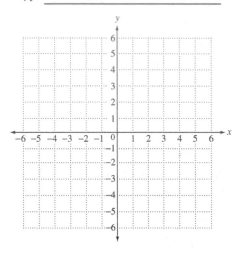

8. $2x - 3y = 6$ 8. _____

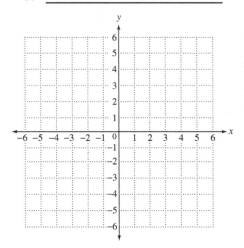

Objective 4   Recognize equations of horizontal and vertical lines.

Video Examples

Review these examples for Objective 4:

4. Graph each equation.

    a. $y = -2$

For any value of x, y is always –2. Three ordered pairs that satisfy the equation are $(-4, -2)$, $(0, -2)$ and $(2, -2)$. Drawing a line through these points gives the horizontal line. The y-intercept is $(0, -2)$. There is no x-intercept.

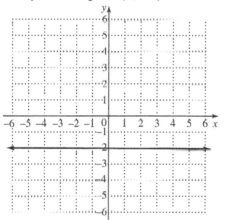

    b. $x + 4 = 0$

First we subtract 4 from each side of the equation to get the equivalent equation $x = -4$. All ordered-pair solutions of this equation have x-coordinate –4.

Three ordered pairs that satisfy the equation are $(-4, -1)$, $(-4, 0)$, and $(-4, 3)$. The graph is a vertical line. The x-intercept is $(-4, 0)$. There is no y-intercept.

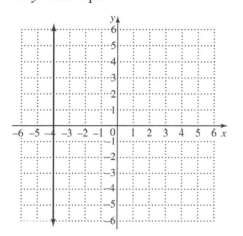

Now Try:

4. Graph each equation.

    a. $y = 4$

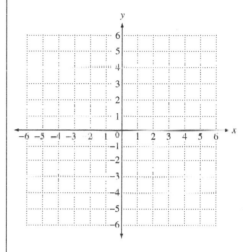

    b. $x = 0$

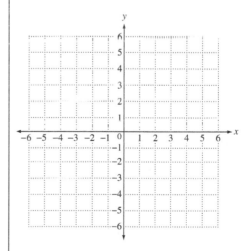

Name: Date:
Instructor: Section:

Objective 4 Practice Exercises

For extra help, see Example 4 on pages 632–633 of your text.

Find the intercepts, and graph the line.

9. $x - 1 = 0$

9.

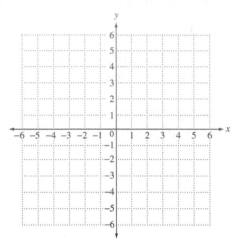

10. $y + 3 = 0$

10.

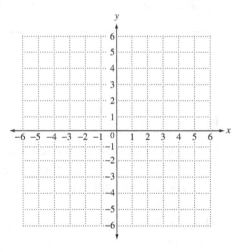

Objective 5   **Find the slope of a line.**

Video Examples

Review this example for Objective 5:

5.   Find the slope of the line passing through $(-5, 4)$ and $(2, -6)$

    Apply the slope formula.

$$(x_1,\ y_1) = (-5,\ 4) \text{ and } (x_2,\ y_2) = (2, -6)$$

$$\text{slope } m = \frac{y_2 - y_1}{x_2 - x_1} = \frac{-6 - 4}{2 - (-5)}$$

$$= \frac{-10}{7}, \text{ or } -\frac{10}{7}$$

Now Try:

5.   Find the slope of the line passing through $(-6, 7)$ and $(3, -9)$

    _____

Objective 5 Practice Exercises

For extra help, see Examples 5–7 on pages 634–635 of your text.

Find the slope of the line through the given points.

11.   (4, 3) and (3, 5)   11. _____

12.   (5, –2) and (2, 7)   12. _____

13.   (7, 2) and (–7, 3)   13. _____

Objective 6   Graph a line given its slope and a point on the line.

Video Examples

Review this example for Objective 6:

8.   Graph the line passing through the point $(1, -3)$, with slope $-\frac{5}{2}$.

First, locate the point $(1, -3)$. Then write the slope $-\frac{5}{2}$ as

$$\text{slope } m = \frac{\text{change in } y \text{ (rise)}}{\text{change in } x \text{ (run)}} = \frac{5}{-2}.$$

Locate another point on the line by counting up 5 units from $(1, -3)$, and then to the left 2 units. Finally, draw the line through this new point, $(-1, 2)$.

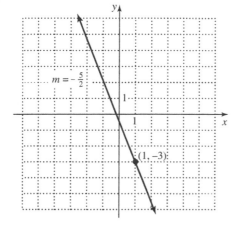

Now Try:

8.   Graph the line passing through the point $(2, 2)$, with slope $\frac{1}{3}$.

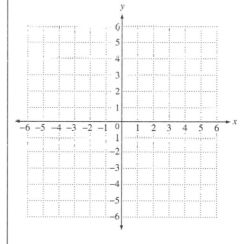

Name: Date:

Instructor: Section:

Objective 6 Practice Exercises

For extra help, see Example 8 on page 636 of your text.

Graph the line passing through the given point and having the given slope.

14. $(4,-2);\ m=-1$

14.

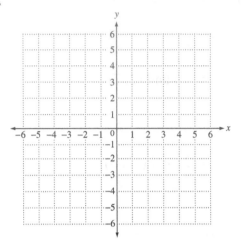

15. $(-3,-2);\ m=\frac{2}{3}$

15.

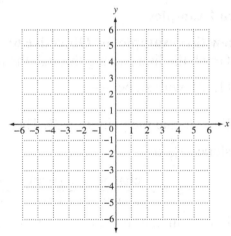

16. $(-3,-1);$ undefined slope

16.

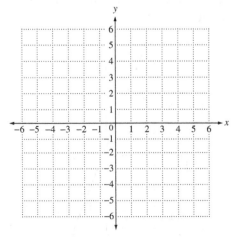

          Copyright © 2018 Pearson Education, Inc.

Chapter 8   EQUATIONS, INEQUALITIES, GRAPHS, AND SYSTEMS REVISITED

8.5   Review of Systems of Linear Equations in Two Variables

Learning Objectives	
1	Solve linear systems with two equations and two variables.
2	Solve special systems.

Key Terms

Use the vocabulary terms listed below to complete each statement in exercises 1–7.

system of equations	**linear system**
solution set of a system	**consistent system** **inconsistent system**
independent equations	**dependent equations**

1. Equations of a system that have different graphs are called

   _____.

2. A system of equations with at least one solution is a

   _____   _____.

3. Two or more equations that are to be solved at the same time form a

   _____.

4. The _____ of linear equations includes all the
   ordered pairs that make all the equations of the system true at the same time.

5. Equations of a system that have the same graph (because they are different forms of
   the same equation) are called _____.

6. A system with no solution is called a(n) _____ _____.

7. A(n) _____ consists of two or more linear
   equations with the same variables.

Objective 1 Solve linear systems with two equations and two variables.

Video Examples

Review these examples for Objective 1:

1. Solve the system of equations by graphing.

   $$x - 2y = 6 \quad (1)$$
   $$2x + y = 2 \quad (2)$$

   To graph these linear equations, we plot several points for each line.

   $x - 2y = 6$

x	y
0	−3
6	0
4	−1

   $2x + y = 2$

x	y
0	2
1	0
−1	4

   From the graph we see that the solution is (2,–2).

   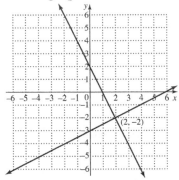

   To check, substitute 2 for x and –2 for y in each equation.

   $$x - 2y = 6 \qquad\qquad 2x + y = 2$$
   $$\overset{?}{2 - 2(-2) = 6} \quad \overset{?}{2(2) + (-2) = 2}$$
   $$\overset{?}{2 + 4 = 6} \qquad\quad \overset{?}{4 - 2 = 2}$$
   $$\text{True } 6 = 6 \qquad\quad \text{True } 2 = 2$$

   The solution set is {(2,–2)}.

2. Solve the system by substitution.

   $$x - 2y = -1 \quad (1)$$
   $$y = x - 2 \quad (2)$$

   Since equation (2) is solved for y, substitute $x - 2$ for y in equation (1).

   $$x - 2(x - 2) = -1$$
   $$x - 2x + 4 = -1$$
   $$-x + 4 = -1$$
   $$-x = -5$$
   $$x = 5$$

Now Try:

1. Solve the system of equations by graphing.

   $$2x - 5y = 8 \quad (1)$$
   $$5x - 4y = 3 \quad (2)$$

   _____

   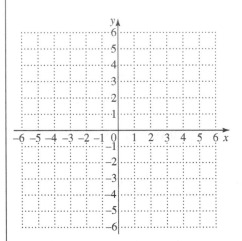

2. Solve the system by substitution.

   $$5x - 2y = 4 \qquad (1)$$
   $$x = y - 1 \quad (2)$$

   _____

Now find x by substituting 5 for x in equation (2).

$$y = 5 - 2 = 3$$

Thus, $x = 5$ and $y = 3$, giving the ordered pair (5, 3).

Check

$x - 2y = -1$	$y = x - 2$
$5 - 2(3) \stackrel{?}{=} -1$	$3 \stackrel{?}{=} 5 - 2$
$5 - 6 \stackrel{?}{=} -1$	$3 = 3$ True
$-1 = -1$ True	

The solution set is $\{(5, 3)\}$.

5. Solve the system by elimination.

$$4x + 3y = 41 \quad (1)$$
$$3x - 2y = 1 \quad (2)$$

Step 1 Both equations are in standard form.

Step 2 Multiply equation (1) by 2 and equation (2) by 3.

$$8x + 6y = 82 \quad \text{Multiply (1) by 2.}$$
$$9x - 6y = 3 \quad \text{Multiply (2) by 3.}$$

Step 3 Now add.

$$\begin{array}{r} 8x + 6y = 82 \\ 9x - 6y = 3 \\ \hline 17x \quad\quad = 85 \end{array}$$

Step 4 Solve for x. $x = 5$

Step 5 To find y, substitute 5 for y in either equation (1) or (2).

$$3x - 2y = 1 \quad (2)$$
$$3(5) - 2y = 1$$
$$15 - 2y = 1$$
$$-2y = -14$$
$$y = 7$$

Step 6 Check the solution (5, 7) in both equations (1) and (2).

$4x + 3y = 41$	$3x - 2y = 1$
$4(5) + 3(7) \stackrel{?}{=} 41$	$3(5) - 2(7) \stackrel{?}{=} 1$
$20 + 21 \stackrel{?}{=} 41$	$15 - 14 \stackrel{?}{=} 1$
$41 = 41$ True	$1 = 1$ True

The solution set is $\{(5, 7)\}$.

5. Solve the system by elimination.

$$4x + 5y = 11 \quad (1)$$
$$3x - 2y = -9 \quad (2)$$

_____

Name: Date:

Instructor: Section:

Objective 1 Practice Exercises

For extra help, see Examples 1–5 on pages 640–645 of your text.

Solve the system by graphing.

1. $y - 2 = 0$ 1.

      $3x - 4y = -17$

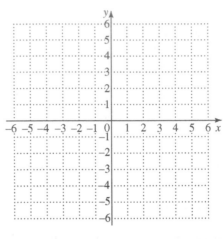

_____

Solve the system by substitution.

2. $2x + y = 6$ 2. _____

          $y = 5 - 3x$

Solve the system by elimination.

3. $3x + 4y = -13$ 3. _____

      $5x - 2y = -13$

Objective 2   Solve special systems.

Video Examples

Review this example for Objective 2:	**Now Try:**

Review this example for Objective 2:

6. Solve the system.

$$x + 5y = 9 \quad (1)$$

$$3x + 15y = 27 \quad (2)$$

We multiply equation (1) by 3 and then add the result to equation (2).

$$-3x - 15y = -27 \quad \text{Multiply (1) by } -3.$$

$$\underline{3x + 15y = 27 \quad (2)}$$

$$0 = 0 \quad \text{True}$$

The equations are dependent.
Since the statement is true, the solution set is the set of all points on the line with equation $x + 5y = 9$, written as $\{(x, y) \mid x + 5y = 9\}$.

Now Try:

6. Solve the system.

$$12x + 9y = 36 \quad (1)$$

$$4x + 3y = 12 \quad (2)$$

_____

Objective 2 Practice Exercises

For extra help, see Examples 6–7 on pages 645–646 of your text.

Solve each system of equations using any method.

4. $8x + 4y = -1$

  $4x + 2y = 3$

4. _____

5. $x + 2y = 4$

  $8y = -4x + 16$

5. _____

6. $-3x + 2y = 6$

  $-6x + 4y = 12$

6. _____

Chapter 8   EQUATIONS, INEQUALITIES, GRAPHS, AND SYSTEMS REVISITED

8.6   Systems of Linear Equations in Three Variables; Applications

Learning Objectives
1   Understand the geometry of systems of three equations in three variables.
2   Solve linear systems with three equations and three variables by elimination.
3   Solve linear systems with three equations and three variables in which some of the equations have missing terms.
4   Solve special systems.
5   Solve application problems with three variables using a system of three equations.

Key Terms

Use the vocabulary terms listed below to complete each statement in exercises 1−3.

ordered triple   inconsistent system   dependent system

1.   The solution of a linear system of equations in three variables is written as a(n)
      _____.

2.   A system of equations in which all solutions of the first equation are also solutions of
      the second equation is a(n) _____.

3.   A system of equations that has no common solution is called a(n)
      _____.

Objective 1   Understand the geometry of systems of three equations in three variables.

Objective 1 Practice Exercises

For extra help, see pages 649–540 of your text.

Answer each question.

1.   If a system of linear equations in three variables has   1. _____
      a single solution, how do the planes that are the
      graphs of the equations intersect?

2.   If a system of linear equations in three variables has   2. _____
      no solution, how do the planes that are the graphs of
      the equations intersect?

Objective 2   Solve linear systems with three equations and three variables by elimination.

Video Examples

Review this example for Objective 2:

1. Solve the system.

$$x - 2y + 5z = -7 \quad (1)$$
$$2x + 3y - 4z = 14 \quad (2)$$
$$3x - 5y + z = 7 \quad (3)$$

Step 1: Since x in equation (1) has coefficient 1, choose x as the focus variable and (1) as the working equation.

Step 2: Multiply working equation (1) by -2 and add the result to equation (2) to eliminate focus variable x.

$$\begin{array}{l} -2x + 4y - 10z = 14 \quad \text{Multiply (1) by } -2 \\ \underline{2x + 3y - 4z = 14 \quad (2)} \\ 7y - 14z = 28 \quad (4) \end{array}$$

Step 3: Multiply working equation (1) by -3 and add the result to equation (3) to eliminate focus variable x.

$$\begin{array}{l} -3x + 6y - 15z = 21 \quad \text{Multiply (1) by } -3 \\ \underline{3x - 5y + z = 7 \quad (3)} \\ y - 14z = 28 \quad (5) \end{array}$$

Step 4: Write equations (4) and (5) as a system, then solve the system.

$$7y - 14z = 28 \quad (4)$$
$$y - 14z = 28 \quad (5)$$

We will eliminate z.

$$\begin{array}{l} -7y + 14z = -28 \quad \text{Multiply (4) by } -1 \\ \underline{y - 14z = 28 \quad (5)} \\ -6y = 0 \quad \text{Add.} \\ y = 0 \quad \text{Divide by } -6. \end{array}$$

Substitute 0 for y in either equation to find z.

$$y - 14z = 28 \quad (5)$$
$$0 - 14z = 28 \quad \text{Let } y = 0.$$
$$-14z = 28$$
$$z = -2$$

Now Try:

1. Solve the system.

$$2x + y + 2z = -1$$
$$3x - y + 2z = -6$$
$$3x + y - z = -10$$

_____

Step 5: Now substitute $y = 0$ and $z = -2$ in working equation (1) to find the value of the remaining variable, focus variable x.

$$x - 2y + 5z = -7 \quad (1)$$

$$x - 2(0) + 5(-2) = -7 \quad \text{Let } y = 0 \text{ and } z = -2.$$

$$x - 10 = -7$$

$$x = 3$$

Step 6: It appears that the ordered triple $(3, 0, -2)$ is the solution of the system. We must check that the solution satisfies all three original equations of the system.

$$x - 2y + 5z = -7 \quad (1)$$

$$\overset{?}{3 - 2(0) + 5(-2)} = -7$$

$$\overset{?}{3 - 0 - 10} = -7$$

$$-7 = -7$$

Because $(3, 0, -2)$ also satisfies equations (2) and (3), the solution set is $\{(3, 0, -2)\}$.

Objective 2 Practice Exercises

For extra help, see Example 1 on pages 651–652 of your text.

Solve each system of equations.

3. $x + y + z = 2$ 3. _____
     $x - y + z = -2$
     $x - y - z = -4$

Name: Date:
Instructor: Section:

4. $2x + y - z = 9$

$x + 2y + z = 3$

$3x + 3y - z = 14$

4. ______________

5. $2x - 5y + 2z = 30$

$x + 4y + 5z = -7$

$\frac{1}{2}x - \frac{1}{4}y + z = 4$

5. ______________

Objective 3   **Solve linear systems with three equations and three variables in which some of the equations have missing terms.**

Video Examples

Review this example for Objective 3:

2.   Solve the system.

$$3x \quad\quad -4z = -23 \quad (1)$$
$$\quad\quad y + 5z = \quad 24 \quad (2)$$
$$x - 3y \quad\quad = \quad 2 \quad (3)$$

Since equation (1) is missing the variable y, one way to begin is to eliminate y again, using equations (2) and (3).

$$3y + 15z = 72 \quad \text{Multiply (2) by 3}$$
$$\underline{x - 3y \quad\quad = \quad 2 \quad (3)}$$
$$x \quad\quad + 15z = 74 \quad (4)$$

Now solve the system composed of equations (1) and (4).

$$3x \; -4z = \; -23 \quad (1)$$
$$\underline{-3x - 45z = -222 \quad \text{Multiply (4) by } -3}$$
$$-49z = -245$$
$$z = 5$$

Substitute 5 for z in (1) and solve for x.

$$3x - 4z = -23 \quad (1)$$
$$3x - 4(5) = -23 \quad \text{Let } z = 5.$$
$$3x - 20 = -23$$
$$3x = -3$$
$$x = -1$$

Substitute 5 for z in (2) and solve for y.

$$y + 5z = 24 \quad (2)$$
$$y + 5(5) = 24 \quad \text{Let } z = 5.$$
$$y + 25 = 24$$
$$y = -1$$

A check verifies that the solution set is $\{(-1, -1, 5)\}$.

Now Try:

2.   Solve the system.

$$x + 5y \quad\quad = -23$$
$$\quad\quad 4y - 3z = -29$$
$$2x \quad\quad + 5z = \quad 19$$

_____

Name: Date:
Instructor: Section:

Objective 3 Practice Exercises

For extra help, see Example 2 on pages 652–653 of your text.

Solve each system of equations.

6. $\begin{aligned} 7x \quad\;\; + z &= -1 \\ 3y - 2z &= 8 \\ 5x + y \quad\;\; &= 2 \end{aligned}$

6. _____

7. $\begin{aligned} 2x + 5y \quad\;\; &= 18 \\ 3y + 2z &= 4 \\ \tfrac{1}{4}x - y \quad\;\; &= -1 \end{aligned}$

7. _____

8. $\begin{aligned} 5x \quad\;\; - 2z &= 8 \\ 4y + 3z &= -9 \\ \tfrac{1}{2}x + \tfrac{2}{3}y \quad\;\; &= -1 \end{aligned}$

8. _____

Name: Date:

Instructor: Section:

Objective 4   Solve special systems.

Video Examples

Review these examples for Objective 4:

4. Solve the system.
$$3x - 2y + 5z = 4 \quad (1)$$
$$-6x + 4y - 10z = -8 \quad (2)$$
$$\frac{3}{2}x - y + \frac{5}{2}z = 2 \quad (3)$$

Multiplying each side of equation (1) by -2 gives equation (2). Multiplying each side of equation (1) by $\frac{1}{2}$ gives equation (3). Thus, the equations are dependent, and all three equations have the same graph as shown below. The solution set is written
$$\{(x, y, z) \mid 3x - 2y + 5z = 4\}.$$

3. Solve the system.
$$x - y + z = 7 \quad (1)$$
$$2x + 5y - 4z = 2 \quad (2)$$
$$-x + y - z = 4 \quad (3)$$

Since x in equation (1) has coefficient 1, choose x as the focus variable and (1) as the working equation. Using equations (1) and (3), we have
$$x - y + z = 7 \quad (1)$$
$$\underline{-x + y - z = 4 \quad (3)}$$
$$0 = 11 \quad \text{False}$$

The resulting false statement indicates that equations (1) and (3) have no common solution. Thus, the system is inconsistent and the solution set is ∅. The graph of this system would show that three planes are parallel to each other as shown below.

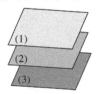

Now Try:

4. Solve the system.
$$x - 5y + 2z = 0$$
$$-x + 5y - 2z = 0$$
$$\frac{1}{2}x - \frac{5}{2}y + z = 0$$

_____

3. Solve the system.
$$-4x - 2y + z = -19$$
$$-6x + 2y - 6z = -8$$
$$-4x + 2y - 5z = -6$$

_____

Objective 4 Practice Exercises

For extra help, see Examples 3–5 on pages 653–654 of your text.

Solve each system of equations.

9.　　$8x - 7y + 2z = 1$
　　　$3x + 4y - z = 6$
　　　$-8x + 7y - 2z = 5$

9. _____

10.　　$3x - 2y + 4z = 5$
　　　$-3x + 2y - 4z = -5$
　　　$\frac{3}{2}x - y + 2z = \frac{5}{2}$

10. _____

11.　　$-x + 5y - 2z = 3$
　　　$2x - 10y + 4z = -6$
　　　$-3x + 15y - 6z = 9$

11. _____

Objective 5 **Solve application problems with three variables using a system of three equations.**

Video Examples

Review this example for Objective 5:

7. Lee has some $5, $10, and $20-bills. He has a total of 51 bills, worth $795. The number of $5-bills is 25 less than the number of $20-bills. Find the number of each type of bill he has.

Step 1: Read the problem again. There are three unknowns.

Step 2: Assign variables.
Let x = the number of $5 bills,
let y = the number of $10 bills,
let z = the number of $20 bills.

Step 3: Write a system of three equations.
There are a total of 51 bills, so
$$x + y + z = 51 \quad (1)$$
The bills amounted to $795, so
$$5x + 10y + 20z = 795 \quad (2)$$
The number of $5-bills is 25 less than the number of $20-bills, so $x = z - 25$ or
$$x - z = -25 \quad (3)$$
The system is
$$x + y + z = 51 \quad (1)$$
$$5x + 10y + 20z = 795 \quad (2)$$
$$x \qquad - z = -25 \quad (3)$$

Step 4: Solve the system.
Eliminate y.

$$-10x - 10y - 10z = -510 \quad \text{Multiply (1) by } -10$$
$$\underline{5x + 10y + 20z = 795 \quad (2)}$$
$$-5x + 10z = 285 \quad (4)$$

Solve the system consisting of equations (3) and (4).

$$5x - 5z = -125 \quad \text{Multiply (3) by 5}$$
$$\underline{-5x + 10z = 285 \quad (4)}$$
$$5z = 160$$
$$z = 32$$

Substitute 32 for z in equation (3) and solve for x.

Now Try:

7. The manager of the Sweet Candy Shop wishes to mix candy worth $4 per pound, $6 per pound, and $10 per pound to get 100 pounds of a mixture worth $7.60 per pound. The amount of $10 candy must equal the total amounts of the $4 and the $6 candy. How many pounds of each must be used?

$4 candy _____

$6 candy _____

$10 candy _____

$$x - z = -25 \quad (3)$$
$$x - 32 = -25 \quad \text{Let } z = 32.$$
$$x = 7$$

Substitute 7 for x and 32 for z in equation (1) and solve for y.

$$x + y + z = 51 \quad (1)$$
$$7 + y + 32 = 51 \quad \text{Let } x = 7, \, z = 32.$$
$$y + 39 = 51$$
$$y = 12$$

Step 5: State the answer. Lee has 7 $5-bills, 12 $10-bills, and 32 $20-bills.

Step 6: Check that the total value of the bills is $795 and that the number of $5-bills is 25 less than the number of $20-bills.

Objective 5 Practice Exercises

For extra help, see Examples 6–7 on pages 655–657 of your text.

Solve each problem involving three unknowns.

12. Julie has $80,000 to invest. She invests part at 5%, one fourth this amount at 6%, and the balance 7%. Her total annual income from interest is $4700. Find the amount invested at each rate.

12. 5% _____

    6% _____

    7% _____

13. A merchant wishes to mix gourmet coffee selling for   13. $8/lb _____
     $8 per pound, $10 per pound, and $15 per pound to
     get 50 pounds of a mixture that can be sold for $10/lb _____
     $11.70 per pound. The amount of the $8 coffee must
     be 3 pounds more than the amount of the $10 coffee. $15/lb _____
     Find the number of pounds of each that must be
     used.

14. A boy scout troop is selling popcorn. There are three   14. I _____
     different kinds of popcorn in three different
     arrangements. Arrangement I contains 1 bag of   II _____
     cheddar cheese popcorn, 2 bags of caramel popcorn,
     and 3 bags of microwave popcorn. Arrangement II   III _____
     contains 3 bags of cheddar cheese popcorn, 1 bag of
     caramel popcorn, and 2 bags of microwave popcorn.
     Arrangement III contains 2 bags of cheddar cheese
     popcorn, 3 bags of caramel popcorn, and 1 bag of
     microwave popcorn. Jim needs 28 bags of cheddar
     cheese popcorn, 22 bags of caramel popcorn, and 22
     bags of microwave popcorn to give as stocking
     stuffers for Christmas. How many of each
     arrangement should he buy?

Copyright © 2018 Pearson Education, Inc.

Chapter 9   RELATIONS AND FUNCTIONS

9.1   Introduction to Relations and Functions

Learning Objectives
1   Define and identify relations and functions.
2   Find domain and range.
3   Identify functions defined by graphs and equations.

Key Terms

Use the vocabulary terms listed below to complete each statement in exercises 1–6.

dependent variable   **independent variable**   **relation**

function   **domain**   **range**

1. The _____ of a relation is the set of second components (y-values) of the ordered pairs of the relation.

2. A _____ is a set of ordered pairs of real numbers.

3. If the quantity y depends on x, then y is called the _____ in a relation between x and y.

4. The _____ of a relation is the set of first components (x-values) of the ordered pairs of the relation.

5. A _____ is a set of ordered pairs in which each value of the first component, x, corresponds to exactly one value of the second component, y.

6. If the quantity y depends on x, then x is called the _____ in a relation between x and y.

Objective 1   Define and identify relations and functions.

Video Examples

Review these examples for Objective 1:

2. Determine whether each relation defines a function.

  a. $F = \{(3, 10), (-4, 6), (7,-5)\}$

For $x = 3$, there is only one value of y, 10.
For $x = -4$, there is only one value of y, 6.
For $x = 7$, there is only one value of y, –5.
Relation F is a function.

  c. $H = \{(-9, 2), (-6, 2), (-6, 9)\}$

In relation H, the last two ordered pairs have the same x-value pair with different y-values. H is a relation, but not a function.

Now Try:

2. Determine whether each relation defines a function.

  a. $\{(3,-3), (5,-5), (7,-7)\}$

_____

  c. $\{(0, 2), (2, 6), (6, 3), (0, 7)\}$

_____

Objective 1 Practice Exercises

For extra help, see Examples 1–2 on pages 678–679 of your text.

Decide whether each relation is a function.

1. $\{(1,\ 3),\ (1,\ 4),\ (2,-1),\ (3,\ 7)\}$

    1. _____

2. $\{(2,-2,),\ (3,-3),\ (4,-4)\}$

    2. _____

3. $\{(3,\ 4),\ (5,\ 2),\ (4,\ 3),\ (5,\ 3),\ (-2,\ 2)\}$

    3. _____

Objective 2   Find domain and range.

Video Examples

Review these examples for Objective 2:

3a. Give the domain and range of the relation. Tell whether the relation defines a function.

$\{(15, 2),\ (20, 3),\ (6, 10),\ (-1, 2)\}$

The domain is the set of *x*-values $\{-1, 6, 15, 20\}$.
The range is the set of *y*-values $\{2, 3, 10\}$.
The relation is a function, because each *x*-value corresponds to exactly one *y*-value.

Now Try:

3a. Give the domain and range of the relation. Tell whether the relation defines a function.
$\{(13,-1),\ (13,-2),\ (13, 4)\}$

domain: _____

range: _____

_____

4. Give the domain and range of each relation.

c.

The arrowheads on the graphed line indicate that the line extends indefinitely left and right, as well as up and down. Therefore the domain and range include all real numbers.
The domain is $(-\infty, \infty)$. The range is $(-\infty, \infty)$.

4. Give the domain and range of each relation.

c.

_____

_____

d.

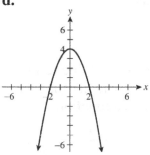

d.

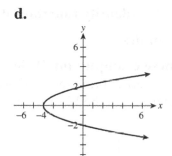

The graph extends indefinitely left and right, as well as downward. The domain is $(-\infty, \infty)$. Because there is a greatest y-value, 4, the range includes all numbers less than or equal to 4, written $(-\infty, 4]$.

_____

_____

Objective 2 Practice Exercises

For extra help, see Examples 3–4 on pages 681–682 of your text.

Decide whether the relation is a function, and give the domain and range of the relation.

4. $\{(5, 2), (3, -1), (1, -3), (-1, 5)\}$

4. _____

domain:_____

range: _____

5.

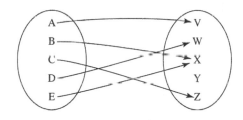

5. _____

domain:_____

range: _____

6.

x	y
1	3
2	-1
-1	4
1	4

6. _____

domain:_____

range: _____

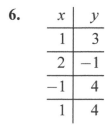

Objective 3   Identify functions defined by graphs and equations.

Video Examples

Review these examples for Objective 3:

5b. Use the vertical line test to determine whether the relation graphed is a function.

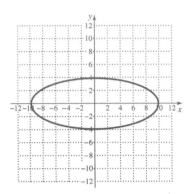

The graph is not a function.

Now Try:

5b. Use the vertical line test to determine whether the relation graphed is a function.

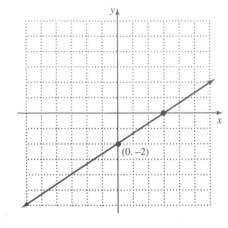

_____

6a. Decide whether the relation defines y as a function of x. Give the domain.

$$y = 2x - 4$$

Each x value corresponds to just one y-value and the relation defines a function. Since x can be any real number, the domain is $(-\infty, \infty)$.

6a. Decide whether the relation defines y as a function of x. Give the domain.

$$y = 2x - 4$$

_____

_____

Objective 3 Practice Exercises

For extra help, see Examples 5–6 on pages 683–685 of your text.

Use the vertical line test to determine whether the relation graphed is a function.

7.

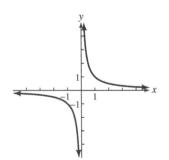

7. _____

Name: Date:
Instructor: Section:

Decide whether each equation defines y as a function of x. Give the domain.

8. $y^2 = x + 1$

8. _____

9. $y = \dfrac{3}{x+6}$

9. _____

Chapter 9 RELATIONS AND FUNCTIONS

9.2 Function Notation and Linear Functions

Learning Objectives
1   Use function notation.
2   Graph linear and constant functions.

Key Terms

Use the vocabulary terms listed below to complete each statement in exercises 1–3.

function notation   **linear function**   **constant function**

1.   A function defined by an equation of the form $f(x) = ax + b$, for real numbers a and b, is a _____.

2.   _____ $f(x)$ represents the value of the function at x, that is, the y-value that corresponds to x.

3.   A _____ is a linear function of the form $f(x) = b$, for a real number b.

Objective 1   Use function notation.

Video Examples

Review these examples for Objective 1:

1a. Let $f(x) = 7x - 3$. Evaluate the function f for the following.

$x = 4$

Start with the given function. Replace x with 4.

$$f(x) = 7x - 3$$
$$f(4) = 7(4) - 3$$
$$f(4) = 28 - 3$$
$$f(4) = 25$$

Thus, $f(4) = 25$.

3. Let $g(x) = 5x + 6$. Find and simplify $g(n+8)$.

Replace x with $n + 8$.

$$g(x) = 5x + 6$$
$$g(n+8) = 5(n+8) + 6$$
$$g(n+8) = 5n + 40 + 6$$
$$g(n+8) = 5n + 46$$

Now Try:

1a. Let $f(x) = 8x - 7$. Evaluate the function f for the following.

$x = 3$

_____

3. Let $g(x) = 4x - 7$. Find and simplify $g(a-1)$.

_____

Name: _____ Date: _____

Instructor: _____ Section: _____

4c. For the function, find $f(5)$.

$$f = \{(7, -27),\ (5, -25),\ (3, -23),\ (1, -21)\}$$

From the ordered pair $(5, -25)$, we have $f(5) = -25$.

4c. For the function, find $f(-6)$.

$$f = \{(-2, 11),\ (-4, 17),\ (-6, 21),\ (-8, 24)\}$$

_____

6b. Write the equation using function notation $f(x)$. Then find $f(-5)$.

$$x - 5y = 8$$

Step 1 $x - 5y - 8$

$$-5y = -x + 8$$

$$y = \frac{1}{5}x - \frac{8}{5}$$

Step 2 $f(x) - \frac{1}{5}x - \frac{8}{5}$

$$f(-5) - \frac{1}{5}(-5) - \frac{8}{5}$$

$$f(-5) = -\frac{13}{5}$$

6b. Write the equation using function notation $f(x)$. Then find $f(-3)$.

$$2x + 3y = 7$$

_____

Objective 1 Practice Exercises

For extra help, see Examples 1–6 on pages 689–691 of your text.

For each function f, find (a) $f(-2)$, (b) $f(0)$, and (c) $f(-x)$.

1. $f(x) = 3x - 7$

1. a. _____

    b. _____

    c. _____

2. $f(x) = 2x^2 + x - 5$

2. a. _____

    b. _____

    c. _____

3. $f(x) = 9$

3. a. _____

b. _____

c. _____

Objective 2   Graph linear and constant functions.

Video Examples

Review these examples for Objective 2:

7. Graph each function. Give the domain and range.

a. $f(x) = -2x - 3$

The graph of the function has slope -2 and y-intercept -3. To graph this function, plot the y-intercept $(0, -3)$ and use the definition of slope as $\dfrac{\text{rise}}{\text{run}}$ to find a second point on the line. Since the slope is -2, move down two units and right one unit to the point $(1, -5)$. Draw the straight line through the points to obtain the graph. The domain and range are both $(-\infty, \infty)$.

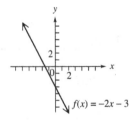

Now Try:

7. Graph each function. Give the domain and range.

a. $f(x) = \dfrac{1}{2}x + \dfrac{1}{2}$

_____

_____

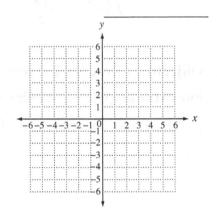

b. $f(x) = -2$

The graph of this constant function is a horizontal line containing all points with y-coordinate -2.

The domain is $(-\infty, \infty)$ and the range is $\{-2\}$.

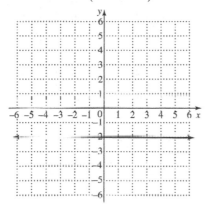

b. $f(x) = -3$

_____

_____

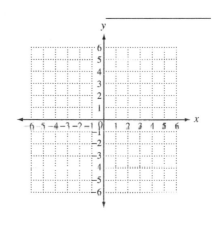

Objective 2 Practice Exercises

For extra help, see Example 7 on page 692 of your text.

Graph each function. Give the domain and range.

4. $2x - y = -2$

4. domain _____

        range _____

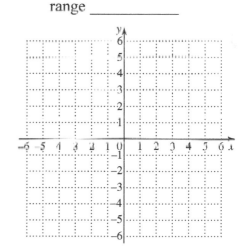

5. $y + \dfrac{1}{2}x = -2$

5. domain _____

range _____

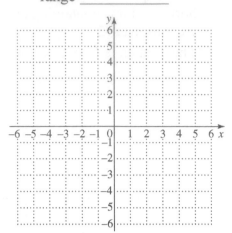

6. $y = 2$

6. domain _____

range _____

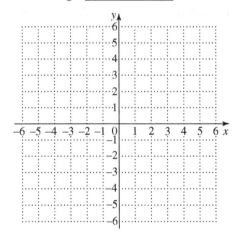

Chapter 9 RELATIONS AND FUNCTIONS

9.3   Polynomial Functions, Operations, and Graphs

Learning Objectives	
1	Recognize and evaluate polynomial functions.
2	Perform operations on polynomial functions.
3	Graph polynomial functions.

Key Terms

Use the vocabulary terms listed below to complete each statement in exercises 1–4.

polynomial function of degree n identity function

squaring function   cubing function

1.   The polynomial function $f(x) = x^2$ is the _____.

2.   A function defined by $f(x) = a_n x^n + a_{n-1} x^{n-1} + \cdots + a_1 x + a_0$, where $a_n \neq 0$ and n is a whole number is a _____.

3.   The polynomial function $f(x) = x^3$ is the _____.

4.   The polynomial function $f(x) = x$ is the _____.

Objective 1   Recognize and evaluate polynomial functions.

Video Examples

Review these examples for Objective 1:

1.   Let $f(x) = 6x^3 - 6x + 1$. Find each value.

   a. $f(2)$

   Substitute 2 for x.
   $$f(2) = 6(2)^3 - 6(2) + 1$$
   $$= 6(8) - 6(2) + 1$$
   $$= 48 - 12 + 1$$
   $$= 37$$

   b. $f(-3)$

   Substitute –3 for x.
   $$f(-3) = 6(-3)^3 - 6(-3) + 1$$
   $$= 6(-27) - 6(-3) + 1$$
   $$= -162 + 18 + 1$$
   $$= -143$$

Now Try:

1.   Let $f(x) = -x^4 + 3x^2 - x + 7$. Find each value.

   a. $f(2)$

   _____

   b. $f(-3)$

   _____

Objective 1 Practice Exercises

For extra help, see Example 1 on page 697 of your text.

For each polynomial function, find (a) $f(-2)$ and (b) $f(3)$.

1. $f(x) = -x^2 - x - 5$

    1. (a) _____

       (b) _____

2. $f(x) = 2x^2 + 3x - 5$

    2. (a) _____

       (b) _____

3. $f(x) = 3x^4 - 5x^2$

    3. (a) _____

       (b) _____

Objective 2   Perform operations on polynomial functions.

Video Examples

Review these examples for Objective 2:

2. For $f(x) = 2x^2 + 4x - 5$ and
    $g(x) = -x^2 + 3x - 8$, find each of the following.

    a. $(f+g)(x)$

    $(f+g)(x) = f(x) + g(x)$
    $$= (2x^2 + 4x - 5) + (-x^2 + 3x - 8)$$
    $$= x^2 + 7x - 13$$

    b. $(f-g)(x)$

    $(f-g)(x) = f(x) - g(x)$
    $$= (2x^2 + 4x - 5) - (-x^2 + 3x - 8)$$
    $$= (2x^2 + 4x - 5) + (x^2 - 3x + 8)$$
    $$= 3x^2 + x + 3$$

Now Try:

2. For $f(x) = 6x^2 - 7x + 12$ and
    $g(x) = -3x^2 + x + 9$, find each
    of the following.

    a. $(f+g)(x)$

    _____

    b. $(f-g)(x)$

    _____

4. For $f(x)=5x+2$ and $g(x)=3x^2+4x$, find $(fg)(x)$ and $(fg)(-2)$.

$(fg)(x)=f(x)\cdot g(x)$

$\quad =(5x+2)(3x^2+4x)$

$\quad =15x^3+20x^2+6x^2+8x$

$\quad =15x^3+26x^2+8x$

$(fg)(-2)=15(-2)^3+26(-2)^2+8(-2)$

$\quad --120+104-16$

$\quad =-32$

4. For $f(x)=6x+5$ and $g(x)=7x^2+2x$, find $(fg)(x)$ and $(fg)(-3)$.

_____

Objective 2 Practice Exercises

For extra help, see Examples 2–5 on pages 699–700 of your text.

For the pair of functions, find (a) $(f+g)(x)$ and (b) $(f-g)(x)$.

4. $f(x)=5x^2+7x-1,\ g(x)=-4x^2+2x+1$

4. a._____

b._____

For the pair of functions, find (a) $(fg)(x)$ and (b) $(fg)(-1)$.

5. $f(x)-3x^2+2,\ g(x)=-5x$

5. a._____

b._____

For the pair of functions, find the quotient $\left(\frac{f}{g}\right)(x)$ and give any x-values that are not in the domain of the quotient function.

6. $f(x)=4x^2-11x-45,\ g(x)=x-5$

6. _____

Name:

Date:

Instructor:

Section:

Objective 3   Graph basic polynomial functions.

Video Examples

Review these examples for Objective 2:

6.   Graph each function. Give the domain and range.

   a. $f(x) = -3x + 2$

   Plot the points and join them with a straight line.
   The domain and range are both $(-\infty, \infty)$.

x	$f(x) = -3x + 2$
-1	5
0	2
1	-1
2	-4
3	-7

   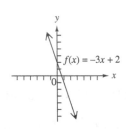

   b. $g(x) = -2x^2$

   The graph of $g(x)$ has the same shape as that of
   $f(x) = x^2$ but is narrower and opens downward.
   The domain is $(-\infty, \infty)$. The range is
   $(-\infty, 0]$.

x	$g(x) = -2x^2$
-2	-8
-1	-2
0	0
1	-2
2	-8

   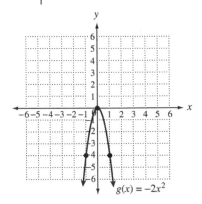

Now Try:

6.   Graph each function. Give the domain and range.

   a. $f(x) = -2x - 3$

   _____

   _____

   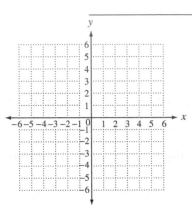

   b. $f(x) = x^2 - 1$

   _____

   _____

   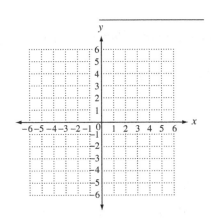

c. $f(x) = x^3 + 3$.

For this function, cube the input and add 3 to the result. The graph is the cubing function shifted 3 units up.

x	$f(x) = x^3 + 3$
-2	-5
-1	2
0	3
1	4
2	11

The domain and range is $(-\infty, \infty)$.

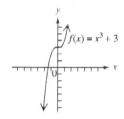

c. $f(x) = -x^3 + 1$.

_____

_____

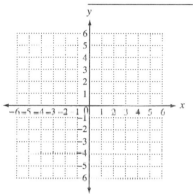

Objective 3 Practice Exercises

For extra help, see Example 6 on page 702 of your text.

Graph each function. Give the domain and range.

7. $f(x) = \dfrac{1}{2}x + \dfrac{1}{2}$

7. domain _____

   range _____

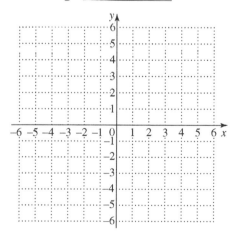

8. $f(x) = -2x^2$

8. domain _____

   range _____

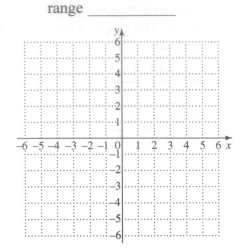

9. $f(x) = x^3 + 2$

9. domain _____

   range _____

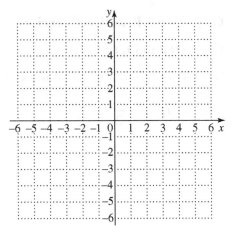

Chapter 9   RELATIONS AND FUNCTIONS

9.4   Variation

Learning Objectives

1	Write an equation expressing direct variation.
2	Find the constant of variation, and solve direct variation problems.
3	Solve inverse variation problems.
4	Solve joint variation problems.
5	Solve combined variation problems.

Key Terms

Use the vocabulary terms listed below to complete each statement in exercises 1–3.

varies directly   **varies inversely**   **constant of variation**

1.   In the equations for direct and inverse variation, k is the _____.

2.   If there exists a real number k such that $y = \dfrac{k}{x}$, then y _____ as x.

3.   If there exists a real number k such that $y = kx$, then y _____ as x.

Objective 1   Write an equation expressing direct variation.

For extra help, see page 706 of your text.

Objective 2   Find the constant of variation, and solve direct variation problems.

Video Examples

Review these examples for Objective 2:

2.   A person's weight on the moon varies directly with the person's weight on Earth. A 120-pound person would weigh about 20 pounds on the moon. How much would a 150-pound person weigh on the moon?

If m represents the person's weight on the moon and w represents the person's weight on Earth.

$$m = kw$$

$$20 = k \cdot 120 \quad \text{Let } m = 20 \text{ and } w = 120.$$

$$\frac{20}{120} = \frac{1}{6} = k \quad \text{Solve for } k; \text{ lowest terms}$$

Now, substitute $\dfrac{1}{6}$ for k and 150 for w in the

Now Try:

2.   The pressure exerted by a certain liquid at a given point varies directly as the depth of the point beneath the surface of the liquid. The pressure at 10 feet is 50 pounds per square inch (psi). What is the pressure at 25 feet?

_____

variation equation.

$m = kw$

$m = \dfrac{1}{6} \cdot 150 = 25$

A 150-pound person will weigh 25 pounds on the moon.

3. The surface area of a sphere varies directly as the square of its radius. If the surface area of a sphere with a radius of 12 inches is 576π square inches, find the surface area of a sphere with a radius of 3 inches.

Step 1 A represents the surface area and r represents the radius.

$A = kr^2$

Step 2 Find the value of k when A is 576π and r is 12. $A = kr^2$

$576\pi = k \cdot 12^2$

$576\pi = 144k$

$\dfrac{576\pi}{144} = 4\pi = k$

Step 3 Rewrite the variation equation.

$A = 4\pi r^2$

Step 4 Let $r = 3$ to find the surface area.

$A = 4\pi \cdot 3^2 = 4\pi \cdot 9 = 36\pi$

The surface area of a sphere with a radius of 3 inches is 36π square inches.

3. The area of a circle varies directly as the square of the radius. A circle with a radius of 5 centimeters has an area of 78.5 square centimeters. Find the area if the radius changes to 7 centimeters.

_____

Objective 2 Practice Exercises

For extra help, see Examples 1–3 on pages 706–708 of your text.

Find the constant of variation, and write a direct variation equation.

1. $y = 13.75$ when $x = 55$

1. _____

Solve each problem.

2. The circumference of a circle varies directly as the radius. A circle with a radius of 7 centimeters has a circumference of 43.96 centimeters. Find the circumference of the circle if the radius changes to 11 centimeters.

2. _____

3. The force required to compress a spring varies directly as the change in length of the spring. If a force of 20 newtons is required to compress a spring 2 centimeters in length, how much force is required to compress a spring of length 10 centimeters?

3. _____

Objective 3   Solve inverse variation problems.

Video Examples

Review this example for Objective 3:

4. For a specified distance, time varies inversely with speed. If Ramona walks a certain distance on a treadmill in 40 minutes at 4.2 miles per hour, how long will it take her to walk the same distance at 3.5 miles per hour?

   Let t = time and s = speed.
   Since t varies inversely as s, there is a constant k
   such that $t = \dfrac{k}{s}$. Recall that $40 \min = \dfrac{40}{60}$ hr .

   $$t = \frac{k}{s}$$
   $$\frac{40}{60} = \frac{k}{4.2}$$
   $$2.8 = k$$

Now Try:

4. The length of a violin string varies inversely with the frequency of its vibrations. A 10-inch violin string vibrates at a frequency of 512 cycles per second. Find the frequency of an 8-inch string.

   _____

Now use $t = \dfrac{k}{s}$ to find the value of t

when $s = 3.5$.

$$t = \dfrac{2.8}{3.5} = \dfrac{4}{5}$$

It takes $\dfrac{4}{5}$ hr, or 48 min to walk the same

distance.

Objective 3 Practice Exercises

For extra help, see Examples 4–5 on pages 709–710 of your text.

Solve each problem.

4. The illumination produced by a light source varies
    inversely as the square of the distance from the
    source. If the illumination produced 4 feet from a
    light source is 75 footcandles, find the illumination
    produced 9 feet from the same source.

4.

5. The weight of an object varies inversely as the
    square of its distance from the center of Earth. If an
    object 8000 miles from the center of Earth weighs 90
    pounds, find its weight when it is 12,000 miles from
    the center of Earth.

5.

6. The speed of a pulley varies inversely as its
    diameter. One kind of pulley, with a diameter of 3
    inches, turns at 150 revolutions per minute. Find the
    speed of a similar pulley with diameter of 5 inches.

6. _____

Objective 4   Solve joint variation problems.

Video Examples

Review this example for Objective 4:

6. For a fixed interest rate, interest varies jointly as the principal and the time in years. If $5000 invested for 4 years earns $900, how much interest will $6000 invested for 3 years earn at the same interest rate?

Let I = the interest, p = the principal, and t = the time in years. Then, $I = kpt$.

$$I = kpt$$
$$900 = k \cdot 5000 \cdot 4 \quad \text{Substitute given values.}$$
$$\frac{900}{20,000} = \frac{9}{200} = k$$

Now use $k = \frac{9}{200}$.

$$I = \frac{9}{200} \cdot 6000 \cdot 3 = 810$$

$6000 invested for three years will earn $810 in interest.

Now Try:

6. The strength of a rectangular beam varies jointly as its width and the square of its depth. If the strength of a beam 2 inches wide by 10 inches deep is 1000 pounds per square inch, what is the strength of a beam 4 inches wide and 8 inches deep?

_____

Objective 4 Practice Exercises

For extra help, see Example 6 on page 711 of your text.

Solve each problem.

7. Suppose d varies jointly as f^2 and g^2, and $d = 384$ when $f = 3$ and $g = 8$. Find d when $f = 6$ and $g = 2$.

7. _____

8. The work w (in joules) done when lifting an object is
    jointly proportional to the product of the mass m (in
    kg) of the object and the height h (in meter) the
    object is lifted. If the work done when a 120 kg
    object is lifted 1.8 meters above the ground is 2116.8
    joules, how much work is done when lifting a 100kg
    object 1.5 meters above the ground?

8. _____

9. The absolute temperature of an ideal gas varies
    jointly as its pressure and its volume. If the absolute
    temperature is 250° when the pressure is 25 pounds
    per square centimeter and the volume is 50 cubic
    centimeters, find the absolute temperature when the
    pressure is 50 pounds per square centimeter and the
    volume is 75 cubic centimeters.

9. _____

Objective 5   Solve combined variation problems.

Video Examples

Review this example for Objective 5:

7. The number of hours h that it takes w workers to
    assemble x machines varies directly as the
    number of machines and inversely as the number
    of workers. If four workers can assemble 12
    machines in four hours, how many workers are
    needed to assemble 36 machines in eight hours?

    The variation equation is $h = \dfrac{kx}{w}$.

    To find k, let $h = 4$, $x = 12$, and $w = 4$.

Now Try:

7. The volume of a gas varies
    directly as its temperature and
    inversely as its pressure. The
    volume of a gas at 85° C at a
    pressure of 12 kg/cm^2 is 300
    cm^3. What is the volume when
    the pressure is 20 kg/cm^2 and
    the temperature is 30° C?

    _____

$$4 = \frac{k \cdot 12}{4}$$

$$k = \frac{4 \cdot 4}{12}$$

$$k = \frac{4}{3}$$

Now find w when $x = 36$ and $h = 8$.

$$8 = \frac{\frac{4}{3} \cdot 36}{w}$$

$$8w = 48$$

$$w = 6$$

Eight workers are needed to assemble 36 machines in eight hours.

Objective 5 Practice Exercises

For extra help, see Example 7 on page 711 of your text.

Solve each problem.

10. The volume of a gas varies inversely as the pressure and directly as the temperature. If a certain gas occupies a volume of 1.3 liters at 300 K and a pressure of 18 kilograms per square centimeter, find the volume at 340 K and a pressure of 24 kilograms per square centimeter.

10. _____

11. The time required to lay a sidewalk varies directly as its length and inversely as the number of people who are working on the job. If three people can lay a sidewalk 100 feet long in 15 hours, how long would it take two people to lay a sidewalk 40 feet long?

11. _____

12. When an object is moving in a circular path, the centripetal force varies directly as the square of the velocity and inversely as the radius of the circle. A stone that is whirled at the end of a string 50 centimeters long at 900 centimeters per second has a centripetal force of 3,240,000 dynes. Find the centripetal force if the stone is whirled at the end of a string 75 centimeters long at 1500 centimeters per second.

12. _____

Chapter 10 ROOTS, RADICALS, AND ROOT FUNCTIONS

10.1 Radical Expressions and Graphs

Learning Objectives
1   Find square roots.
2   Decide whether a given root is rational, irrational, or not a real number.
3   Find cube, fourth, and other roots.
4   Graph functions defined by radical expressions.
5   Find nth roots of nth powers.
6   Use a calculator to find roots.

Key Terms

Use the vocabulary terms listed below to complete each statement in exercises 1−10.

        square root   **principal square root**   **radicand**   **index (order)**

        radical   **radical expression**   **perfect square**

        irrational number   **cube root**   **perfect cube**

1.   The number or expression inside a radical sign is called the _____.

2.   A number with a rational square root is called a _____.

3.   In a radical of the form $\sqrt[n]{a}$, the number n is the _____.

4.   The number b is a _____ of a if $b^2 = a$.

5.   The expression $\sqrt[n]{a}$ is called a _____.

6.   The positive square root of a number is its _____.

7.   A real number that is not rational is called an _____. .

8.   A number with a rational cube root is called a _____.

9.   A _____ is a radical sign and the number or expression in it.

10.   The number b is a _____ of a if $b^3 = a$.

Objective 1   Find square roots.

Video Examples

Review these examples for Objective 1:

1. Find the square roots of 64.

    What number multiplied by itself equals 64?
    $8^2 = 64$ and $(-8)^2 = 64$.
    Thus, 64 has two square roots: 8 and –8.

2. Find each square root.

    a. $\sqrt{121}$

    $11^2 = 121$, so $\sqrt{121} = 11$.

    d. $-\sqrt{\dfrac{16}{25}}$

    $-\sqrt{\dfrac{16}{25}} = -\dfrac{4}{5}$

3. Find the square of each radical expression.

    b. $-\sqrt{31}$

    $\left(-\sqrt{31}\right)^2 = 31$

    c. $\sqrt{w^2 + 3}$

    $\left(\sqrt{w^2 + 3}\right)^2 = w^2 + 3$

Now Try:

1. Find the square roots of 81.

    _____

2. Find each square root.

    a. $\sqrt{169}$

    _____

    d. $-\sqrt{\dfrac{9}{49}}$

    _____

3. Find the square of each radical expression.

    b. $-\sqrt{37}$

    _____

    c. $\sqrt{n^2 + 5}$

    _____

Objective 1 Practice Exercises

For extra help, see Examples 1–3 on pages 726–727 of your text.

Find all square roots of each number.

1. 625

2. $\dfrac{121}{196}$

Find the square root.

3. $\sqrt{\dfrac{900}{49}}$

1. _____

2. _____

3. _____

Objective 2   Decide whether a given root is rational, irrational, or not a real number.

Video Examples

Review these examples for Objective 2:

4.   Tell whether each square root is rational, irrational, or not a real number.

    a. $\sqrt{5}$

    Because 5 is not a perfect square, $\sqrt{5}$ is irrational.

    b. $\sqrt{81}$

    81 is a perfect square, 9^2, so $\sqrt{81} = 9$ is a rational number.

    c. $\sqrt{-16}$

    There is no real number whose square is –16. Therefore, $\sqrt{-16}$ is not a real number.

Now Try:

4.   Tell whether each square root is rational, irrational, or not a real number.

    a. $\sqrt{11}$

    _____

    b. $\sqrt{100}$

    _____

    c. $\sqrt{-9}$

    _____

Objective 2 Practice Exercises

For extra help, see Example 4 on page 728 of your text.

Tell whether each square root is rational, irrational, or not a real number.

4. $\sqrt{72}$ 4.   _____

5. $\sqrt{-36}$ 5.   _____

6. $\sqrt{6400}$ 6.   _____

Objective 3   Find cube, fourth, and other roots.

Video Examples

Review these examples for Objective 3:

5.   Find each cube root.

    c. $\sqrt[3]{729}$

    $\sqrt[3]{729} = 9$, because $9^3 = 729$.

    b. $\sqrt[3]{-64}$

    $\sqrt[3]{-64} = -4$, because $(-4)^3 = -64$.

Now Try:

5.   Find each cube root.

    c. $\sqrt[3]{343}$

    _____

    b. $\sqrt[3]{-125}$

    _____

6. Find each root.

c. $\sqrt[4]{-81}$

For a real number fourth root, the radicand must be nonnegative. There is no real number that equals $\sqrt[4]{-81}$.

a. $\sqrt[4]{81}$

$\sqrt[4]{81} = 3$, because 3 is positive and $3^4 = 81$.

6. Find each root.

c. $\sqrt[4]{-625}$

_____

a. $\sqrt[4]{625}$

_____

Objective 3 Practice Exercises

For extra help, see Examples 5–6 on pages 729–730 of your text.

Find each root.

7. $\sqrt[3]{-27}$

7. _____

8. $\sqrt[4]{256}$

8. _____

9. $\sqrt[7]{-1}$

9. _____

Objective 4   Graph functions defined by radical expressions.

Video Examples

Review these examples for Objective 4:

7. Graph each function by creating a table of values. Give the domain and the range.

a. $f(x) = \sqrt{x-1}$

Create a table of values.

x	$f(x) = \sqrt{x-1}$
1	$\sqrt{1-1} = 0$
5	$\sqrt{5-1} = 2$
10	$\sqrt{10-1} = 3$

For the radicand to be nonnegative, we must have $x-1 \geq 0$ or $x \geq 1$. Therefore, the domain is $[1, \infty)$. Function values are nonnegative, so

Now Try:

7. Graph each function by creating a table of values. Give the domain and the range.

a. $f(x) = \sqrt{x} - 1$

the range is $[0, \infty)$.

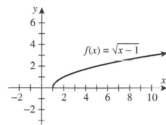

b. $f(x) = \sqrt[3]{x} + 1$

Create a table of values.

x	$f(x) = \sqrt[3]{x} + 1$
-8	$\sqrt[3]{-8} + 1 = -1$
-1	$\sqrt[3]{-1} + 1 = 0$
0	$\sqrt[3]{0} + 1 - 1$
1	$\sqrt[3]{1} + 1 = 2$
8	$\sqrt[3]{8} + 1 = 3$

Both the domain and range are $(-\infty, \infty)$.

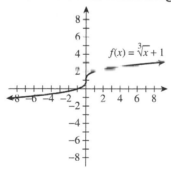

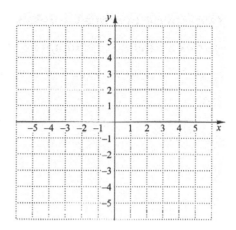

domain: _____

range: _____

b. $f(x) = \sqrt[3]{x} + 1$

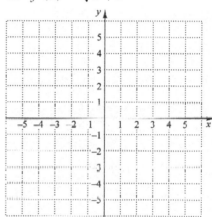

domain: _____

range: _____

Name: _____   Date: _____

Instructor: _____   Section: _____

Objective 4 Practice Exercises

For extra help, see Example 7 on page 731 of your text.

Graph each function and give its domain and its range.

10. $f(x) = \sqrt{x} + 2$

10.

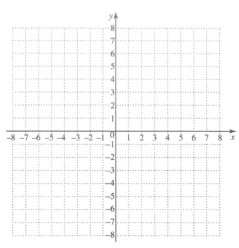

domain: _____

range: _____

11. $f(x) = \sqrt[3]{x} - 2$

11.

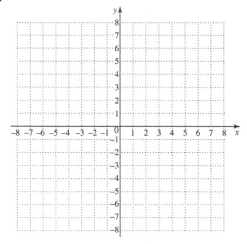

domain: _____

range: _____

12. $f(x) = \sqrt[3]{x} + 2$ **12.**

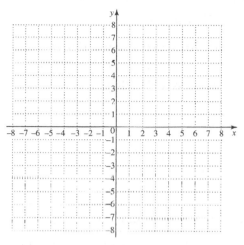

domain: _____

range: _____

Objective 5   Find *n*th roots of *n*th powers.

Video Examples

Review these examples for Objective 5:
9.   Simplify each root.

   c. $-\sqrt[6]{(-8)^6}$

   n is even. Use absolute value.

   $-\sqrt[6]{(-8)^6} = -|-8| = -8$

   e. $\sqrt[5]{s^{20}}$

   $\sqrt[5]{s^{20}} = s^4$, because $s^{20} = (s^4)^5$.

Now Try:
9.   Simplify each root.

   c. $-\sqrt[4]{(-2)^4}$

   _____

   e. $\sqrt[3]{w^{30}}$

   _____

Objective 5 Practice Exercises

For extra help, see Examples 8–9 on page 732 of your text.

Simplify each root.

13. $\sqrt{(-9)^2}$ **13.** _____

14. $-\sqrt[5]{x^5}$ **14.** _____

15. $-\sqrt[4]{x^{16}}$ **15.** _____

Objective 6   Use a calculator to find roots.

Video Examples

Review this example for Objective 6:

10a. Use a calculator to approximate the radical to three decimal places.

$$\sqrt{12}$$

Use the square root key on a calculator.
$$\sqrt{12} \approx 3.464$$

Now Try:

10a. Use a calculator to approximate the radical to three decimal places.

$$\sqrt{14}$$

_____

Objective 6 Practice Exercises

For extra help, see Examples 10–11 on page 733 of your text.

Use a calculator to find a decimal approximation for each radical. Give the answer to the nearest thousandth.

16. $\sqrt[3]{701}$ 16. _____

17. $-\sqrt{990}$ 17. _____

18.   The time t in seconds for one complete swing of a simple pendulum, where L is the length of the pendulum in feet is $t = 2\pi\sqrt{\dfrac{L}{32}}$. Find the time of a complete swing of a 4-ft pendulum to the nearest tenth of a second.

18. _____

Chapter 10 ROOTS, RADICALS, AND ROOT FUNCTIONS

10.2   Rational Exponents

Learning Objectives
1   Use exponential notation for nth roots.
2   Define and use expressions of the form $a^{m/n}$.
3   Convert between radicals and rational exponents.
4   Use the rules for exponents with rational exponents.

Key Terms

Use the vocabulary terms listed below to complete each statement in exercises 1–3

   product rule for exponents   **quotient rule for exponents**

   power rule for exponents

1. $\left(x^2 y^3\right)^4 = x^8 y^{12}$ is an example of the _____.

2. $w^5 w^3 = w^8$ is an example of the _____.

3. $\dfrac{z^6}{z^4} = z^2$ is an example of the _____.

Objective 1   Use exponential notation for nth roots.

Video Examples

Review these examples for Objective 1:   Now Try:
1.   Evaluate each exponential.   1.   Evaluate each exponential.

   b. $25^{1/2}$ **b.** $121^{1/2}$

   $25^{1/2} = \sqrt{25} = 5$

                                               _____

   e. $(-8)^{1/3}$ **e.** $(-243)^{1/5}$

   $(-8)^{1/3} = \sqrt[3]{-8} = -2$

                                               _____

   d. $(-81)^{1/4}$ **d.** $(-1024)^{1/10}$

   $(-81)^{1/4} = \sqrt[4]{-81}$ is not a real number.

                                               _____

Objective 1 Practice Exercises

For extra help, see Example 1 on page 739 of your text.

Evaluate each exponential.

1. $-625^{1/4}$

2. $16^{1/2}$

3. $(-3375)^{1/3}$

1. _____

2. _____

3. _____

Objective 2   Define and use expressions of the form $a^{m/n}$.

Video Examples

Review these examples for Objective 2:

2. Evaluate each polynomial.

   a. $100^{3/2}$

   $100^{3/2} = (100^{1/2})^3 = 10^3 = 1000$

   c. $-729^{5/6}$

   $-729^{5/6} = -(729^{1/6})^5 = -(3)^5 = -243$

3a. Evaluate the exponential.

   $625^{-3/4}$

   $625^{-3/4} = \dfrac{1}{625^{3/4}} = \dfrac{1}{(625^{1/4})^3} = \dfrac{1}{\left(\sqrt[4]{625}\right)^3}$

   $\quad = \dfrac{1}{5^3} = \dfrac{1}{125}$

Now Try:

2. Evaluate each polynomial.

   a. $27^{2/3}$

   _____

   c. $-36^{3/2}$

   _____

3a. Evaluate the exponential.

   $32^{-2/5}$

   _____

Name: Date:

Instructor: Section:

Objective 2 Practice Exercises

For extra help, see Examples 2–3 on pages 740–741 of your text.

Evaluate each exponential.

4. $-81^{5/4}$ 4. _____

5. $36^{5/2}$ 5. _____

6. $\left(\dfrac{125}{27}\right)^{-2/3}$ 6. _____

Objective 3   Convert between radicals and rational exponents.

Video Examples

Review these examples for Objective 3:	**Now Try:**
4. Write each radical as an exponential and simplify. Assume that all variables represent positive real numbers. Use the definition that takes the root first.	4. Write each radical as an exponential and simplify. Assume that all variables represent positive real numbers. Use the definition that takes the root first.
a. $17^{1/2}$	**a.** $23^{1/3}$
$17^{1/2} = \sqrt{17}$	
	_____
d. $2x^{2/5} - (4x)^{5/6}$	**d.** $(2x)^{4/3} - 3x^{2/5}$
$2x^{2/5} - (4x)^{5/6} = 2\left(\sqrt[5]{x}\right)^2 - \left(\sqrt[6]{4x}\right)^5$	
	_____
5c. Write the radical as an exponential and simplify. Assume that all variables represent positive real numbers.	**5c.** Write the radical as an exponential and simplify. Assume that all variables represent positive real numbers.
$\sqrt[3]{3^6}$	$\sqrt{10^4}$
$\sqrt[3]{3^6} = 3^{6/3} = 3^2 = 9$	
	_____

Objective 3 Practice Exercises

For extra help, see Examples 4–5 on page 742 of your text.

Write with radicals. Assume that all variables represent positive real numbers.

7. $4y^{2/5} + (5x)^{1/5}$

7. _____

8. $\left(2x^4 - 3y^2\right)^{-4/3}$

8. _____

Simplify the radical by rewriting it with a rational exponent. Write answers in radical form if necessary. Assume that variables represent positive real numbers.

9. $\sqrt[8]{a^2}$

9. _____

Objective 4   Use the rules for exponents with rational exponents.

Video Examples

Review these examples for Objective 4:

6. Write with only positive exponents. Assume that all variables represent positive real numbers.

a. $13^{4/5} \cdot 13^{1/2}$

$13^{4/5} \cdot 13^{1/2} = 13^{4/5+1/2} = 13^{13/10}$

d. $\left(\dfrac{c^6 x^3}{c^{-2} x^{1/2}}\right)^{-3/4}$

$\left(\dfrac{c^6 x^{1/2}}{c^{-2} x^3}\right)^{-3/4} = \left(c^{6-(-2)} x^{1/2-3}\right)^{-3/4}$

$= \left(c^8 x^{-5/2}\right)^{-3/4}$

$= \left(c^8\right)^{-3/4} \left(x^{-5/2}\right)^{-3/4}$

$= c^{-6} x^{15/8}$

$= \dfrac{x^{15/8}}{c^6}$

Now Try:

6. Write with only positive exponents. Assume that all variables represent positive real numbers.

a. $5^{3/4} \cdot 5^{7/4}$

_____

d. $\left(\dfrac{x^{-1} y^{2/3}}{x^{1/3} y^{1/2}}\right)^{-3/2}$

_____

7. Write all radicals as exponentials, and then apply the rules for rational exponents. Leave answers in exponential form. Assume that all variables represent positive real numbers.

    a. $\sqrt[4]{x^3} \cdot \sqrt[5]{x}$

$$\sqrt[4]{x^3} \cdot \sqrt[5]{x} = x^{3/4} \cdot x^{1/5}$$
$$= x^{3/4+1/5}$$
$$= x^{15/20+4/20}$$
$$= x^{19/20}$$

    c. $\sqrt{\sqrt[4]{y^3}}$

$$\sqrt{\sqrt[4]{y^3}} = \sqrt{y^{3/4}} = \left(y^{3/4}\right)^{1/2} = y^{3/8}$$

7. Write all radicals as exponentials, and then apply the rules for rational exponents. Leave answers in exponential form. Assume that all variables represent positive real numbers.

    a. $\sqrt[6]{x^3} \cdot \sqrt[3]{x^2}$

    _____

    c. $\sqrt[3]{\sqrt[4]{x^3}}$

    _____

Objective 4 Practice Exercises

For extra help, see Examples 6–7 on pages 743–744 of your text.

Use the rules of exponents to simplify each expression. Write all answers with positive exponents. Assume that variables represent positive real numbers.

10. $y^{7/3} \cdot y^{-4/3}$ 10. _____

11. $\dfrac{a^{2/3} \cdot a^{-1/3}}{\left(a^{-1/6}\right)^3}$ 11. _____

12. $\dfrac{\left(x^{-3}y^2\right)^{2/3}}{\left(x^2y^{-5}\right)^{2/5}}$ 12. _____

Name: Date:
Instructor: Section:

Chapter 10 ROOTS, RADICALS, AND ROOT FUNCTIONS

10.3   Simplifying Radical Expressions

Learning Objectives
1   Use the product rule for radicals.
2   Use the quotient rule for radicals.
3   Simplify radicals.
4   Simplify products and quotients of radicals with different indexes.
5   Use the Pythagorean theorem.
6   Use the distance formula.

Key Terms

Use the vocabulary terms listed below to complete each statement in exercises 1–3.

       index   **radicand**   **hypotenuse**   **legs**

1. In a right triangle, the side opposite the right angle is called the

    _____.

2. In the expression $\sqrt[4]{x^2}$, the "4" is the _____ and x^2 is the

    _____.

3. In a right triangle, the sides that form the right angle are called the

    _____.

Objective 1   Use the product rule for radicals.

Video Examples

Review these examples for Objective 1:

1. Multiply. Assume that all variables represent positive real numbers.

    a. $\sqrt{13} \cdot \sqrt{5}$

    $\sqrt{13} \cdot \sqrt{5} = \sqrt{13 \cdot 5} = \sqrt{65}$

    c. $\sqrt{5x} \cdot \sqrt{2yz}$

    $\sqrt{5x} \cdot \sqrt{2yz} = \sqrt{10xyz}$

Now Try:

1. Multiply. Assume that all variables represent positive real numbers.

    a. $\sqrt{2} \cdot \sqrt{7}$

    _____

    c. $\sqrt{3} \cdot \sqrt{11mn}$

    _____

2. Multiply. Assume that all variables represent positive real numbers.

   a. $\sqrt[4]{2} \cdot \sqrt[4]{2x}$

   $\sqrt[4]{2} \cdot \sqrt[4]{2x} = \sqrt[4]{2 \cdot 2x} = \sqrt[4]{4x}$

   b. $\sqrt[3]{8x} \cdot \sqrt[3]{2y^2}$

   $\sqrt[3]{8x} \cdot \sqrt[3]{2y^2} = \sqrt[3]{8x \cdot 2y^2} = \sqrt[3]{16xy^2}$

   c. $\sqrt[5]{6r^2} \cdot \sqrt[5]{4r^2}$

   $\sqrt[5]{6r^2} \cdot \sqrt[5]{4r^2} = \sqrt[5]{6r^2 \cdot 4r^2} = \sqrt[3]{24r^4}$

   d. $\sqrt[5]{2} \cdot \sqrt[4]{6}$

   $\sqrt[5]{2} \cdot \sqrt[4]{6}$ cannot be simplified using the product rule for radicals, because the indexes (5 and 4) are different.

2. Multiply. Assume that all variables represent positive real numbers.

   a. $\sqrt[3]{3} \cdot \sqrt[3]{7}$

   _____

   b. $\sqrt[3]{7x} \cdot \sqrt[3]{5y}$

   _____

   c. $\sqrt[5]{4w} \cdot \sqrt[5]{2w^3}$

   _____

   d. $\sqrt{3} \cdot \sqrt[3]{64}$

   _____

Objective 1 Practice Exercises

For extra help, see Examples 1–2 on page 749 of your text.

Multiply. Assume that variables represent positive real numbers.

1. $\sqrt{7x} \cdot \sqrt{6t}$

1. _____

2. $\sqrt[5]{6r^2t^3} \cdot \sqrt[5]{4r^2t}$

2. _____

3. $\sqrt{3} \cdot \sqrt[3]{7}$

3. _____

Objective 2   Use the quotient rule for radicals.

Video Examples

Review these examples for Objective 2:

3.   Simplify. Assume that all variables represent positive real numbers.

a. $\sqrt{\dfrac{64}{9}}$

$\sqrt{\dfrac{64}{9}} = \dfrac{\sqrt{64}}{\sqrt{9}} = \dfrac{8}{3}$

b. $\sqrt{\dfrac{5}{16}}$

$\sqrt{\dfrac{5}{16}} = \dfrac{\sqrt{5}}{\sqrt{16}} = \dfrac{\sqrt{5}}{4}$

c. $\sqrt[3]{-\dfrac{27}{8}}$

$\sqrt[3]{-\dfrac{27}{8}} = \dfrac{\sqrt[3]{-27}}{\sqrt[3]{8}} = \dfrac{-3}{2} = -\dfrac{3}{2}$

e. $\sqrt[5]{-\dfrac{a^3}{243}}$

$\sqrt[5]{-\dfrac{a^3}{243}} = \dfrac{\sqrt[5]{a^3}}{\sqrt[5]{-243}} = \dfrac{\sqrt[5]{a^3}}{-3} = -\dfrac{\sqrt[5]{a^3}}{3}$

f. $\sqrt{\dfrac{z^4}{36}}$

$\sqrt{\dfrac{z^4}{36}} = \dfrac{\sqrt{z^4}}{\sqrt{36}} = \dfrac{z^2}{6}$

Now Try:

3.   Simplify. Assume that all variables represent positive real numbers.

a. $\sqrt{\dfrac{36}{49}}$

_____

b. $\sqrt{\dfrac{13}{81}}$

_____

c. $\sqrt[3]{-\dfrac{343}{125}}$

_____

e. $\sqrt[3]{-\dfrac{a^6}{125}}$

_____

f. $\sqrt[4]{\dfrac{m}{81}}$

_____

Objective 2 Practice Exercises

For extra help, see Example 3 on page 750 of your text.

Simplify each radical. Assume that variables represent positive real numbers.

4. $\sqrt[3]{\dfrac{27}{8}}$

4.   _____

5. $\sqrt[5]{\dfrac{7x}{32}}$ 5. _____

6. $\sqrt[3]{-\dfrac{a^6}{125}}$ 6. _____

Objective 3   Simplify radicals.

Video Examples

Review these examples for Objective 3: | **Now Try:**

4. Simplify.

a. $\sqrt{90}$

$\sqrt{90} = \sqrt{9 \cdot 10}$

$\qquad - \sqrt{9} \cdot \sqrt{10}$

$\qquad = 3\sqrt{10}$

b. $\sqrt{288}$

$\sqrt{288} = \sqrt{144 \cdot 2}$

$\qquad = \sqrt{144} \cdot \sqrt{2}$

$\qquad = 12\sqrt{2}$

c. $\sqrt{35}$

No perfect square (other than 1) divides into 35, so $\sqrt{35}$ cannot be simplified further.

d. $\sqrt[3]{81}$

$\sqrt[3]{81} = \sqrt[3]{27 \cdot 3} = \sqrt[3]{27} \cdot \sqrt[3]{3} = 3\sqrt[3]{3}$

e. $-\sqrt[4]{3125}$

$-\sqrt[4]{3125} = -\sqrt[4]{5^5} = \sqrt[4]{5^4 \cdot 5}$

$\qquad = -\sqrt[4]{5^4} \cdot \sqrt[4]{5}$

$\qquad = -5\sqrt[4]{5}$

Now Try:

4. Simplify.

a. $\sqrt{84}$

_____

b. $\sqrt{162}$

_____

c. $\sqrt{95}$

_____

d. $\sqrt[3]{256}$

_____

e. $-\sqrt[5]{512}$

_____

5. Simplify. Assume that all variables represent positive real numbers.

a. $\sqrt{81x^3}$

$$\sqrt{81x^3} = \sqrt{9^2 \cdot x^2 \cdot x} = 9x\sqrt{x}$$

b. $\sqrt{56x^7 y^6}$

$$\sqrt{56x^7 y^6} = \sqrt{4 \cdot 14 \cdot (x^3)^2 \cdot x \cdot (y^3)^2}$$
$$= 2x^3 y^3 \sqrt{14x}$$

c. $\sqrt[3]{-270b^4 c^8}$

$$\sqrt[3]{-270b^4 c^8} = \sqrt[3]{(-27b^3 c^6)(10bc^2)}$$
$$= \sqrt[3]{-27b^3 c^6} \cdot \sqrt[3]{10bc^2}$$
$$= -3bc^2 \sqrt[3]{10bc^2}$$

d. $-\sqrt[6]{448a^7 b^7}$

$$-\sqrt[6]{448a^7 b^7} = -\sqrt[6]{(64a^6 b^6)(7ab)}$$
$$= -\sqrt[6]{64a^6 b^6} \cdot \sqrt[6]{7ab}$$
$$= -2ab \sqrt[6]{7ab}$$

6. Simplify. Assume that all variables represent positive real numbers.

a. $\sqrt[24]{5^4}$

$$\sqrt[24]{5^4} = (5^4)^{1/24} = 5^{4/24} = 5^{1/6} = \sqrt[6]{5}$$

c. $\sqrt[8]{x^{12}}$

$$\sqrt[8]{x^{12}} = (x^{12})^{1/8} = x^{12/8} = x^{3/2} = \sqrt{x^3}$$
$$= \sqrt{x^2 \cdot x} = \sqrt{x^2} \cdot \sqrt{x} = x\sqrt{x}$$

5. Simplify. Assume that all variables represent positive real numbers.

a. $\sqrt{100y^3}$

_____

b. $\sqrt{48m^5 r^9}$

_____

c. $\sqrt[3]{-32n^7 t^5}$

_____

d. $-\sqrt[4]{405x^3 y^9}$

_____

6. Simplify. Assume that all variables represent positive real numbers.

a. $\sqrt[12]{11^9}$

_____

c. $\sqrt[24]{z^{30}}$

_____

Objective 3 Practice Exercises

For extra help, see Examples 4–6 on pages 751–752 of your text.

Simplify each radical. Assume that variables represent positive real numbers.

7. $\sqrt[42]{x^{28}}$ 7. _____

8. $\sqrt{8x^3y^6z^{11}}$ 8. _____

9. $\sqrt[3]{1250a^5b^7}$ 9. _____

Objective 4   **Simplify products and quotients of radicals with different indexes.**

Video Examples

Review this example for Objective 4:

7.   Simplify $\sqrt{3} \cdot \sqrt[5]{6}$.

   Because the different indexes, 2 and 5, have least common multiple index of 10, we use rational exponents to write each radical as a tenth root.

$$\sqrt{3} = 3^{1/2} = 3^{5/10} = \sqrt[10]{3^5} = \sqrt[10]{243}$$

$$\sqrt[5]{6} = 6^{1/5} = 6^{2/10} = \sqrt[10]{6^2} = \sqrt[10]{36}$$

$$\sqrt{3} \cdot \sqrt[5]{6} = \sqrt[10]{243} \cdot \sqrt[10]{36}$$

$$= \sqrt[10]{243 \cdot 36}$$

$$= \sqrt[10]{8748}$$

Now Try:

7.   Simplify $\sqrt[3]{3} \cdot \sqrt[6]{7}$.

                     _____

Objective 4 Practice Exercises

For extra help, see Example 7 on page 753 of your text.

Simplify each radical. Assume that variables represent positive real numbers.

10. $\sqrt{r} \cdot \sqrt[3]{r}$ 10. _____

11. $\sqrt[4]{2} \cdot \sqrt[8]{7}$

11. _____

12. $\sqrt{3} \cdot \sqrt[5]{64}$

12. _____

Objective 5   Use the Pythagorean theorem.

Video Examples

Review this example for Objective 5:

8. Use the Pythagorean theorem to find the length of the unknown side of the triangle.

$$a^2 + b^2 = c^2$$
$$12^2 + b^2 = 25^2$$
$$144 + b^2 = 625$$
$$b^2 = 481$$
$$b = \sqrt{481}$$

The length of the side is $\sqrt{481}$.

Now Try:

8. Use the Pythagorean theorem to find the length of the unknown side of the triangle.

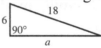

_____

Objective 5 Practice Exercises

For extra help, see Example 8 on page 754 of your text.

Find the unknown length in each right triangle. Simplify the answer if necessary.

13.

13. _____

14.

14. _____

15.

15. _____

Objective 6   Use the distance formula.

Video Examples

Review this example for Objective 6:

9. Find the distance between the points $(2,-2)$ and $(-6, 1)$.

   Use the distance formula. Let $(x_1, y_1) = (2,-2)$ and $(x_2, y_2) = (-6, 1)$.

   $$d = \sqrt{(x_2 - x_1)^2 + (y_2 - y_1)^2}$$
   $$= \sqrt{(-6-2)^2 + [1-(-2)]^2}$$
   $$= \sqrt{(-8)^2 + 3^2}$$
   $$= \sqrt{64+9}$$
   $$= \sqrt{73}$$

Now Try:

9. Find the distance between the points $(-1,-2)$ and $(-4, 3)$.

   _____

Objective 6 Practice Exercises

For extra help, see Example 9 on page 755 of your text.

Find the distance between each pair of points.

16. (3, 4) and (−1,−2)   16. _____

17. (−2,−3) and (−5, 1)   17. _____

18. (4, 2) and (3,−1)   18. _____

Chapter 10 ROOTS, RADICALS, AND ROOT FUNCTIONS

10.4   Adding and Subtracting Radical Expressions

Learning Objectives
1   Simplify radical expressions involving addition and subtraction.

Key Terms

Use the vocabulary terms listed below to complete each statement in exercises 1–2.

like radicals   unlike radicals

1.   The expressions $2\sqrt{2}$ and $6\sqrt[3]{2}$ are _____.

2.   The expressions $2\sqrt{2}$ and $7\sqrt{2}$ are _____.

Objective 1   Simplify radical expressions involving addition and subtraction.

Video Examples

Review these examples for Objective 1:

1.   Add or subtract to simplify each radical expression.

c. $3\sqrt{13} + 5\sqrt{52}$

$$
\begin{aligned}
3\sqrt{13} + 5\sqrt{52} &= 3\sqrt{13} + 5\sqrt{4}\sqrt{13} \\
&= 3\sqrt{13} + 5\cdot 2\sqrt{13} \\
&= 3\sqrt{13} + 10\sqrt{13} \\
&= (3+10)\sqrt{13} \\
&= 13\sqrt{13}
\end{aligned}
$$

d. $\sqrt{48x} - \sqrt{12x},\ x \ge 0$

$$
\begin{aligned}
\sqrt{48x} - \sqrt{12x} &= \sqrt{16}\cdot\sqrt{3x} - \sqrt{4}\cdot\sqrt{3x} \\
&= 4\sqrt{3x} - 2\sqrt{3x} \\
&= (4-2)\sqrt{3x} \\
&= 2\sqrt{3x}
\end{aligned}
$$

e. $7\sqrt{3} - 6\sqrt{21}$

The radicands differ and are already simplified, so this expression cannot be simplified further.

Now Try:

1.   Add or subtract to simplify each radical expression.

c. $3\sqrt{54} - 5\sqrt{24}$

_____

d. $3\sqrt{18x} + 2\sqrt{8x},\ x \ge 0$

_____

e. $3\sqrt{7} + 2\sqrt{6}$

_____

2. Simplify. Assume that all variables represent positive real numbers.

    a. $7\sqrt[4]{32} - 9\sqrt[4]{2}$

    $$7\sqrt[4]{32} - 9\sqrt[4]{2} = 7\sqrt[4]{16} \cdot \sqrt[4]{2} - 9\sqrt[4]{2}$$
    $$= 7 \cdot 2 \cdot \sqrt[4]{2} - 9\sqrt[4]{2}$$
    $$= 14\sqrt[4]{2} - 9\sqrt[4]{2}$$
    $$= (14 - 9)\sqrt[4]{2}$$
    $$= 5\sqrt[4]{2}$$

    b. $6\sqrt[3]{27x^5 r} + 2x\sqrt[3]{x^2 r}$

    $$6\sqrt[3]{27x^5 r} + 2x\sqrt[3]{x^2 r}$$
    $$= 6 \cdot \sqrt[3]{27x^3} \cdot \sqrt[3]{x^2 r} + 2x\sqrt[3]{x^2 r}$$
    $$= 18x\sqrt[3]{x^2 r} + 2x\sqrt[3]{x^2 r}$$
    $$= (18x + 2x)\sqrt[3]{x^2 r}$$
    $$= 20x\sqrt[3]{x^2 r}$$

2. Simplify. Assume that all variables represent positive real numbers.

    a. $7\sqrt[3]{54} - 6\sqrt[3]{128}$

    _____

    b. $\sqrt[4]{32y^2 z^5} + 3z\sqrt[4]{2y^2 z}$

    _____

3. Simplify. Assume that all variables represent positive real numbers.

    a. $\dfrac{\sqrt{32}}{3} + \dfrac{\sqrt{8}}{\sqrt{18}}$

    $$\frac{\sqrt{32}}{3} + \frac{\sqrt{8}}{\sqrt{18}} = \frac{\sqrt{16 \cdot 2}}{3} + \frac{\sqrt{4 \cdot 2}}{\sqrt{9 \cdot 2}}$$
    $$= \frac{4\sqrt{2}}{3} + \frac{2\sqrt{2}}{3\sqrt{2}}$$
    $$= \frac{4\sqrt{2}}{3} + \frac{2}{3}$$
    $$= \frac{4\sqrt{2} + 2}{3}$$

3. Simplify. Assume that all variables represent positive real numbers.

    a. $\sqrt{\dfrac{10}{18}} + \dfrac{\sqrt{15}}{\sqrt{27}}$

    _____

b. $\sqrt[3]{\dfrac{81}{y^6}} + 5\sqrt[3]{\dfrac{27}{y^3}}$

$$\sqrt[3]{\dfrac{81}{y^6}} + 5\sqrt[3]{\dfrac{27}{y^3}} = \dfrac{\sqrt[3]{81}}{\sqrt[3]{y^6}} + 5\dfrac{\sqrt[3]{27}}{\sqrt[3]{y^3}}$$

$$= \dfrac{3\sqrt[3]{3}}{y^2} + 5\left(\dfrac{3}{y}\right)$$

$$= \dfrac{3\sqrt[3]{3}}{y^2} + \dfrac{15}{y}$$

$$= \dfrac{3\sqrt[3]{3} + 15y}{y^2}$$

b. $\sqrt[3]{\dfrac{216}{w^6}} + \sqrt{\dfrac{121}{w^4}}$

_____

Objective 1 Practice Exercises

For extra help, see Examples 1–3 on pages 762–764 of your text.

Add or subtract. Assume that all variables represent positive real numbers.

1. $\sqrt{100x} - \sqrt{9x} + \sqrt{25x}$

1. _____

2. $2\sqrt[3]{16r} + \sqrt[3]{54r} - \sqrt[3]{16r}$

2. _____

3. $\sqrt[3]{\dfrac{y^7}{125}} + y^2\sqrt[3]{\dfrac{y}{27}}$

3. _____

Chapter 10 ROOTS, RADICALS, AND ROOT FUNCTIONS

10.5   Multiplying and Dividing Radical Expressions

Learning Objectives	
1	Multiply radical expressions.
2	Rationalize denominators with one radical term.
3	Rationalize denominators with binomials involving radicals.
4	Write radical quotients in lowest terms.

Key Terms

Use the vocabulary terms listed below to complete each statement in exercises 1−2.

rationalizing the denominator   **conjugate**

1.   The _____ of $a + b$ is $a - b$.

2.   The process of removing radicals from the denominator so that the denominator contains only rational quantities is called _____.

Objective 1   Multiply radical expressions.

Video Examples

Review this example for Objective 1:
2a. Multiply, using the FOIL method.

$$\left(2\sqrt{3} - \sqrt{7}\right)\left(\sqrt{5} + \sqrt{2}\right)$$

$$\left(2\sqrt{3} - \sqrt{7}\right)\left(\sqrt{5} + \sqrt{2}\right)$$
$$= 2\sqrt{3}\cdot\sqrt{5} + 2\sqrt{3}\cdot\sqrt{2} - \sqrt{7}\cdot\sqrt{5} - \sqrt{7}\sqrt{2}$$
$$= 2\sqrt{15} + 2\sqrt{6} - \sqrt{35} - \sqrt{14}$$

Now Try:
2a. Multiply, using the FOIL method.
$$\left(\sqrt{5} + \sqrt{3}\right)\left(\sqrt{2} - \sqrt{11}\right)$$

_____

Objective 1 Practice Exercises

For extra help, see Examples 1−2 on pages 767−768 of your text.

Multiply each product, then simplify. Assume that variables represent positive real numbers.

1. $\left(\sqrt{5} + \sqrt{6}\right)\left(\sqrt{2} - 4\right)$

1.   _____

2. $\left(\sqrt{2}-\sqrt{12}\right)^2$

2. _____

3. $\left(2+\sqrt[3]{5}\right)\left(2-\sqrt[3]{5}\right)$

3. _____

Objective 2   Rationalize denominators with one radical term.

Video Examples

Review these examples for Objective 2:
3b. Rationalize the denominator.

$$\frac{\sqrt{3}}{8\sqrt{5}}$$

$$\frac{\sqrt{3}}{8\sqrt{5}}=\frac{\sqrt{3}\cdot\sqrt{5}}{8\sqrt{5}\cdot\sqrt{5}}=\frac{\sqrt{15}}{8\cdot5}=\frac{\sqrt{15}}{40}$$

4a. Simplify the radical.

$$-\sqrt{\frac{27}{98}}$$

$$-\sqrt{\frac{27}{98}}=-\frac{\sqrt{27}}{\sqrt{98}}$$

$$=-\frac{\sqrt{9\cdot3}}{\sqrt{49\cdot2}}$$

$$=-\frac{3\sqrt{3}}{7\sqrt{2}}$$

$$=-\frac{3\sqrt{3}\cdot\sqrt{2}}{7\sqrt{2}\cdot\sqrt{2}}$$

$$=-\frac{3\sqrt{6}}{7\cdot2}$$

$$=-\frac{3\sqrt{6}}{14}$$

Now Try:
3b. Rationalize the denominator.

$$\frac{6\sqrt{3}}{\sqrt{14}}$$

_____

4a. Simplify the radical.

$$-\sqrt{\frac{45}{32}}$$

_____

Name: _____ Date: _____

Instructor: _____ Section: _____

5a. Simplify.

$$\sqrt[3]{\frac{16}{9}}$$

$$\sqrt[3]{\frac{16}{9}} = \frac{\sqrt[3]{8 \cdot 2}}{\sqrt[3]{9}} = \frac{2\sqrt[3]{2}}{\sqrt[3]{9}}$$

$$= \frac{2\sqrt[3]{2} \cdot \sqrt[3]{3}}{\sqrt[3]{9} \cdot \sqrt[3]{3}}$$

$$= \frac{2\sqrt[3]{6}}{\sqrt[3]{27}}$$

$$= \frac{2\sqrt[3]{6}}{3}$$

5a. Simplify.

$$\sqrt[3]{\frac{8}{100}}$$

_____

Objective 2 Practice Exercises

For extra help, see Examples 3–5 on pages 769–770 of your text.

Simplify. Assume that variables represent positive real numbers.

4. $\sqrt{\dfrac{5a^2b^3}{6}}$

4. _____

5. $\sqrt{\dfrac{7y^2}{12b}}$

5. _____

6. $\sqrt[3]{\dfrac{5}{49x}}$

6. _____

Name: Date:

Instructor: Section:

Objective 3   Rationalize denominators with binomials involving radicals.

Video Examples

Review these examples for Objective 3:

6.   Rationalize each denominator.

a. $\dfrac{5}{5+\sqrt{2}}$

$$\dfrac{5}{5+\sqrt{2}} = \dfrac{5(5-\sqrt{2})}{(5+\sqrt{2})(5-\sqrt{2})}$$

$$= \dfrac{5(5-\sqrt{2})}{25-2}$$

$$= \dfrac{5(5-\sqrt{2})}{23}$$

c. $\dfrac{\sqrt{2}-\sqrt{3}}{\sqrt{2}-\sqrt{5}}$

$$\dfrac{\sqrt{2}-\sqrt{3}}{\sqrt{2}-\sqrt{5}} = \dfrac{(\sqrt{2}-\sqrt{3})(\sqrt{2}+\sqrt{5})}{(\sqrt{2}-\sqrt{5})(\sqrt{2}+\sqrt{5})}$$

$$= \dfrac{2-\sqrt{10}-\sqrt{6}-\sqrt{15}}{2-5}$$

$$= \dfrac{2-\sqrt{10}-\sqrt{6}-\sqrt{15}}{-3}$$

Now Try:

6.   Rationalize each denominator.

a. $\dfrac{4}{\sqrt{3}+2}$

_____

c. $\dfrac{\sqrt{3}+\sqrt{2}}{\sqrt{5}-\sqrt{2}}$

_____

Objective 3 Practice Exercises

For extra help, see Example 6 on page 772 of your text.

Rationalize each denominator. Write quotients in lowest terms. Assume that variables represent positive real numbers.

7. $\dfrac{4}{\sqrt{3}+2}$

7. _____

8. $\dfrac{5}{\sqrt{3}-\sqrt{10}}$

8. _____

9. $\dfrac{\sqrt{6}+2}{4-\sqrt{2}}$

9. _____

Objective 4   Write radical quotients in lowest terms.

Video Examples

Review this example for Objective 4:

7b. Write the quotient in lowest terms.

$$\dfrac{7y-\sqrt{98y^5}}{14y},\ y>0$$

$$\dfrac{7y-\sqrt{98y^5}}{14y}=\dfrac{7y-\sqrt{49y^4\cdot 2y}}{14y}$$

$$=\dfrac{7y-7y^2\sqrt{2y}}{14y}$$

$$=\dfrac{7y\left(1-y\sqrt{2y}\right)}{14y}$$

$$=\dfrac{1-y\sqrt{2y}}{2}$$

Now Try:

7b. Write the quotient in lowest terms.

$$\dfrac{2x-\sqrt{8x^3}}{4x},\ x>0$$

_____

Objective 4 Practice Exercises

For extra help, see Example 7 on page 773 of your text.

Write each quotient in lowest terms. Assume that variables represent positive real numbers.

10. $\dfrac{7-\sqrt{98}}{14}$

10. _____

11. $\dfrac{16-12\sqrt{72}}{24}$

11. _____

12. $\dfrac{2x-\sqrt{8x^2}}{4x}$

12. _____

Chapter 10 ROOTS, RADICALS, AND ROOT FUNCTIONS

10.6 Solving Equations with Radicals

Learning Objectives
1   Solve radical equations using the power rule.
2   Solve radical equations that require additional steps.
3   Solve radical equations with indexes greater than 2.
4   Use the power rule to solve a formula for a specified variable.

Key Terms

Use the vocabulary terms listed below to complete each statement in exercises 1−3.

         radical equation   **proposed solution**   **extraneous solution**

1.   A(n) _____ is a potential solution to an equation that does not satisfy the equation.

2.   An equation with a variable in the radicand is a(n) _____.

3.   A value of a variable that appears to be a solution of an equation is a(n) _____.

Objective 1   Solve radical equations using the power rule.

Video Examples

Review these examples for Objective 1:

1.   Solve $\sqrt{3w+4} = 7$.

$$\left(\sqrt{3w+4}\right)^2 = 7^2$$
$$3w+4 = 49$$
$$3w = 45$$
$$w = 15$$

Check $\sqrt{3w+4} = 7$
$$\sqrt{3(15)+4} \overset{?}{=} 7$$
$$\sqrt{49} \overset{?}{=} 7$$
$$7 = 7 \quad \text{True}$$

Since 15 satisfies the original equation, the solution set is {15}.

Now Try:

1.   Solve $\sqrt{7x-6} = 8$.

_____

2. Solve $\sqrt{12p+1}+7=0$.

$$\text{Step 1} \qquad \sqrt{12p+1}=-7$$

$$\text{Step 2} \quad \left(\sqrt{12p+1}\right)^2=(-7)^2$$

$$\text{Step 3} \qquad 12p+1=49$$
$$12p=48$$
$$p=4$$

Step 4 Check $\quad \sqrt{12p+1}+7=0$

$$\sqrt{12(4)+1}+7\overset{?}{=}0$$

$$\sqrt{49}+7\overset{?}{=}0$$

$$14=0 \quad \text{False}$$

The false result shows that the proposed solution 4 is not a solution of the original equation. It is extraneous. The solution set is $\varnothing$.

2. Solve $\sqrt{4x-19}+5=0$.

_____

Objective 1 Practice Exercises

For extra help, see Examples 1–2 on pages 779–780 of your text.

Solve each equation.

1. $\sqrt{4x-19}=5$

1. _____

2. $\sqrt{7p+3}+5=0$

2. _____

3. $\sqrt{5r-4}-9=0$

3. _____

Objective 2   Solve radical equations that require additional steps.

Video Examples

Review these example for Objective 2:

3.   Solve $\sqrt{x+3}=x-3$.

Step 1 The radical is isolated on the left side of the equation.

Step 2 Square each side.
$$\left(\sqrt{x+3}\right)^2=(x-3)^2$$
$$x+3=x^2-6x+9$$

Step 3 Write the equation in standard form and solve.
$$0=x^2-7x+6$$
$$0=(x-1)(x-6)$$
$$x-1=0 \quad \text{or} \quad x-6=0$$
$$x=1 \quad \text{or} \quad\quad x=6$$

Step 4 Check each proposed solution in the original equation.

$$\sqrt{x+3}=x-3 \qquad\qquad \sqrt{x+3}=x-3$$
$$\sqrt{1+3}\overset{?}{=}1-3 \qquad\qquad \sqrt{6+3}\overset{?}{=}6-3$$
$$\sqrt{4}\overset{?}{=}-2 \qquad\qquad\qquad \sqrt{9}\overset{?}{=}3$$
$$2=-2 \quad \text{False} \qquad\qquad 3=3 \quad \text{True}$$

The solution set is $\{6\}$. The other proposed solution, 1, is extraneous.

5.   Solve $\sqrt{3x}-4=\sqrt{x-2}$.

$$\left(\sqrt{3x}-4\right)^2=\left(\sqrt{x-2}\right)^2$$
$$3x-8\sqrt{3x}+16=x-2$$
$$-8\sqrt{3x}=-2x-18$$
$$\left(-8\sqrt{3x}\right)^2=(-2x-18)^2$$
$$192x=4x^2+72x+324$$
$$0=4x^2-120x+324$$
$$0=4\left(x^2-30x+81\right)$$
$$0=4(x-3)(x-27)$$
$$x-3=0 \quad \text{or} \quad x-27=0$$
$$x=3 \quad \text{or} \quad\quad x=27$$

Now Try:

3.   Solve $\sqrt{x+11}=x-1$.

_____

5.   Solve $\sqrt{3x+4}=\sqrt{9x}-2$.

_____

Check

$$\sqrt{3x} - 4 = \sqrt{x-2}$$

$$\sqrt{3(3)} - 4 \overset{?}{=} \sqrt{3-2}$$

$$3 - 4 \overset{?}{=} \sqrt{1}$$

$$-1 = 1 \quad \text{False}$$

$$\sqrt{3x} - 4 = \sqrt{x-2}$$

$$\sqrt{3(27)} - 4 \overset{?}{=} \sqrt{27-2}$$

$$9 - 4 \overset{?}{=} \sqrt{25}$$

$$5 = 5 \quad \text{True}$$

The proposed solution, 27, is valid, but 3 is extraneous and must be rejected. The solution set is {27}.

Objective 2 Practice Exercises

For extra help, see Examples 3–5 on pages 781–782 of your text.

Solve each equation.

4. $\sqrt{27 - 18v} = 2v - 3$

4. _____

5. $\sqrt{7r + 8} - \sqrt{r + 1} = 5$

5. _____

6. $\sqrt{k + 10} + \sqrt{2k + 19} = 2$

6. _____

Objective 3   Solve radical equations with indexes greater than 2.

Video Examples

Review this example for Objective 3:

6. Solve $\sqrt[3]{5r-6}=\sqrt[3]{3r+4}$.

$$\left(\sqrt[3]{5r-6}\right)^3=\left(\sqrt[3]{3r+4}\right)^3$$

$$5r-6=3r+4$$

$$2r=10$$

$$r=5$$

Check $\sqrt[3]{5r-6}=\sqrt[3]{3r+4}$

$$\sqrt[3]{5(5)-6}\overset{?}{=}\sqrt[3]{3(5)+4}$$

$$\sqrt[3]{19}=\sqrt[3]{19}\quad\text{True}$$

The solution set is $\{5\}$.

Now Try:

6. Solve $\sqrt[4]{8x+5}=\sqrt[4]{7x+7}$.

_____

Objective 3 Practice Exercises

For extra help, see Example 6 on page 783 of your text.

Solve each equation.

7. $\sqrt[3]{2a-63}+5=0$

7. _____

8. $\sqrt[5]{5a+1}-\sqrt[5]{2a-11}=0$

8. _____

9. $\sqrt[4]{15x-12} = \sqrt[4]{24-9x}$

9. _____

Objective 4 Use the power rule to solve a formula for a specified variable.

Video Examples

Review this example for Objective 4:

7. Solve the formula $d = \sqrt{\dfrac{H}{1.6n}}$ for n.

$$d^2 = \left(\sqrt{\dfrac{H}{1.6n}}\right)^2$$

$$d^2 = \dfrac{H}{1.6n}$$

$$1.6d^2 n = H$$

$$n = \dfrac{H}{1.6d^2}$$

Now Try:

7. Solve the formula $r = \sqrt{\dfrac{3v}{\pi h}}$ for h.

_____

Objective 4 Practice Exercises

For extra help, see Example 7 on page 783 of your text.

Solve each equation for the indicated variable.

10. $Z = \sqrt{\dfrac{L}{C}}$, for L

10. _____

11. $f = \dfrac{1}{2\pi\sqrt{LC}}$, for C

12. $N = \dfrac{1}{2\pi}\sqrt{\dfrac{a}{r}}$, for r

Chapter 10 ROOTS, RADICALS, AND ROOT FUNCTIONS

10.7   Complex Numbers

Learning Objectives
1   Simplify numbers of the form $\sqrt{-b}$, where $b > 0$.
2   Recognize subsets of the complex numbers.
3   Add and subtract complex numbers.
4   Multiply complex numbers.
5   Divide complex numbers.
6   Simplify powers of i.

Key Terms

Use the vocabulary terms listed below to complete each statement in exercises 1–7.

> **complex number**   **real part**   **imaginary part**
>
> **pure imaginary number**   **standard form (of a complex number)**
>
> **nonreal complex number**   **complex conjugate**

1. A _____ is a number that can be written in the form $a + bi$, where a and b are real numbers.

2. The _____ of $a + bi$ is $a - bi$.

3. The _____ of $a + bi$ is bi.

4. The _____ of $a + bi$ is a.

5. A complex number is in _____ if it is written in the form $a + bi$.

6. A complex number $a + bi$ with $a = 0$ and $b \neq 0$ is called a

   _____.

7. A complex number $a + bi$ $b \neq 0$ is called a _____.

Objective 1   Simplify numbers of the form $\sqrt{-b}$, where $b > 0$.

Video Examples

Review these examples for Objective 1:	Now Try:
1a. Write the number as a product of a real number and i.	1a. Write the number as a product of a real number and i.
$\sqrt{-36}$	$\sqrt{-16}$
$\sqrt{-36} = i\sqrt{36} = 6i$	_____

2b. Multiply.

$$\sqrt{-6} \cdot \sqrt{-7}$$

$$\sqrt{-6} \cdot \sqrt{-7} = i\sqrt{6} \cdot i\sqrt{7}$$

$$= i^2\sqrt{6 \cdot 7}$$

$$= (-1)\sqrt{42}$$

$$= -\sqrt{42}$$

3a. Divide.

$$\frac{\sqrt{-125}}{\sqrt{-5}}$$

$$\frac{\sqrt{-125}}{\sqrt{-5}} = \frac{i\sqrt{125}}{i\sqrt{5}}$$

$$= \sqrt{\frac{125}{5}}$$

$$= \sqrt{25}$$

$$= 5$$

2b. Multiply.

$$\sqrt{-5} \cdot \sqrt{-6}$$

_____

3a. Divide.

$$\frac{\sqrt{-200}}{\sqrt{-8}}$$

_____

Objective 1 Practice Exercises

For extra help, see Examples 1–3 on pages 788–789 of your text.

Write the number as a product of a real number and i. Simplify all radical expressions.

1. $-\sqrt{-162}$

1. _____

Multiply or divide as indicated.

2. $\sqrt{-5} \cdot \sqrt{-3} \cdot \sqrt{-7}$

2. _____

3. $\dfrac{\sqrt{-42} \cdot \sqrt{-6}}{\sqrt{-7}}$

3. _____

Name: Date:
Instructor: Section:

Objective 2   Recognize subsets of the complex numbers.

For extra help, see page 790 of your text.

Objective 3   Add and subtract complex numbers.

Video Examples

Review these examples for Objective 3:

4a. Add.

$$(2+9i)+(10-3i)$$

$$(2+9i)+(10-3i) = (2+10)+(9-3)i$$
$$= 12+6i$$

5b. Subtract.

$$(8+3i)-(5+3i)$$

$$(8+3i)-(5+3i) = (8-5)+(3i-3i)$$
$$= 3$$

Now Try:

4a. Add.

$$(4-7i)+(6-2i)$$

_____

5b. Subtract.

$$(2-5i)-(2-3i)$$

_____

Objective 3 Practice Exercises

For extra help, see Examples 4–5 on pages 791 of your text.

Add or subtract as indicated. Write answers in standard form.

4. $(-7-2i)-(-3-3i)$

4. _____

5. $4i-(9+5i)+(2+3i)$

5. _____

6. $(7-9i)-(5-6i)$

6. _____

Objective 4   **Multiply complex numbers.**

Video Examples

Review this example for Objective 4:

6a. Multiply.

     $6i(2-7i)$

     $6i(2-7i) = 6i(2) + 6i(-7i)$

                  $= 12i - 42i^2$

                  $= 12i - 42(-1)$

                  $= 42 + 12i$

Now Try:

6a. Multiply.

     $2i(4+7i)$

             _____

Objective 4 Practice Exercises

For extra help, see Example 6 on pages 791–792 of your text.

Multiply.

     7. $(2-5i)(2+5i)$

                                                              7. _____

     8. $(1+3i)^2$

                                                              8. _____

     9. $(12+2i)(-1+i)$

                                                              9. _____

Objective 5   Divide complex numbers.

Video Examples

Review this example for Objective 5:

7a. Find the quotient.

$$\frac{6-i}{2-3i}$$

Multiply the numerator and denominator by $2 + 3i$, the conjugate of the denominator.

$$\frac{6-i}{2-3i} = \frac{(6-i)(2+3i)}{(2-3i)(2+3i)}$$

$$= \frac{12+18i-2i-3i^2}{2^2+3^2}$$

$$= \frac{12+16i-3(-1)}{4+9}$$

$$= \frac{15+16i}{13}, \text{ or } \frac{15}{13} + \frac{16}{13}i$$

Now Try:

7a. Find the quotient.

$$\frac{4+i}{5-2i}$$

_____

Objective 5 Practice Exercises

For extra help, see Example 7 on pages 792–793 of your text.

Write each quotient in the form a + bi.

10. $\dfrac{3-2i}{2+i}$

10. _____

11. $\dfrac{5+2i}{9-4i}$

11. _____

12. $\dfrac{6-i}{2-3i}$ 12. _____

Objective 6   Simplify powers of i.

Video Examples

Review this example for Objective 6:
8a. Find the power of i.

i^{100}

$i^{100} = (i^4)^{25} = 1^{25} = 1$

Now Try:
8a. Find the power of i.

i^{48}

_____

Objective 6 Practice Exercises

For extra help, see Example 8 on page 793 of your text.

Find each power of i.

13. i^{14} 13. _____

14. i^{113} 14. _____

15. i^{-21} 15. _____

Chapter 11 QUADRATIC EQUATIONS, INEQUALITIES, AND FUNCTIONS

11.1   Solving Quadratic Equations by the Square Root Property

Learning Objectives
1   Review the zero-factor property.
2   Solve quadratic equations of the form $x^2 = k$, where $k > 0$.
3   Solve quadratic equations of the form $(ax + b)^2 = k$, where $k > 0$.
4   Solve quadratic equations with nonreal complex solutions.

Key Terms

Use the vocabulary terms listed below to complete each statement in exercises 1–2.

        quadratic equation   **zero-factor property**

1.   An equation that can be written in the form $ax^2 + bx + c = 0$ is a

       _____.

2.   The _____ states that if a product equals 0, then at least one of the factors of the product also equals zero.

Objective 1   Review the zero-factor property.

Video Examples

Review this example for Objective 1:

1a. Solve $2x^2 + 5x - 3 = 0$.

    Use the zero factor property.
$$2x^2 + 5x - 3 = 0$$
$$(2x - 1)(x + 3) = 0$$
$$2x - 1 = 0 \quad \text{or} \quad x + 3 = 0$$
$$2x = 1 \quad \text{or} \quad x = -3$$
$$x = \frac{1}{2}$$

    The solution set is $\left\{ -3, \frac{1}{2} \right\}$.

Now Try:

1a. Solve $3x^2 - 5x - 28 = 0$.

_____

Objective 1 Practice Exercises

For extra help, see Example 1 on page 810 of your text.

Solve each equation using the zero-factor property.

1. $15s^2 - 2 = s$

1. _____

2. $z^2 = 6z - 9$

2. _____

3. $16m^2 - 64 = 0$

3. _____

Objective 2   Solve quadratic equations of the form $x^2 = k$, **where** $k > 0$.

Video Examples

Review these examples for Objective 2:

2.   Solve each equation.

b. $m^2 - 52 = 0$

$m^2 - 52 = 0$

$m^2 = 52$

$m = \sqrt{52}$ or $m = -\sqrt{52}$

$m = 2\sqrt{13}$ or $m = -2\sqrt{13}$

The solution set is $\{2\sqrt{13}, -2\sqrt{13}\}$, or $\{\pm 2\sqrt{13}\}$.

c. $5p^2 - 100 = 0$

$5p^2 - 100 = 0$

$5p^2 = 100$

$p^2 = 20$

$p = \sqrt{20}$ or $p = -\sqrt{20}$

$p = 2\sqrt{5}$ or $p = -2\sqrt{5}$

The solution set is $\{2\sqrt{5}, -2\sqrt{5}\}$, or $\{\pm 2\sqrt{5}\}$.

Now Try:

2.   Solve each equation.

b. $n^2 - 40 = 0$

_____

c. $3x^2 - 54 = 0$

_____

Objective 2 Practice Exercises

For extra help, see Examples 2–3 on pages 811–812 of your text.

Solve each equation by using the square root property. Express all radicals in simplest form.

4. $r^2 = 900$

4. _____

5. $x^2 - 98 = 0$

5. _____

6. $3d^2 - 500 = 250$

6. _____

Objective 3   **Solve quadratic equations of the form** $(ax + b)^2 = k$, **where** $k > 0$.

Video Examples

Review these examples for Objective 3:

5.   Solve $(3x + 4)^2 = 40$.

$$(3x + 4)^2 = 40$$
$$3x + 4 = \sqrt{40} \qquad \text{or} \quad 3x + 4 = -\sqrt{40}$$
$$3x = -4 + \sqrt{40} \quad \text{or} \qquad 3x = -4 - \sqrt{40}$$
$$x = \frac{-4 + 2\sqrt{10}}{3} \quad \text{or} \qquad x = \frac{-4 - 2\sqrt{10}}{3}$$

Now Try:

5.   Solve $(2x + 5)^2 = 32$.

_____

Check

$$(3x+4)^2 = 40$$

$$\left[3\left(\frac{-4-2\sqrt{10}}{3}\right)+4\right]^2 \overset{?}{=} 40$$

$$\left(-4-2\sqrt{10}+4\right)^2 \overset{?}{=} 40$$

$$\left(-2\sqrt{10}\right)^2 \overset{?}{=} 40$$

$$40 = 40$$

The check for the second solution is similar.

The solution sets is $\left\{\dfrac{-4-2\sqrt{10}}{3}, \dfrac{-4+2\sqrt{10}}{3}\right\}$.

4b. Solve $(x+3)^2 = 10$.

$$(x+3)^2 = 10$$

$$x+3 = \sqrt{10} \quad \text{or} \quad x+3 = -\sqrt{10}$$

$$x = -3+\sqrt{10} \quad \text{or} \quad x = -3-\sqrt{10}$$

Check

$$\left(-3+\sqrt{10}+3\right)^2 = \left(\sqrt{10}\right)^2 = 10$$

$$\left(-3-\sqrt{10}+3\right)^2 = \left(-\sqrt{10}\right)^2 = 10$$

The solution set is $\left\{-3-\sqrt{10},\ -3+\sqrt{10}\right\}$.

4b. Solve $(x+5)^2 = 14$.

_____

Objective 3 Practice Exercises

For extra help, see Examples 4–5 on pages 812–813 of your text.

Solve each equation by using the square root property. Express all radicals in simplest form.

7. $(y+2)^2 = 16$

7. _____

8. $(q-4)^2 - 7 = 0$ **8.** _____

9. $(3f+4)^2 = 32$ **9.** _____

Objective 4   Solve quadratic equations with nonreal complex solutions.

Video Examples

Review these examples for Objective 4: | **Now Try:**

6.   Solve each equation.

a. $x^2 = -48$

$$x^2 = -48$$
$$x = \sqrt{-48} \quad \text{or} \quad x = -\sqrt{-48}$$
$$x = 4i\sqrt{3} \quad \text{or} \quad x = -4i\sqrt{3}$$

The solution set is $\{-4i\sqrt{3},\ 4i\sqrt{3}\}$.

b. $(x-3)^2 = -25$

$$(x-3)^2 = -25$$
$$x-3 = \sqrt{-25} \quad \text{or} \quad x-3 = -\sqrt{-25}$$
$$x-3 = 5i \quad \text{or} \quad x-3 = -5i$$
$$x = 3+5i \quad \text{or} \quad x = 3-5i$$

The solution set is $\{3-5i,\ 3+5i\}$.

6.   Solve each equation.

a. $y^2 = -32$

_____

b. $(x+2)^2 = -49$

_____

Objective 4 Practice Exercises

For extra help, see Example 6 on page 813 of your text.

Find the complex solutions of each equation.

10. $(10m-5)^2 + 9 = 0$ 10. _____

11. $(m+1)^2 = -36$ 11. _____

12. $(x-1)^2 + 2 = 0$ 12. _____

Chapter 11   QUADRATIC EQUATIONS, INEQUALITIES, AND FUNCTIONS

11.2   Solving Quadratic Equations by Completing the Square

Learning Objectives
1   Solve quadratic equations by completing the square when the coefficient of the second-degree term is 1.
2   Solve quadratic equations by completing the square when the coefficient of the second-degree term is not 1.
3   Simplify the terms of an equation before solving.

Key Terms

Use the vocabulary terms listed below to complete each statement in exercises 1–3.

completing the square   **perfect square trinomial**   **square root property**

1.   A _____ can be written in the form $x^2 + 2kx + k^2$ or $x^2 - 2kx + k^2$

2.   The _____ says that, if k is positive and $a^2 = k$, then $a = \pm\sqrt{k}$.

3.   Use the process called _____ in order to rewrite an equation so it can be solved using the square root property.

Objective 1   Solve quadratic equations by completing the square when the coefficient of the second-degree term is 1.

Video Examples

Review these examples for Objective 1:	Now Try:
3. Solve $x^2 + 5x + 2 = 0$.	3. Solve $x^2 - 11x + 8 = 0$.

Since the coefficient of the second degree term is 1, begin with Step 2.

Step 2 $x^2 + 5x = -2$

Step 3 Take half the coefficient of the first-degree term and square the result.

$$\left[\frac{1}{2}(5)\right]^2 = \left(\frac{5}{2}\right)^2 = \frac{25}{4}$$

Step 4 $x^2 + 5x + \frac{25}{4} = -2 + \frac{25}{4}$

Step 5 $\left(x + \frac{5}{2}\right)^2 = \frac{17}{4}$

Step 6

$$x + \frac{5}{2} = \sqrt{\frac{17}{4}} \qquad \text{or} \quad x + \frac{5}{2} = -\sqrt{\frac{17}{4}}$$

$$x + \frac{5}{2} = \frac{\sqrt{17}}{2} \qquad \text{or} \quad x + \frac{5}{2} = -\frac{\sqrt{17}}{2}$$

$$x = -\frac{5}{2} + \frac{\sqrt{17}}{2} \quad \text{or} \qquad x = -\frac{5}{2} - \frac{\sqrt{17}}{2}$$

A check shows the solution set is

$$\left\{ -\frac{5}{2} - \frac{\sqrt{17}}{2}, -\frac{5}{2} + \frac{\sqrt{17}}{2} \right\}.$$

2. Solve $x^2 - 8x = 30$.

Take half the coefficient of x and square it.

$$\left[\frac{1}{2}(-8) \right]^2 = (-4)^2 = 16$$

Add the result, 16, to each side of the equation.

$$x^2 - 8x + 16 = 30 + 16$$

$$(x - 4)^2 = 46$$

$$x - 4 = \sqrt{46} \qquad \text{or} \quad x - 4 = -\sqrt{46}$$

$$x = 4 + \sqrt{46} \quad \text{or} \qquad x = 4 - \sqrt{46}$$

A check indicates the solution set is

$$\left\{ 4 - \sqrt{46},\ 4 + \sqrt{46} \right\}.$$

2. Solve $x^2 - 10x = 10$.

2. _____

Objective 1 Practice Exercises

For extra help, see Examples 1–3 on pages 817–819 of your text.

Solve each equation by completing the square.

1. $r^2 + 8r = -4$

1. _____

2. $x^2 - 4x = 2$

2. _____

3. $x^2 - 9x + 8 = 0$ 3. _____

Objective 2 **Solve quadratic equations by completing the square when the coefficient of the second-degree term is not 1.**

Video Examples

Review these examples for Objective 2:	**Now Try:**
4. Solve $4n^2 + 4n = 15$.	**4.** Solve $9x^2 + 6x - 8 = 0$.

4. Solve $4n^2 + 4n = 15$.

Step 1 Divide each side by 4, to get a 1 as the coefficient of x^2.

$$n^2 + n = \frac{15}{4}$$

Step 2 The equation is already in the correct form.

Step 3 Complete the square.

$$\left[\frac{1}{2}(1)\right]^2 = \left(\frac{1}{2}\right)^2 = \frac{1}{4}$$

Step 4 $n^2 + n + \frac{1}{4} = \frac{15}{4} + \frac{1}{4}$

Step 5 $\left(n + \frac{1}{2}\right)^2 = 4$

Step 6 Solve using the square root property.

$$n + \frac{1}{2} = \sqrt{4} \quad \text{or} \quad n + \frac{1}{2} = -\sqrt{4}$$

$$n + \frac{1}{2} = 2 \quad \text{or} \quad n + \frac{1}{2} = -2$$

$$n = \frac{3}{2} \quad \text{or} \quad n = -\frac{5}{2}$$

Check by substituting each value of n in to the original equation. The solution set is $\left\{-\frac{5}{2}, \frac{3}{2}\right\}$.

5. Solve $3x^2 - 6x - 2 = 0$.

$$x^2 - 2x - \frac{2}{3} = 0 \qquad \text{Step 1}$$

$$x^2 - 2x = \frac{2}{3} \qquad \text{Step 2}$$

$$\left[\frac{1}{2}(-2)\right]^2 = (-1)^2 = 1 \qquad \text{Step 3}$$

$$x^2 - 2x + 1 = \frac{2}{3} + 1 \qquad \text{Step 4}$$

$$(x-1)^2 = \frac{5}{3} \qquad \text{Step 5}$$

$$x - 1 = \sqrt{\frac{5}{3}} \quad \text{or} \quad x - 1 = -\sqrt{\frac{5}{3}} \qquad \text{Step 6}$$

$$x = 1 + \sqrt{\frac{5}{3}} \quad \text{or} \quad x = 1 - \sqrt{\frac{5}{3}}$$

$$x = 1 + \frac{\sqrt{15}}{3} \quad \text{or} \quad x = 1 - \frac{\sqrt{15}}{3}$$

$$x = \frac{3 + \sqrt{15}}{3} \quad \text{or} \quad x = \frac{3 - \sqrt{15}}{3}$$

The solution set is $\left\{\dfrac{3 - \sqrt{15}}{3},\ \dfrac{3 + \sqrt{15}}{3}\right\}$.

5. Solve $2p^2 + 6p - 1 = 0$.

_____

Objective 2 Practice Exercises

For extra help, see Examples 4–6 on pages 820–822 of your text.

Solve each equation by completing the square.

4. $3t^2 + t - 2 = 0$

4. _____

5. $4q^2 + 8q = -5$

5. _____

6. $6q^2 + 4q = 1$

6. _____

Objective 3   Simplify the terms of an equation before solving.

Video Examples

Review this example for Objective 3:

7.   Solve $(x - 7)(x - 5) = 32$.

$$(x - 7)(x - 5) = 32$$
$$x^2 - 12x + 35 = 32$$
$$x^2 - 12x = -3$$
$$x^2 - 12x + 36 = -3 + 36$$
$$(x - 6)^2 = 33$$
$$x - 6 = \sqrt{33} \quad \text{or} \quad x - 6 = -\sqrt{33}$$
$$x - 6 + \sqrt{33} \quad \text{or} \quad x = 6 - \sqrt{33}$$

The solution set is $\{6 + \sqrt{33}, \ 6 - \sqrt{33}\}$.

Now Try:

7.   Solve $(s + 3)(s + 1) = 1$.

_____

Objective 3 Practice Exercises

For extra help, see Example 7 on page 822 of your text.

Simplify each of the following equations and then solve by completing the square.

7. $2z^2 = 6z + 3 - 4z^2$

7. _____

8. $(b-1)(b+7)=9$

9. $(5m+2)^2-9(5m+2)-36=0$

Chapter 11 QUADRATIC EQUATIONS, INEQUALITIES, AND FUNCTIONS

11.3 Solving Quadratic Equations by the Quadratic Formula

Learning Objectives	
1	Derive the quadratic formula.
2	Solve quadratic equations by using the quadratic formula.
3	Use the discriminant to determine the number and type of solutions.

Key Terms

Use the vocabulary terms listed below to complete each statement in exercises 1–2.

quadratic formula   **discriminant**

1. The expression under the radical in the quadratic formula is called the

    _____ .

2. The formula $x = \dfrac{-b \pm \sqrt{b^2 - 4ac}}{2a}$ is called the _____ .

Objective 1   Derive the quadratic formula.

For extra help, see pages 825–826 of your text.

Objective 2   Solve quadratic equations by using the quadratic formula.

Video Examples

Review these examples for Objective 2:

3. Solve $4x^2 = -4x + 1$.

    First write the equation in standard
    form as $4x^2 + 4x - 1 = 0$.
    $a = 4$, $b = 4$, $c = -1$

    $$x = \frac{-b \pm \sqrt{b^2 - 4ac}}{2a}$$

    $$x = \frac{-4 \pm \sqrt{(4)^2 - 4(4)(-1)}}{2(4)}$$

    $$x = \frac{-4 \pm \sqrt{16 + 16}}{8}$$

    $$x = \frac{-4 \pm \sqrt{32}}{8}$$

    $$x = \frac{-4 \pm 4\sqrt{2}}{8}$$

Now Try:

3. Solve $2x^2 = 2x + 3$.

    _____

$$x = \frac{4(-1 \pm \sqrt{2})}{4(2)}$$

$$x = \frac{-1 \pm \sqrt{2}}{2}$$

The solution set is $\left\{\dfrac{-1-\sqrt{2}}{2}, \dfrac{-1+\sqrt{2}}{2}\right\}$.

4. Solve $(5x-2)(x+2) = -9$.

$$(5x-2)(x+2) = -9$$

$$5x^2 + 8x - 4 = -9$$

$$5x^2 + 8x + 5 = 0$$

From the standard form, we identify
$a = 5$, $b = 8$, and $c = 5$.

$$x = \frac{-b \pm \sqrt{b^2 - 4ac}}{2a}$$

$$x = \frac{-8 \pm \sqrt{(8)^2 - 4(5)(5)}}{2(5)}$$

$$x = \frac{-8 \pm \sqrt{-36}}{10}$$

$$x = \frac{-8 \pm 6i}{10}$$

$$x = \frac{2(-4 \pm 3i)}{2(5)}$$

$$x = \frac{-4 \pm 3i}{5} = -\frac{4}{5} \pm \frac{3}{5}i$$

The solution set is $\left\{-\dfrac{4}{5}-\dfrac{3}{5}i, -\dfrac{4}{5}+\dfrac{3}{5}i\right\}$.

4. Solve $(2x-6)(x+1) = -16$.

_____

Objective 2 Practice Exercises

For extra help, see Examples 1–4 on pages 826–829 of your text.

Use the quadratic formula to solve each equation.

1. $(z+2)^2 = 2(5z-2)$

1. _____

2. $5k^2 + 4k - 2 = 0$

2. _____

3. $34 - 10x = x^2$

3. _____

Objective 3   Use the discriminant to determine the number and type of solutions.

Video Examples

Review these examples for Objective 3:

5. Find the discriminant. Use it to predict the number and type of solutions for each equation. Then tell whether the equation can be solved by factoring or whether the quadratic formula should be used.

c. $16x^2 + 25 = 40x$

Write in standard form: $16x^2 - 40x + 25 = 0$.
$a = 16$, $b = -40$, and $c = 25$

$$b^2 - 4ac = (-40)^2 - 4(16)(25)$$
$$= 1600 - 1600$$
$$= 0$$

Because the discriminant is 0, this quadratic equation will have one distinct rational solution. The equation can be solved by factoring.

Now Try:

5. Find the discriminant. Use it to predict the number and type of solutions for each equation. Then tell whether the equation can be solved by factoring or whether the quadratic formula should be used.

c. $25x^2 + 9 = 30x$

_____

b. $6x^2 - 3x = 10$

Write in standard form: $6x^2 - 3x - 10 = 0$.
$a = 6$, $b = -3$, and $c = -10$

$$b^2 - 4ac = (-3)^2 - 4(6)(-10)$$
$$= 9 + 240$$
$$= 249$$

Because 249 is positive but not the square of an integer and a, b, and c are integers, the equation will have two irrational solutions and is best solved using the quadratic formula.

a. $3x^2 + x - 2 = 0$

First identify the values of a, b, and c.
$a = 3$, $b = 1$, and $c = -2$.
Then find the discriminant.

$$b^2 - 4ac = 1^2 - 4(3)(-2)$$
$$= 1 + 24$$
$$= 25, \text{ or } 5^2$$

Since a, b, and c are integers and the discriminant 25 is a perfect square, there will be two rational solutions. The equation can be solved by factoring.

d. $5y^2 - 5y + 2 = 0$

$a = 5$, $b = -5$, and $c = 2$
$$b^2 - 4ac = (-5)^2 - 4(5)(2)$$
$$= 25 - 40$$
$$= -15$$

Because the discriminant is negative and a, b, and c are integers, this quadratic equation will have two nonreal complex solutions. The quadratic equation should be used to solve it.

b. $2x^2 - 4x = 8$

_____

a. $10x^2 + 21x + 9 = 0$

_____

d. $2y^2 + 4y + 8 = 0$

_____

Objective 3 Practice Exercises

For extra help, see Example 5 on page 830 of your text.

Use the discriminant to determine whether the solutions for each equation are

    A. *two rational numbers*   B. *one rational number,*
    C. *two irrational numbers*   D. *two imaginary numbers.*

Do not actually solve.

4. $m^2 - 4m + 4 = 0$ 4. _____

5. $z^2 + 6z + 3 = 0$ 5. _____

6. $16x^2 - 12x + 9 = 0$ 6. _____

Chapter 11 QUADRATIC EQUATIONS, INEQUALITIES, AND FUNCTIONS

11.4 Equations Quadratic in Form

Learning Objectives
1   Solve rational equations that lead to quadratic equations.
2   Solve applied problems involving quadratic equations.
3   Solve radical equations that lead to quadratic equations.
4   Solve equations that are quadratic in form.

Key Terms

Use the vocabulary terms listed below to complete each statement in exercises 1–2.

quadratic in form   **standard form**

1. A quadratic equation written in the form $ax^2 + bx + c = 0$, $a \neq 0$ is written in

   _____.

2. A nonquadratic equation that can be written as a quadratic equation is called

   _____.

Objective 1   Solve rational equations that lead to quadratic equations.

Video Examples

Review this example for Objective 1:

1. Solve $5 + \dfrac{6}{m+1} = \dfrac{14}{m}$.

   Multiply each side by the least common denominator, $m(m + 1)$. The domain must be restricted to $m \neq 0$, $m \neq -1$.

   $$5 + \frac{6}{m+1} = \frac{14}{m}$$

   $$m(m+1)\left(5 + \frac{6}{m+1}\right) = m(m+1)\frac{14}{m}$$

   $$m(m+1)(5) + m(m+1)\frac{6}{m+1} = m(m+1)\frac{14}{m}$$

   $$5m^2 + 5m + 6m = 14m + 14$$

   $$5m^2 + 11m = 14m + 14$$

   $$5m^2 - 3m - 14 = 0$$

   $$(5m+7)(m-2) = 0$$

   $$5m + 7 = 0 \quad \text{or} \quad m - 2 = 0$$

   $$m = -\frac{7}{5} \quad \text{or} \quad m = 2$$

   The solution set is $\left\{-\dfrac{7}{5},\ 2\right\}$.

Now Try:

1. Solve $4 - \dfrac{8}{x-1} = -\dfrac{35}{x}$.

   _____

Objective 1 Practice Exercises

For extra help, see Example 1 on page 833 of your text.

Solve each equation. Check your solutions.

1. $\dfrac{5}{x} + \dfrac{1}{2x+7} = -\dfrac{2}{3}$

    1. _____

2. $\dfrac{2m}{m-5} + \dfrac{7}{m+1} = 0$

    2. _____

3. $1 + \dfrac{49}{2x} - \dfrac{15}{x+1}$

    3. _____

Objective 2   Solve applied problems involving quadratic equations.

Video Examples

Review these examples for Objective 2:

2.   Amy rows her boat 6 miles upstream and then returns in $2\frac{6}{7}$ hours. The speed of the current is 2 miles per hour. How fast can she row?

Step 1   Read the problem carefully.

Step 2   Assign a variable. Let x = the rate that Amy rows in still water. The current slows Amy when she is going upstream, so Amy's rate going upstream is her rate in still water less the rate of the current, or $x - 2$. Similarly, the current makes Amy row faster when she is going downstream, so her downstream rate is $x + 2$.

Complete a table. Recall that $d = rt$.

	d	r	t
Upstream	6	$x - 2$	$\dfrac{6}{x-2}$
Downstream	6	$x + 2$	$\dfrac{6}{x+2}$

Step 3   Write an equation. The time upstream plus the time downstream equals the total time, $2\frac{6}{7}$ hours, or $\frac{20}{7}$ hours.

$$\frac{6}{x-2} + \frac{6}{x+2} = \frac{20}{7}$$

Step 4   Solve the equation. The LCD is $7(x-2)(x+2)$.

$$7(x-2)(x+2)\left(\frac{6}{x-2} + \frac{6}{x+2}\right)$$
$$= 7(x-2)(x+2)\left(\frac{20}{7}\right)$$

$$7(x+2)6 + 7(x-2)6 = (x-2)(x+2)20$$
$$42(x+2) + 42(x-2) = 20(x^2 - 4)$$
$$42x + 84 + 42x - 84 = 20x^2 - 80$$
$$20x^2 - 84x - 80 = 0$$
$$4(5x^2 - 21x - 20) = 0$$
$$4(5x + 4)(x - 5) = 0$$
$$5x + 4 = 0 \quad \text{or} \quad x - 5 = 0$$
$$x = -\frac{4}{5} \quad \text{or} \quad x = 5$$

Now Try:

2.   Mike can row 3 miles per hour in still water. It takes him 3 hours and 36 minutes to row 3 miles round trip, upstream and back. Find the speed of the current.

           _____

Step 5 State the answer. The rate cannot be $-\dfrac{4}{5}$ mph, so the answer is Amy rows at 5 mph.

Step 6 Check that this value satisfies the original equation.

3. It takes two painters working together $4\dfrac{4}{5}$ hours to paint a house. If each worked alone, one of them could do the job in 4 hours less than the other. How long would it take each painter to complete the job alone?

Step 1 Read the problem carefully. There will be two answers.

Step 2 Assign a variable. Let $x =$ the number of hours for the slower painter to complete the job alone. Then the faster painter could do the entire job in $x - 4$ hours. The faster painter's rate is $\dfrac{1}{x-4}$ and the slower painter's rate is $\dfrac{1}{x}$.

Together, they do the job in $4\dfrac{4}{5}$ hours.

Complete a table.

	Rate	Time working together	Fractional part of the job done
Slower painter	$\dfrac{1}{x}$	$\dfrac{24}{5}$	$\dfrac{1}{x}\left(\dfrac{24}{5}\right)$
Faster painter	$\dfrac{1}{x-4}$	$\dfrac{24}{5}$	$\dfrac{1}{x-4}\left(\dfrac{24}{5}\right)$

Step 3 Write an equation. The sum of the two fractional parts is 1.

$$\frac{24}{5x} + \frac{24}{5(x-4)} = 1$$

Step 4 Solve the equation. The LCD is $5x(x-4)$.

$$5x(x-4)\left(\frac{24}{5x} + \frac{24}{5(x-4)}\right) = 5x(x-4)(1)$$

$$24(x-4) + 24x = 5x(x-4)$$

$$24x - 96 + 24x = 5x^2 - 20x$$

$$48x - 96 = 5x^2 - 20x$$

$$5x^2 - 68x + 96 = 0$$

$$(5x - 8)(x - 12) = 0$$

3. Working together, Tom and Huck painted a fence in 8 hours. If each worked alone, Tom could paint the fence 12 hours faster than Huck could. How long would it take each to paint the fence alone?

_____

$$5x - 8 = 0 \quad \text{or} \quad x - 12 = 0$$
$$x = \frac{8}{5} \quad \text{or} \quad x = 12$$

Step 5 State the answer.

If the slower painter can do the job in $\frac{8}{5}$ hours

(or 1.6 hours), then the faster painter's time to do the job is 1.6 – 4 = –2.4 hr, which cannot represent the time for the slower painter.
If the slower painter can do the job in 12 hours, then the faster painter's time is 12 – 4 = 8 hours.

Step 6 Check that these results satisfy the original problem.

Objective 2 Practice Exercises

For extra help, see Examples 2–3 on pages 833–836 of your text.

Solve each problem. Round answers to the nearest tenth, if necessary.

4. Two pipes together can fill a large tank in 10 hours. One of the pipes, used alone, takes 15 hours longer than the other to fill the tank. How long would each pipe used alone take to fill the tank?

4. pipe 1 _____

pipe 2 _____

5. A jet plane traveling at a constant speed goes 1200 miles with the wind, then turns around and travels for 1000 miles against the wind. If the speed of the wind is 50 miles per hour and the total flight takes 4 hours, find the speed of the plane.

5. _____

6. A man rode a bicycle for 12 miles and then hiked an additional 8 miles. The total time for the trip was 5 hours. If his rate when he was riding the bicycle was 10 miles per hour faster than his rate walking, what was each rate?

6. bike _____

hike _____

Objective 3   Solve radical equations that lead to quadratic equations.

Video Examples

Review this example for Objective 3:

4a. Solve the equation.

$$y = \sqrt{y+42}$$

Start by squaring each side.

$$y = \sqrt{y+42}$$
$$y^2 = \left(\sqrt{y+42}\right)^2$$
$$y^2 = y+42$$
$$y^2 - y - 42 = 0$$
$$(y+6)(y-7) = 0$$
$$y+6 = 0 \quad \text{or} \quad y-7 = 0$$
$$y = -6 \quad \text{or} \quad y = 7$$

We must check all proposed solutions in the original equation because squaring each side of an equation can introduce extraneous solutions.
Check

$$y = \sqrt{y+42} \qquad\qquad y = \sqrt{y+42}$$
$$-6 \overset{?}{=} \sqrt{-6+42} \qquad 7 \overset{?}{=} \sqrt{7+42}$$
$$-6 \overset{?}{=} \sqrt{36} \qquad\qquad 7 \overset{?}{=} \sqrt{49}$$
$$-6 = 6 \quad \text{False} \qquad 7 = 7 \quad \text{True}$$

The solution set is $\{7\}$.

Now Try:

4a. Solve the equation.

$$x = \sqrt{x+2}$$

_____

Objective 3 Practice Exercises

For extra help, see Example 4 on pages 836–837 of your text.

Solve each equation. Check your solutions.

7. $\sqrt{7y-10} = y$ 7. _____

8. $x = \sqrt{\dfrac{x+3}{2}}$ 8. _____

9. $\sqrt{4x} + 3 = x$ 9. _____

Name: Date:

Instructor: Section:

Objective 4   **Solve equations that are quadratic in form.**

Video Examples

Review this example for Objective 4:

6b. Solve the equation.

$$x^4 - 5x^2 + 4 = 0$$

$$x^4 - 5x^2 + 4 = 0$$

$$u^2 - 5u + 4 = 0 \quad \text{Let } u = x^2.$$

$$(u - 4)(u - 1) = 0$$

$$u - 4 = 0 \quad \text{or} \quad u - 1 = 0$$

$$u = 4 \quad \text{or} \quad u = 1$$

$$x^2 = 4 \quad \text{or} \quad x^2 = 1$$

$$x = \pm 2 \quad \text{or} \quad x = \pm 1$$

The solution set is $\{-2, -1, 1, 2\}$.

Now Try:

6b. Solve the equation.

$$16x^4 = 25x^2 - 9$$

_____

Objective 4 Practice Exercises

For extra help, see Examples 5–7 on pages 837–840 of your text.

Solve each equation. Check your solutions.

10. $4t^4 = 21t^2 - 5$ **10.** _____

11. $p^{4/3} - 12p^{2/3} + 27 = 0$ **11.** _____

Name:

Instructor:

Date:

Section:

12. $\left(t^2 - 3t\right)^2 = 14\left(t^2 - 3t\right) - 40$

12. _____

Chapter 11 QUADRATIC EQUATIONS, INEQUALITIES, AND FUNCTIONS

11.5   Formulas and Further Applications

Learning Objectives
1   Solve formulas involving squares and square roots for specified variables.
2   Solve applied problems using the Pythagorean theorem.
3   Solve applied problems using area formulas.
4   Solve applied problems using quadratic functions as models.

Key Terms

Use the vocabulary terms listed below to complete each statement in exercises 1–2.

quadratic function   **Pythagorean theorem**

1.   A function defined by $f(x) = ax^2 + bx + c$, for real numbers a, b, and c, with $a \neq 0$, is a _____.

2.   The _____ states that the sum of the squares of the lengths of the legs of a right triangle equals the square of the length of the hypotenuse.

Objective 1   Solve formulas involving squares and square roots for specified variables.

Video Examples

Review these examples for Objective 1:

1a. Solve the formula for the given variable.

$$y = \frac{1}{2}gt^2 \text{ for } t$$

The goal is to isolate t on one side.

$$y = \frac{1}{2}gt^2$$

$$2y = gt^2$$

$$\frac{2y}{g} = t^2$$

$$t = \pm\sqrt{\frac{2y}{g}}$$

$$t = \pm\frac{\sqrt{2y}}{\sqrt{g}} \cdot \frac{\sqrt{g}}{\sqrt{g}}$$

$$t = \pm\frac{\sqrt{2yg}}{g}$$

Now Try:

1a. Solve the formula for the given variable.

$$F = \frac{mx}{t^2} \text{ for } t$$

_____

2. Solve $rk^2 - 3k = -s$ for k.

Write the equation in standard form and then use the quadratic formula to solve for k.

$$rk^2 - 3k = -s$$

$$rk^2 - 3k + s = 0$$

Let $a = r$, $b = -3$, and $c = s$.

$$k = \frac{-(-3) \pm \sqrt{(-3)^2 - 4(r)(s)}}{2r}$$

$$k = \frac{3 \pm \sqrt{9 - 4rs}}{2r}$$

The solutions are

$$\frac{3 + \sqrt{9 - 4rs}}{2r} \text{ and } \frac{3 - \sqrt{9 - 4rs}}{2r}.$$

2. Solve $p^2q^2 + pkq = k^2$ for q.

_____

Objective 1 Practice Exercises

For extra help, see Examples 1–2 on pages 846–847 of your text.

Solve each equation for the indicated variable. (Leave $\pm$ in your answers.)

1. $F = \frac{kl}{\sqrt{d}}$ for d

1. _____

2. $p = \sqrt{\frac{kl}{g}}$ for k

2. _____

3. $b^2a^2 + 2bca = c^2$ for a

3. _____

Name: _____ Date: _____

Instructor: _____ Section: _____

Objective 2   Solve applied problems using the Pythagorean theorem.

Video Examples

Review this example for Objective 2:

3. A 13-foot ladder is leaning against a building. The distance from the bottom of the ladder to the building is 2 feet more than twice the distance from the top of the ladder to the ground. How far is the bottom of the ladder from the building?

Step 1 Read the problem carefully.

Step 2 Assign a variable. Let x – the distance from the top of the ladder to the ground. Then $2x + 2$ = the distance from the bottom of the ladder to the building. Draw a picture to represent the problem.

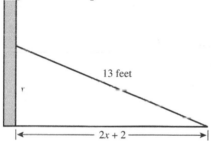

13 feet

$2x + 2$

Step 3 Write an equation. Use the Pythagorean theorem.

$$a^2 + b^2 = c^2$$

$$x^2 + (2x + 2)^2 = 13^2$$

Step 4 Solve.

$$x^2 + 4x^2 + 8x + 4 = 169$$

$$5x^2 + 8x - 165 = 0$$

$$(5x + 33)(x - 5) = 0$$

$$5x + 33 = 0 \quad \text{or} \quad x - 5 = 0$$

$$x = -\frac{33}{5} \quad \text{or} \quad x = 5$$

Step 5 State the answer. Length cannot be negative, so discard the negative solution. The distance from the top of the ladder to the ground is 5 feet. However, we are asked to find the distance from the bottom of the ladder to the building. This distance is $2(5) + 2 = 12$ feet.

Step 6 Check. Since $5^2 + 12^2 = 13^2$, the answer is correct.

Now Try:

3. Two cars left an intersection at the same time, one heading south, the other heading east. Sometime later, the car traveling south had gone 18 miles farther than the car headed east. At that time they were 90 miles apart. How far had each car traveled?

south _____

east _____

Objective 2 Practice Exercises

For extra help, see Example 3 on page 847 of your text.

Solve each problem.

4. A child flying a kite has let out 45 feet of string to the kite. The distance from the kite to the ground is 9 feet more than the distance from the child to a point directly below the kite. How high up is the kite?

4. _____

5. A ladder is leaning against a building so that the top is 8 feet above the ground. The length of the ladder is 2 feet less than twice the distance of the bottom of the ladder from the building. Find the length of the ladder.

5. _____

6. Two cars left an intersection at the same time, one heading north, the other heading west. Later they were exactly 95 miles apart. The car headed west had gone 38 miles less than twice as far as the car headed north. How far had each car traveled?

6. north _____

   west _____

Name:

Date:

Instructor:

Section:

Objective 3   Solve applied problems using area formulas.

Video Examples

Review this example for Objective 3:

4.   A fish pond is 3 feet by 4 feet. How wide a strip of concrete can be poured around the pool if there is enough concrete for 44 square feet?

Step 1 Read the problem carefully.

Step 2 Assign a variable. Let x = the width of the strip of concrete. Then $2x + 3$ = the width of the fish pond with the two strips of concrete and $2x + 4$ = the length of the fish pond with the two strips of concrete.

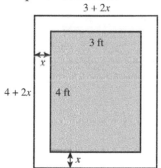

Step 3 Write an equation. The area of the strip is 44 sq ft and the area of the fish pond is $3(4) = 12$ sq ft, so the total area of the outer rectangle is $44 + 12 = 56$ sq ft.

$$(3 + 2x)(4 + 2x) = 56$$

Step 4 Solve.

$$(3 + 2x)(4 + 2x) = 56$$
$$12 + 14x + 4x^2 = 56$$
$$4x^2 + 14x - 44 = 0$$
$$2(2x^2 + 7x - 22) = 0$$
$$2(2x + 11)(x - 2) = 0$$
$$2x + 11 = 0 \quad \text{or} \quad x - 2 = 0$$
$$x = -\frac{11}{2} \quad \text{or} \quad x = 2$$

Step 5 State the answer. Since length cannot be negative, we disregard the negative solution. The concrete strip should be 2 ft wide.

Now Try:

4.   A picture 9 inches by 12 inches is to be mounted on a piece of mat board so that there is an even width of mat all around the picture. How wide will the matted border be if the area of the mounted picture is 238 square inches?

_____

Step 6 Check. If $x = 2$, then the area of the large
rectangle is $(3 + 2 \cdot 2)(4 + 2 \cdot 2) = 7(8) = 56$ sq ft.
The area of the fish pond is $3(4) = 12$ sq ft, so
the area of the concrete strip is $56 - 12 = 44$ sq ft.

Objective 3 Practice Exercises

For extra help, see Example 4 on page 848 of your text.

Solve each problem.

7. A rug is to fit in a room so that a border of even
   width is left on all four sides. If the room is
   16 feet by 20 feet and the area of the rug is 165
   square feet, how wide to the nearest tenth of a foot
   will the border be?

7. _____

8. A rectangular garden has an area of 12 feet by 5 feet.
   A gravel path of equal width is to be built around the
   garden. How wide can the path be if there is enough
   gravel for 138 square feet?

8. _____

9. A doghouse 2 feet by 4 feet is to be built with a
   cement path around it of equal width on all sides.
   The area available for the doghouse and path is 120
   square feet. How wide will the path be?

9. _____

Objective 4   Solve applied problems using quadratic functions as models.

Video Examples

Review this example for Objective 4:

5. A certain projectile is located at a distance of
    $d(t) = 3t^2 - 6t + 1$ feet from its starting point
    after t seconds. How many seconds will it take
    the projectile to travel 10 feet?

    Let $d = 10$ in the formula and solve for t.
    $$d - 3t^2 - 6t + 1$$
    $$10 = 3t^2 - 6t + 1$$
    $$9 = 3t^2 - 6t$$
    $$3 = t^2 \quad 2t$$
    $$3 + 1 - t^2 - 2t + 1$$
    $$4 = (t-1)^2$$
    $$\sqrt{4} = t - 1 \quad \text{or} \quad -\sqrt{4} = t - 1$$
    $$2 = t - 1 \quad \text{or} \quad -2 = t - 1$$
    $$3 = t \quad \text{or} \quad -1 = t$$

    Since t represents time, we reject the negative
    solution. It will take 3 seconds for the projectile
    to travel 10 feet.

Now Try:

5. A baseball is thrown upward
    from a building 20 m high with
    a velocity of 15 m/sec. Its
    distance from the ground after t
    seconds is modeled by the
    function
    $$f(t) = -4.9t^2 + 15t + 20.$$
    When will the ball hit the
    ground? Round your answer to
    the nearest tenth.

    _____

Objective 4 Practice Exercises

For extra help, see Examples 5–6 on pages 849–850 of your text.

Solve each problem. Round answers to the nearest tenth.

10. A population of microorganisms grows according to
     the function $p(x) - 100 + 0.2x + 0.5x^2$, where x is
     given in hours. How many hours does it take to
     reach a population of 250 microorganisms?

10. _____

11.   An object is thrown downward from a tower 280 feet   **11.** _____
      high. The distance the object has fallen at time t in
      seconds is given by $s(t) = 16t^2 + 68t$. How long will
      it take the object to fall 100 feet?

12.   A widget manufacturer estimates that her monthly   **12.** _____
      revenue can be modeled by the function
      $R(x) = -0.006x^2 + 32x - 10,000$. What is the
      minimum number of items that must be sold for the
      revenue to equal \$30,000?

Chapter 11 QUADRATIC EQUATIONS, INEQUALITIES, AND FUNCTIONS

11.6   Graphs of Quadratic Functions

Learning Objectives
1   Graph a quadratic function.
2   Graph parabolas with horizontal and vertical shifts.
3   Use the coefficient of x^2 to predict the shape and direction in which a parabola opens.
4   Find a quadratic function to model data.

Key Terms

Use the vocabulary terms listed below to complete each statement in exercises 1–4.

parabola   vertex   axis   quadratic function

1. The vertical (or horizontal) line through the vertex of a vertical (or horizontal) parabola is its _____.

2. The point on a parabola that has the least y-value (if the parabola opens up) or the greatest y-value (if the parabola opens down) is called the _____ of the parabola.

3. A function defined by $f(x) = ax^2 + bx + c$, for real numbers a, b, and c, with $a \neq 0$, is a _____.

4. The graph of a quadratic function is a _____.

Objective 1   Graph a quadratic function.

For extra help, see page 856 of your text.

Objective 2   Graph parabolas with horizontal and vertical shifts.

Video Examples

Review these examples for Objective 2:

1. Graph $g(x) = x^2 + 2$. Give the vertex, axis, domain, and range.

   The graph of $g(x)$ has the same shape as that of $f(x) = x^2$ but shifted 2 units up with vertex (0, 2). Every function value is 2 more than the corresponding function value of $f(x) = x^2$.

Now Try:

1. Graph $f(x) = x^2 - 1$. Give the vertex, axis, domain, and range.

   Vertex   _____

   Axis   _____

   Domain   _____

x	$f(x) = x^2$	$g(x) = x^2 + 2$
-2	4	6
-1	1	3
0	0	2
1	1	3
2	4	6

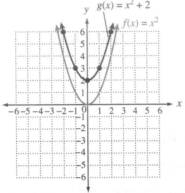

The vertex is (0, 2). The axis is $x = 0$. The domain is $(-\infty, \infty)$. The range is $[2, \infty)$.

2. Graph $g(x) = (x-1)^2$. Give the vertex, axis, domain, and range.

The graph of $g(x)$ has the same shape as that of $f(x) = x^2$ but shifted 1 unit right with vertex (1, 0).

x	$f(x) = x^2$	$g(x) = (x-1)^2$
-2	4	9
-1	1	4
0	0	1
1	1	0
2	4	1
3	9	4

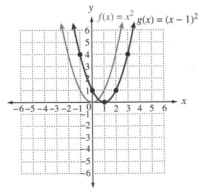

The vertex is (1, 0). The axis is $x = 1$. The domain is $(-\infty, \infty)$. The range is $[0, \infty)$.

Range _____

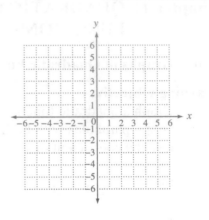

2. Graph $f(x) = (x-3)^2$. Give the vertex, axis, domain, and range.

Vertex _____

Axis _____

Domain _____

Range _____

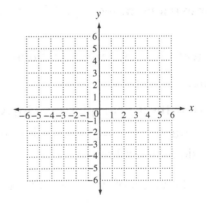

3. Graph $g(x) = (x-1)^2 - 2$. Give the vertex, axis, domain, and range.

The graph of $g(x)$ has the same shape as that of $f(x) = x^2$ but shifted 1 unit right (since $x - 1 = 0$ if $x = 1$) and 2 units down (because of the -2).

x	$g(x) = (x-1)^2 - 2$
-2	7
-1	2
0	-1
1	-2
2	-1
3	2
4	7

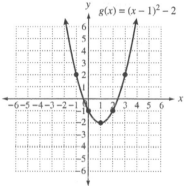

The vertex is $(1, -2)$. The axis is $x = 1$. The domain is $(-\infty, \infty)$. The range is $[-2, \infty)$.

3. Graph $f(x) = (x+2)^2 - 1$. Give the vertex, axis, domain, and range.

Vertex _____

Axis _____

Domain _____

Range _____

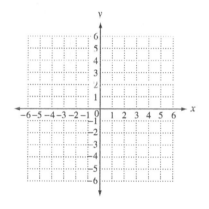

Name: Date:

Instructor: Section:

Objective 2 Practice Exercises

For extra help, see Examples 1–3 on pages 857–858 of your text.

Sketch the graph of each parabola. Give the vertex, axis, domain, and range.

1. $f(x) = x^2 - 4$

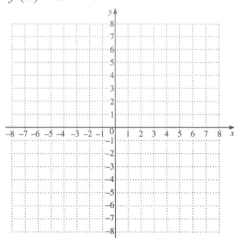

1. vertex _____

     axis _____

     domain _____

     range _____

2. $f(x) = (x + 2)^2 + 1$

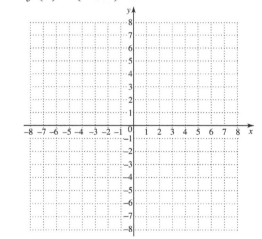

2. vertex _____

     axis _____

     domain _____

     range _____

3. $f(x) = (x - 3)^2 - 1$

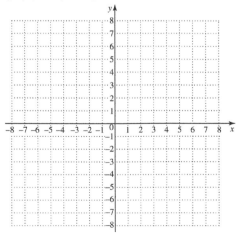

3. vertex _____

     axis _____

     domain _____

     range _____

Name: _____   Date: _____
Instructor: _____   Section: _____

Objective 3   Use the coefficient of x^2 to predict the shape and direction in which a parabola opens.

Video Examples

Review these examples for Objective 3:

4.   Graph $g(x) = -2x^2$. Give the vertex, axis, domain, and range.

The graph of $g(x)$ has the same shape as that of $f(x) = x^2$ but is narrower and opens downward.

x	$g(x) = -2x^2$
-2	-8
-1	-2
0	0
1	-2
2	-8

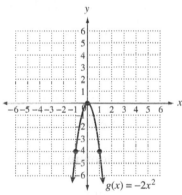

The vertex is (0, 0). The axis is $x = 0$. The domain is $(-\infty, \infty)$. The range is $(-\infty, 0]$.

5.   Graph $g(x) = -\frac{1}{2}(x+1)^2 - 2$. Give the vertex, axis, domain, and range.

The parabola opens down because $a < 0$ and is wider than the graph of $f(x) = x^2$. The parabola has vertex $(-1, -2)$.

x	$g(x) = -\frac{1}{2}(x+1)^2 - 2$
-3	-4
-2	-2.5
-1	-2
0	-2.5
1	-4

Now Try:

4.   Graph $g(x) = -\frac{1}{4}x^2$. Give the vertex, axis, domain, and range.

Vertex   _____

Axis   _____

Domain   _____

Range   _____

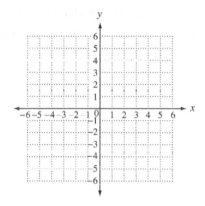

5.   Graph $f(x) = 3(x-1)^2 + 1$. Give the vertex, axis, domain, and range.

Vertex   _____

Axis   _____

Domain   _____

Range   _____

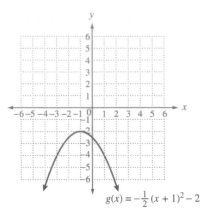

$g(x) = -\frac{1}{2}(x+1)^2 - 2$

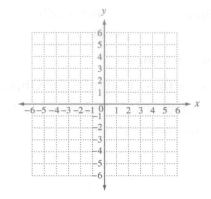

The vertex is $(-1,-2)$. The axis is $x = -1$. The domain is $(-\infty, \infty)$. The range is $(-\infty, -2]$.

Objective 3 Practice Exercises

For extra help, see Examples 4–5 on pages 859–860 of your text.

For each quadratic function, tell whether the graph opens up or down and whether the graph is wider, narrower, or the same shape as the graph of $f(x) = x^2$. Then give the vertex, domain, and range.

4. $f(x) = -\frac{4}{3}x^2 - 1$

4. _____

    vertex _____

    domain _____

    range _____

5. $f(x) = -2(x+1)^2$

5. _____

    vertex _____

    domain _____

    range _____

6. $f(x) = \frac{5}{4}(x-1)^2 + 7$

6. _____

    vertex _____

    domain _____

    range _____

Name: _____ Date: _____
Instructor: _____ Section: _____

Objective 4   Find a quadratic function to model data.

Video Examples

Review this example for Objective 4:

6. The number of ice cream cones sold by an ice cream parlor from 2003–2009 is shown in the following table.

Year	Years since 2003, x	Number of cones sold
2003	0	1775
2004	1	4194
2005	2	5063
2006	3	5161
2007	4	4663
2008	5	4639
2009	6	3710

Use the ordered pairs (x, number of cones sold) to make a scatter diagram of the data. Determine a quadratic function that models these data by using a system of equations. Use the ordered pairs (0, 1775), (3, 5161), and (6, 3710). Round the values of a, b, and c in your model to the nearest tenth, as necessary.

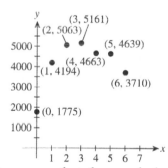

It appears that the parabola opens down, so the coefficient a is negative.
Using the chosen ordered pairs, we substitute the x- and y-values into the quadratic form
$y = ax^2 + bx + c$ to obtain the three equations.

$$a(0)^2 + b(0) + c = 1775 \quad (1)$$

$$a(3)^2 + b(3) + c = 5161 \quad (2)$$

$$a(6)^2 + b(6) + c = 3710 \quad (3)$$

Equation (1) simplifies to $c = 1775$, so substitute 1775 for c in equation (2) and (3).

Now Try:

6. The table lists the average price of a Major League Baseball ticket.

Year	Years since 1990, x	Price
1991	1	$9.14
1994	4	$10.60
1997	7	$12.49
2000	10	$16.81
2004	14	$19.82
2010	20	$26.74

Use the ordered pairs (x, price) to make a scatter diagram of the data. Determine a quadratic function that models these data by using a system of equations. Use the ordered pairs (1, 9.14), (10, 16.81), and (20, 26.74). Round the values of a, b, and c in your model to the nearest hundredth, as necessary.

_____

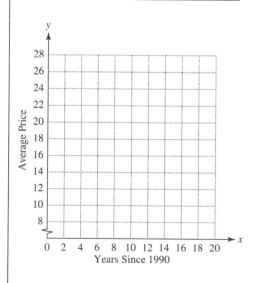

$$9a + 3b + 1775 = 5161 \quad (2)$$

$$36a + 6b + 1775 = 3710 \quad (3)$$

Subtract 1775 from each side of both equations.

$$9a + 3b = 3386 \quad (2)$$

$$36a + 6b = 1935 \quad (3)$$

Solve by elimination. Multiply equation (2)
by -2 and add to equation (3).

$$-18a - 6b = -6772 \quad \text{Multiply (2) by } -2.$$

$$\underline{36a + 6b = 1935 \quad (3)}$$

$$18a = -4837$$

$$a \approx -268.7 \quad \text{Round to the tenth.}$$

Substitute this value for a into equation (2) and
solve for b.

$$9a + 3b = 3386 \quad (2)$$

$$9(-268.7) + 3b = 3386$$

$$-2418.3 + 3b = 3386$$

$$3b = 5804.3$$

$$b \approx 1934.8$$

Therefore, the model

is $y = -268.7x^2 + 1934.8x + 1775$.

Objective 4 Practice Exercises

For extra help, see Example 6 on pages 860–861 of your text.

Tell whether a linear or quadratic function would be a more appropriate model for each set of graphed data. If linear, tell whether the slope should be positive or negative. If quadratic, tell whether the coefficient of x^2 should be positive or negative.

7.

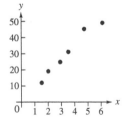

7. _____

8.

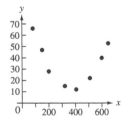

8. _____

Solve the problem.

9. The number of publicly traded companies filing for bankruptcy for selected years between 1990 and 2000 are shown in the table, with 0 representing 1990, 2 representing 1992, etc.

Year	Number of Bankruptcies
0	115
2	91
4	70
6	84
8	120
10	176

Use the ordered pairs to make a scatter diagram of the data.

Use the ordered pairs (0, 115), (4, 70), and (8, 120) to find a function that models the data. Round the values of a, b, and c to three decimal places, if necessary.

Source: Lial, Margaret L., John Hornsby, Terry McGinnis, *Intermediate Algebra* Eighth Edition. Boston: Pearson Education, 2006.

9.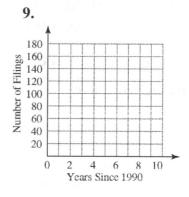

_____

Chapter 11 QUADRATIC EQUATIONS, INEQUALITIES, AND FUNCTIONS

11.7   More about Parabolas and Their Applications

Learning Objectives

1   Find the vertex of a vertical parabola.
2   Graph a quadratic function.
3   Use the discriminant to find the number of x-intercepts of a parabola with a vertical axis.
4   Use quadratic functions to solve problems involving maximum or minimum value.
5   Graph parabolas with horizontal axes.

Key Terms

Use the vocabulary terms listed below to complete each statement in exercises 1–2.

discriminant   **vertex**

1.   The_____ of a quadratic function is found by using the formula $b^2 - 4ac$.

2.   The maximum or minimum value of a quadratic function occurs at the _____ of its graph.

Objective 1   Find the vertex of a vertical parabola.

Video Examples

Review these examples for Objective 1:

2.   Find the vertex of the graph of
$f(x) = 3x^2 + 6x + 10$.

Because the x^2-term has a coefficient other than 1, we factor that coefficient out of the first two terms before completing the square.

$f(x) = 3x^2 + 6x + 10$

$f(x) = 3(x^2 + 2x) + 10$ $\left[\frac{1}{2}(2)\right]^2 = 1$

$f(x) = 3(x^2 + 2x + 1 - 1) + 10$

$f(x) = 3(x^2 + 2x + 1) + 3(-1) + 10$

$f(x) = 3(x + 1)^2 + 7$

The vertex is (−1, 7).

Now Try:

2.   Find the vertex of the graph of
$f(x) = -2x^2 + 4x - 1$.

_____

3. Use the vertex formula to find the vertex of the graph of $f(x) = -4x^2 + 5x + 3$.

The x-coordinate of the vertex of the parabola is given by $\dfrac{-b}{2a}$.

$$\frac{-b}{2a} = \frac{-5}{2(-4)} = \frac{5}{8}$$

The y-coordinate is $f\left(\dfrac{-b}{2a}\right) = f\left(\dfrac{5}{8}\right)$.

$$f\left(\frac{5}{8}\right) = -4\left(\frac{5}{8}\right)^2 + 5\left(\frac{5}{8}\right) + 3 = \frac{73}{16}$$

The vertex is $\left(\dfrac{5}{8},\ \dfrac{73}{16}\right)$.

3. Use the vertex formula to find the vertex of the graph of $f(x) = 2x^2 - 6x + 5$.

_____

Objective 1 Practice Exercises

For extra help, see Examples 1–3 on pages 866–867 of your text.

Find the vertex of each parabola.

1. $f(x) = x^2 - 2x + 4$

1. _____

2. $f(x) = 5x^2 - 4x + 1$

2. _____

3. $f(x) = -\dfrac{1}{4}x^2 - 3x - 9$

3. _____

Objective 2   Graph a quadratic function.

Video Examples

Review this example for Objective 2:

4. Graph the quadratic function defined by

    $f(x) = x^2 - 3x + 2$. Give the vertex, axis, domain, and range.

    Step 1 From the equation, $a = 1$, so the graph opens up.

    Step 2 The x-coordinate of the vertex is $\frac{3}{2}$.

    The y-coordinate of the vertex is

    $$f\left(\frac{3}{2}\right) = \left(\frac{3}{2}\right)^2 - 3\left(\frac{3}{2}\right) + 2 = -\frac{1}{4}.$$

    The vertex is $\left(\frac{3}{2}, -\frac{1}{4}\right)$.

    Step 3 Find any intercepts. Since the vertex is in quadrant IV and the graph opens up, there will be two x-intercepts. Let $f(x) = 0$ and solve.

    $$x^2 - 3x + 2 = 0$$
    $$(x - 1)(x - 2) = 0$$
    $$x - 1 = 0 \quad \text{or} \quad x - 2 = 0$$
    $$x = 1 \quad \text{or} \quad x = 2$$

    The x-intercepts are (1, 0) and (2, 0). Find the y-intercept by evaluating $f(0)$.

    $$f(0) = 0^2 - 3(0) + 2$$

    The y-intercept is (0, 2).

    Step 4 Plot the points found so far and additional points as needed using symmetry about the axis, $x = \frac{3}{2}$.

    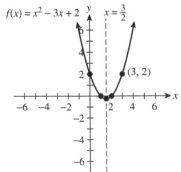

    The domain is $(-\infty, \infty)$.

    The range is $\left[-\frac{1}{4}, \infty\right)$.

Now Try:

4. Graph the quadratic function defined by $f(x) = x^2 + 4x + 5$. Give the vertex, axis, domain, and range.

    Vertex   _____

    Axis   _____

    Domain   _____

    Range   _____

    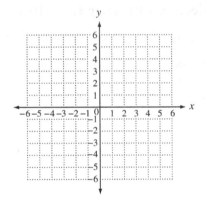

Name: Date:
Instructor: Section:

Objective 2 Practice Exercises

For extra help, see Example 4 on page 868 of your text.

Sketch the graph of each parabola. Give the vertex, axis, domain, and range.

4. $f(x) = -x^2 + 8x - 10$

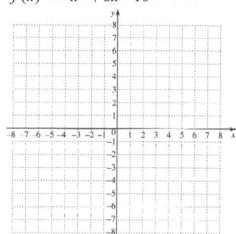

4. vertex _____

   axis _____

   domain _____

   range _____

5. $f(x) = 3x^2 + 6x + 2$

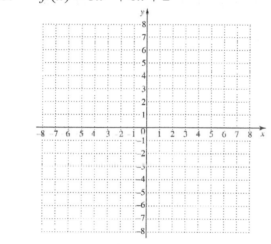

5. vertex _____

   axis _____

   domain _____

   range _____

6. $f(x) = -2x^2 + 4x + 1$

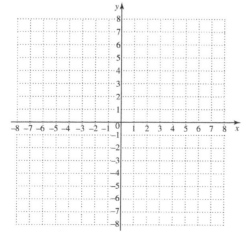

6. vertex _____

   axis _____

   domain _____

   range _____

Objective 3   Use the discriminant to find the number of x-intercepts of a parabola
with a vertical axis.

Video Examples

Review this example for Objective 3:

5b. Find the discriminant and use it to determine
the number of x-intercepts of the graph of the
quadratic function.

$$f(x) = 3x^2 + 3x + 2$$

$$b^2 - 4ac = 3^2 - 4(3)(2) = -15$$

Since the discriminant is negative, the graph has
no x-intercepts.

Now Try:

5b. Find the discriminant and
use it to determine the number
of x-intercepts of the graph of
the quadratic function.

$$f(x) = 9x^2 - 24x + 16$$

_____

Objective 3 Practice Exercises

For extra help, see Example 5 on page 869 of your text.

Use the discriminant to determine the number of x-intercepts of the graph of each function.

7. $f(x) = 2x^2 - 3x + 2$

7. _____

8. $f(x) = -3x^2 - x + 5$

8. _____

9. $f(x) = 3x^2 - 6x + 3$

9. _____

Objective 4   Use quadratic functions to solve problems involving maximum or minimum value.

Video Examples

Review this example for Objective 4:

7. An object is launched directly upward at 64 feet per second from a platform 80 feet high. Its height above the ground is given by

$s(t) = -16t^2 + 64t + 80,$ where t is the number of seconds after launch. What will be the object's maximum height? When will it reach this height?

For this function, $a = -16$, $b = 64$, and $c = 80$.

The vertex formula gives $t = \dfrac{-b}{2a} = \dfrac{-64}{2(-16)} = 2.$

This indicates that the maximum height is reached at 2 seconds. Now calculate $s(2)$ to find the maximum height.

$$s(2) = -16(2)^2 + 64(2) + 80 = 144$$

Thus, it takes 2 seconds to reach the maximum height of 144 feet above the ground.

Now Try:

7. An object is launched directly upward at 48 feet per second from a platform 250 feet high. Its height above the ground is given by

$s(t) = -16t^2 + 48t + 250$

where t is the number of seconds after launch. What will be the object's maximum height? When will it reach this height?

_____

Objective 4 Practice Exercises

For extra help, see Examples 6–7 on pages 870–871 of your text.

Solve each problem.

10. Jean sells ceramic pots. She has weekly costs of

$C(x) = x^2 - 100x + 2700,$ where x is the number of pots she sells each week. How many pots should she sell to minimize her costs? What is the minimum cost?

10. units _____

cost _____

11. The length and width of a rectangle have a sum of 48. What width will produce the maximum area?

11. _____

12. A projectile is fired upward so that its distance (in feet) above the ground t seconds after firing is given by $s(t) = -16t^2 + 80t + 156$. Find the maximum height it reaches and the number of seconds it takes to reach that height.

12. height _____

time _____

Objective 5   Graph parabolas with horizontal axes.

Video Examples

Review this example for Objective 5:

8. Graph $x = (y+1)^2 - 2$. Give the vertex, axis, domain, and range.

The graph has its vertex at $(-2,-1)$ since the roles of x and y are interchanged. It opens to the right since $a > 0$ and has the same shape as $y = x^2$ (but situated horizontally).

x	y
2	-3
-1	-2
-2	-1
-1	0
2	1

The axis is $y = -1$. The domain is $[-2, \infty)$.
The range is $(-\infty, \infty)$.

Now Try:

8. Graph $x = (y-2)^2 + 1$. Give the vertex, axis, domain, and range.

Vertex _____

Axis _____

Domain _____

Range _____

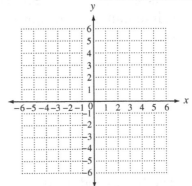

Objective 5 Practice Exercises

For extra help, see Examples 8–9 on page 872 of your text.

Sketch the graph of each parabola. Give the vertex, axis, domain, and range.

13. $x = -y^2 + 2$

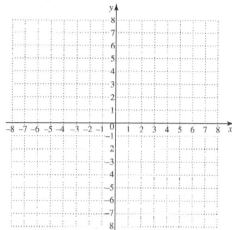

13. vertex _____

      axis _____

      domain _____

      range _____

14. $x = y^2 - 3$

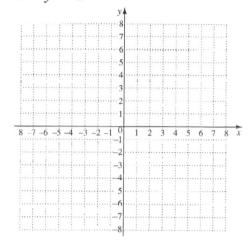

14. vertex _____

      axis _____

      domain _____

      range _____

15. $x = -y^2 - 6y - 10$

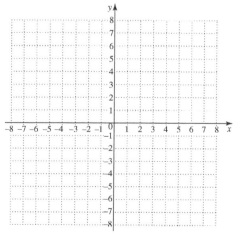

15. vertex _____

      axis _____

      domain _____

      range _____

Chapter 11 QUADRATIC EQUATIONS, INEQUALITIES, AND FUNCTIONS

11.8 Polynomial and Rational Inequalities

Learning Objectives
1   Solve quadratic inequalities.
2   Solve polynomial inequalities of degree 3 or greater.
3   Solve rational inequalities.

Key Terms

Use the vocabulary terms listed below to complete each statement in exercises 1–2.

quadratic inequality   **rational inequality**

1.   An inequality that involves a rational expression is a _____.

2.   An inequality that can be written in the form $ax^2 + bx + c < 0$ or $ax^2 + bx + c > 0$, where a, b, and c are real numbers with $a \neq 0$ is called a

_____.

Objective 1   Solve quadratic inequalities.

Video Examples

Review these examples for Objective 1:

2.   Solve and graph the solution set of
$x^2 + 5x + 4 \geq 0$.

Solve the quadratic equation by factoring.
$(x+1)(x+4) = 0$
$x+1 = 0$ or $x+4 = 0$
$x = -1$ or $x = -4$

The numbers –4 and –1 divide a number line into intervals A, B, and C, as shown below.

```
     A     B    C
  ─────┼────┼┼───────►
      -4   -1 0
```

Since the numbers –4 and –1 are the only numbers that make the quadratic expression

$x^2 + 5x + 4$ equal to 0, all other numbers make the expression either positive or negative. If one number in an interval satisfies the inequality, then all the numbers in that interval will satisfy the inequality.

Choose any number in interval A as a test number; we will choose –5.

Now Try:

2.   Solve and graph the solution set of $x^2 + 7x + 12 > 0$.

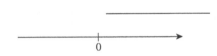

Copyright © 2018 Pearson Education, Inc.

$$x^2 + 5x + 4 \geq 0$$

$$(-5)^2 + 5(-5) + 4 \overset{?}{\geq} 0$$

$$4 \geq 0 \quad \text{True}$$

Because –5 satisfies the inequality, all numbers from interval A are solutions.

Now try –2 from interval B.

$$x^2 + 5x + 4 \geq 0$$

$$(-2)^2 + 5(-2) + 4 \overset{?}{\geq} 0$$

$$-2 \geq 0 \quad \text{False}$$

The numbers in interval B are not solutions.

Finally, try 0 from interval C.

$$x^2 + 5x + 4 \geq 0$$

$$0^2 + 5(0) + 4 \overset{?}{\geq} 0$$

$$4 \geq 0 \quad \text{True}$$

Because 0 satisfies the inequality, all numbers from interval C are solutions.

Because the inequality is greater than or equal to zero, we include the endpoints of the intervals in the solution set. Thus, the solution set is

$$(-\infty, -4] \cup [-1, \infty).$$

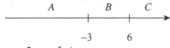

3. Solve and graph the solution set of $x^2 - 3x \leq 18$.

Step 1

$$x^2 - 3x = 18$$

$$x^2 - 3x - 18 = 0$$

$$(x + 3)(x - 6) = 0$$

$$x + 3 = 0 \quad \text{or} \quad x - 6 = 0$$

$$x = -3 \quad \text{or} \quad x = 6$$

Step 2

The numbers –3 and 6 divide a number line into intervals A, B, and C, as shown below.

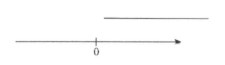

Steps 3 and 4

Substitute a test value form each interval in the *original* inequality $x^2 - 3x \leq 18$ to determine which intervals satisfy the inequality.

3. Solve and graph the solution set of $x^2 - x < 2$.

Interval	Test Number	Test of inequality	True or False?
A	−4	$-28 \leq 18$	F
B	0	$8 \leq 18$	T
C	7	$-8 \leq 18$	F

The numbers in Interval B are solutions.
The solution set is the interval [−3, 6].

$$\begin{array}{ccc} [& &] \\ -3 & 0 & 6 \end{array} \longrightarrow$$

Objective 1 Practice Exercises

For extra help, see Examples 1–4 on pages 878–881 of your text.

Solve each inequality, and graph the solution set.

1. $2y^2 + 5y < 3$

1.   _____

2. $8k^2 + 10k > 3$

2.   _____

3. $(3x - 2)^2 < -1$

3.   _____

Objective 2   Solve polynomial inequalities of degree 3 or greater.

Video Examples

Review this example for Objective 2:

5. Solve and graph the solution set of
$(x+1)(x-2)(x+4)\le 0$.

Set the factored polynomial equal to 0, then use the zero-factor property.

$x+1=0$ or $x-2=0$ or $x+4=0$

$x=-1$ or $x=2$ or $x=-4$

Locate –4, –1, and 2 on a number line to determine the intervals A, B, C, and D.

```
   A     B    C       D
 ───┼───────┼─┼───┼──────►
   -4    -1 0   2
```

Substitute a test number from each interval in the original inequality to determine which intervals satisfy the inequality.

Interval	Test Number	Test of inequality	True or False?
A	-5	$-28\le 0$	T
B	-2	$8\le 0$	F
C	0	$-8\le 0$	T
D	5	$162\le 0$	F

The numbers in intervals A and C are in the solution set. The three endpoints are included in the solution set since the inequality symbol, $\le$, includes equality. Thus, the solution set is $(-\infty,-4]\cup[-1,\ 2]$.

```
 ◄──────┤   ├─┼──┤────────►
       -4   -1 0  2
```

Now Try:

5. Solve and graph the solution sct of $(2x-1)(2x+3)(3x+1)\le 0$.

Objective 2 Practice Exercises

For extra help, see Example 5 on page 882 of your text.

Solve each inequality, and graph the solution set.

4. $(y+2)(y-1)(y-2)<0$

4. _____

```
 ◄─┼─┼─┼─┼─┼─┼─┼─┼─┼─┼─┼─┼─►
```

5. $(k+5)(k-1)(k+3) \leq 0$

5. _____

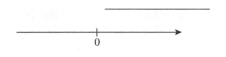

6. $(x-1)(x-3)(x+2) \geq 0$

6. _____

Objective 3   Solve rational inequalities.

Video Examples

Review these examples for Objective 3:

7.   Solve and graph the solution set of $\dfrac{x+1}{x-5} \geq 3$.

Write the inequality so that 0 is on one side.

$$\frac{x+1}{x-5} - 3 \geq 0$$

$$\frac{x+1}{x-5} - \frac{3(x-5)}{x-5} \geq 0$$

$$\frac{x+1}{x-5} - \frac{3x-15}{x-5} \geq 0$$

$$\frac{x+1-3x+15}{x-5} \geq 0$$

$$\frac{-2x+16}{x-5} \geq 0$$

The sign of $\dfrac{-2x+16}{x-5}$ will change from positive to negative or negative to positive only at those numbers that make the numerator or denominator 0. These two numbers, 5 and 8, divide a number line into three intervals.

Now Try:

7.   Solve and graph the solution set of $\dfrac{z+2}{z-3} \leq 2$.

_____

Test a number in each interval using original inequality.

Interval	Test Number	Test of inequality	True or False?
A	0	$-\frac{1}{5} \geq 3$	F
B	6	$7 > 3$	T
C	10	$\frac{11}{5} \geq 3$	F

The solution set is (5, 8]. This interval does not include 5 because it would make the denominator of the original inequality 0. The number 8 is included because the inequality symbol, $\geq$, does includes equality.

Objective 3 Practice Exercises

For extra help, see Examples 6–7 on pages 883 884 of your text.

Solve each inequality, and graph the solution set

7. $\frac{7}{x-1} \leq 1$

7.   _____

8. $\frac{2p-1}{3p+1} \leq 1$

8.   _____

9. $\dfrac{5}{x-3} \le -1$

9. _____

Chapter 12 INVERSE, EXPONENTIAL, AND LOGARITHMIC FUNCTIONS

12.1   Composition of Functions

Learning Objectives
1   Find the composition of functions.

Key Terms

Use the vocabulary terms listed below to complete each statement in exercises 1–2.

> composition   composite function

1.   If f and g are functions, then the _____ of g and f is defined by $(g \circ f)(x) = g(f(x))$ for all x in the domain of f such that $f(x)$ is in the domain of g.

2.   The function $g(f(x))$ is a _____.

Objective 1   Find the composition of functions.

Video Examples

Review these examples for Objective 1:	Now Try:
3. Let $f(x) = 3x - 1$ and $g(x) = x^2 + 2$. Find each of the following	3. Let $f(x) = -3x - 3$ and $g(x) = x^2 - 5$. Find each of the following

b. $(f \circ g)(x)$ **b.** $(f \circ g)(x)$

$$(f \circ g)(x) = f(g(x))$$
$$= 3(g(x)) - 1$$
$$= 3(x^2 + 2) - 1$$
$$= 3x^2 + 6 - 1$$
$$= 3x^2 + 5$$

a. $(f \circ g)(2)$ **a.** $(f \circ g)(2)$

$$(f \circ g)(2) = f(g(2))$$
$$= 3(g(2)) - 1$$
$$= 3(2^2 + 2) - 1$$
$$= 3(6) - 1$$
$$= 17$$

    d. $(g \circ f)(x)$

    $(g \circ f)(x) = g(f(x))$

            $= (f(x))^2 + 2$

            $= (3x - 1)^2 + 2$

            $= 9x^2 - 6x + 3$

    d. $(g \circ f)(x)$

        _____

Objective 1 Practice Exercises

For extra help, see Examples 1–3 on pages 903–904 of your text.

Find the following.

1. Let $f(x) = 4x - 3$ and $g(x) = 2x^2 - 1$. Find the following.

    a. $(f \circ g)(2)$

    b. $(g \circ f)(-1)$

    c. $(f \circ g)(x)$

1. **a.** _____

   b. _____

   c. _____

2. Let $f(x) = \dfrac{1}{x}$ and $g(x) = 3x^2 - 4x + 1$. Find the following.

    a. $(f \circ g)(2)$

    b. $(g \circ f)\left(\dfrac{1}{3}\right)$

    c. $(g \circ f)(x)$

2. **a.** _____

   b. _____

   c. _____

Name: Date:
Instructor: Section:

Chapter 12 INVERSE, EXPONENTIAL, AND LOGARITHMIC FUNCTIONS

12.2   Inverse Functions

Learning Objectives
1   Decide whether a function is one-to-one and, if it is, find its inverse.
2   Use the horizontal line test to determine whether a function is one-to-one.
3   Find the equation of the inverse of a function.
4   Graph f^{-1} from the graph of f.

Key Terms

Use the vocabulary terms listed below to complete each statement in exercises 1–2.

one-to-one function   **inverse of a function f**

1. A function in which each x-value corresponds to just one y-value and each y-value corresponds to just one x-value is a(n) _____.

2. If f is a one-to-one function, the _____ is the set of all ordered pairs of the form (y, x) where (x, y) belongs to f.

Objective 1   Decide whether a function is one-to-one and, if it is, find its inverse.

Video Examples

Review this example for Objective 1:	Now Try:
1b. Find the inverse of the function that is one-to-one.	**1b.** Find the inverse of the function that is one-to-one.
$G = \{(-3,-1), (-2, 0), (-1, 1), (0, 2)\}$	$G = \{(3, 2), (-3,-2), (5,-4), (0, 6)\}$
Every x-value in G corresponds to only one y-value, and every y-value corresponds to only one x-value, so G is a one-to-one function. The inverse function is found by interchanging the x- and y-values in each ordered pair.	_____
$G^{-1} = \{(-1,-3),\ (0,-2),\ (1,-1),\ (2,\ 0)\}$	

Objective 1 Practice Exercises

For extra help, see Example 1 on page 908 of your text.

If the function is one-to-one, find its inverse.

1. $\{(-3,-1), (-2, 2), (-1, 3), (0, 4)\}$

    1. _____

2. {(1, 0), (2, 0), (3, 5), (4, 1)}   2. _____

3. {(0, 0), (1,–1), (–2, 2), (4,–3), (3,–5)}   3. _____

Objective 2   Use the horizontal line test to determine whether a function is one-to-one.

Video Examples

Review these examples for Objective 2:

2. Use the horizontal line test to determine whether each graph is the graph of a one-to-one function.

b.

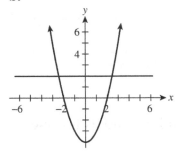

Because a horizontal line intersects the graph in more than one point, the function is not one-to-one.

a.

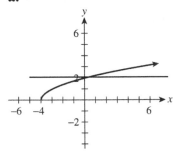

Every horizontal line will intersect the graph in exactly one point. The function is one-to-one.

Now Try:

2. Use the horizontal line test to determine whether each graph is the graph of a one-to-one function.

b.

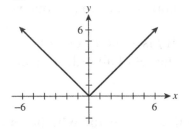

_____

a.

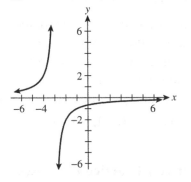

_____

Name: _____ Date: _____

Instructor: _____ Section: _____

Objective 2 Practice Exercises

For extra help, see Example 2 on page 909 of your text.

Use the horizontal line test to determine whether each function is one-to-one.

4.

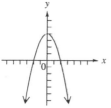

4. _____

5.

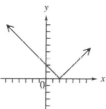

5. _____

6.

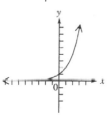

6. _____

Objective 3 Find the equation of the inverse of a function.

Video Examples

Review these examples for Objective 3:

3. Decide whether each equation represents a one-to-one function. If so, find the equation for the inverse.

    a. $f(x) = 3x - 5$

The graph of $y = 3x - 5$ is a nonvertical line, so by the horizontal line test, f is a one-to-one function. To find the inverse, let $y = f(x)$, interchange x and y, then solve for y.

$$y = 3x - 5$$
$$x = 3y - 5 \quad \text{Interchange } x \text{ and } y.$$
$$x + 5 = 3y$$
$$\frac{x+5}{3} = y$$
$$f^{-1}(x) = \frac{x+5}{3} = \frac{x}{3} + \frac{5}{3}$$
$$f^{-1}(x) = \frac{1}{3}x + \frac{5}{3}$$

Now Try:

3. Decide whether each equation represents a one-to-one function. If so, find the equation for the inverse.

    a. $f(x) = 4x - 1$

    _____

c. $f(x) = x^3 + 1$ **c.** $f(x) = 2x^3 - 3$

The graph of $y = x^3 + 1$ is a cubing function.
The function is one-to-one and has an inverse.   _____

$$y = x^3 + 1$$

$$x = y^3 + 1 \quad \text{Interchange } x \text{ and } y.$$

$$x - 1 = y^3$$

$$\sqrt[3]{x-1} = y$$

$$f^{-1}(x) = \sqrt[3]{x-1}$$

Objective 3 Practice Exercises

For extra help, see Example 3 on page 910 of your text.

If the function is one-to-one, find its inverse.

7. $f(x) = 2x - 5$ 7. _____

8. $f(x) = x^3 - 1$ 8. _____

9. $f(x) = x^2 - 1$ 9. _____

Name: Date:
Instructor: Section:

Objective 4   Graph f^{-1} from the graph of f.

Video Examples

Review this example for Objective 4:
4b. Use the given graph to graph the inverse of f.

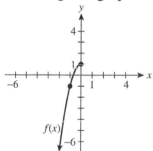

We can find the graph of f^{-1} from the graph of f by locating the mirror image of each point in f with respect to the line $y = x$.

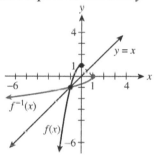

Now Try:
4b. Use the given graph to graph the inverse of f.

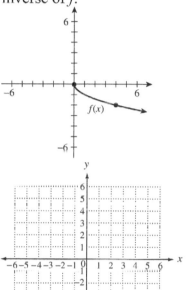

Objective 4 Practice Exercises

For extra help, see Example 4 on page 911 of your text.

If the function is one-to-one, graph the function f and its inverse f^{-1} on the same set of axes.

10.

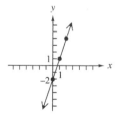

10. _____

11.

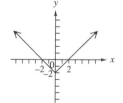

11. _____

12.

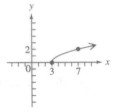

12. _____

Chapter 12 INVERSE, EXPONENTIAL, AND LOGARITHMIC FUNCTIONS

12.3   Exponential Functions

Learning Objectives
1   Define exponential functions.
2   Graph exponential functions.
3   Solve exponential equations of the form $a^x = a^k$ for x.
4   Use exponential functions in applications involving growth or decay.

Key Terms

Use the vocabulary terms listed below to complete each statement in exercises 1–2.

exponential equation   inverse

1.   If f is a one-to-one function, then the _____ of f is the set of all ordered pairs formed by interchanging the coordinates of the ordered pairs of f.

2.   An equation that has a variable as an exponent, is an _____.

Objective 1   Define exponential functions.

For extra help, see page 916 of your text.

Objective 2   Graph exponential functions.

Video Examples

Review these examples for Objective 2:   **Now Try:**

1.   Graph $f(x) = 6^x$.   1.   Graph $f(x) = 3^x$.

Create a table of values, then plot the points and draw a smooth curve through them.

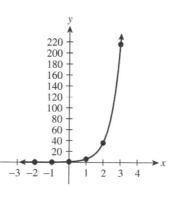

x	$f(x) = 6^x$
-2	$\dfrac{1}{36}$
-1	$\dfrac{1}{6}$
0	1
1	6
2	36
3	216

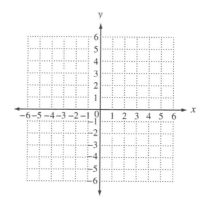

2. Graph $f(x) = \left(\dfrac{1}{6}\right)^x$.

Create a table of values, then plot the points and draw a smooth curve through them.

x	$f(x) = \left(\dfrac{1}{6}\right)^x$
-3	216
-2	36
-1	6
0	1
1	$\dfrac{1}{6}$
2	$\dfrac{1}{36}$

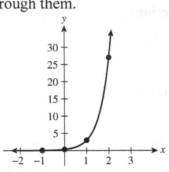

2. Graph $f(x) = \left(\dfrac{1}{3}\right)^x$.

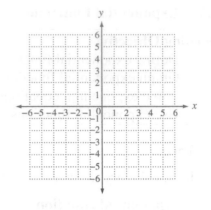

3. Graph $f(x) = 3^{2x-1}$.

Create a table of values, then plot the points and draw a smooth curve through them.

x	$2x - 1$	$f(x) = 3^{2x-1}$
-1	-3	$\dfrac{1}{27}$
0	-1	$\dfrac{1}{3}$
1	1	3
2	3	27

3. Graph $f(x) = 2^{1-x}$.

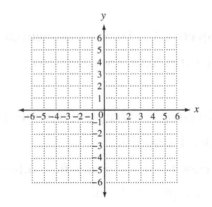

Objective 2 Practice Exercises

For extra help, see Examples 1–3 on pages 916–918 of your text.

Graph each exponential function.

1. $f(x) = 2^{-x}$

1.

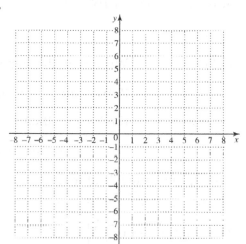

2. $f(x) = \left(\dfrac{1}{8}\right)^x$

2.

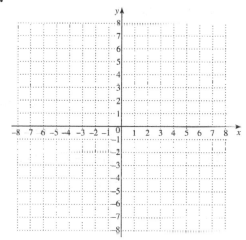

3. $f(x) = 4^{2x-3}$

3.

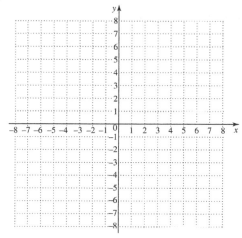

Objective 3   Solve exponential equations of the form $a^x = a^k$ for x.

Video Examples

Review these examples for Objective 3:	Now Try:
4. Solve the equation $16^x = 64$.	4. Solve the equation $25^x = 125$.

$16^x = 64$

$(2^4)^x = 2^6$ Write with the same base.

$2^{4x} = 2^6$ Power rule for exponents

$4x = 6$ If $a^x = a^y$, then $x = y$.

$x = \dfrac{6}{4} = \dfrac{3}{2}$ Solve for x; simplify.

Check: Substitute 3/2 for x.

$16^{3/2} = (16^{1/2})^3 = 4^3 = 64$

The solution set is $\left\{\dfrac{3}{2}\right\}$.

5a. Solve the equation.

$16^{x-2} = 64^x$

5a. Solve the equation.

$4^{x-1} = 8^x$

$16^{x-2} = 64^x$

$(2^4)^{x-2} = (2^6)^x$ Write with the same base.

$2^{4x-8} = 2^{6x}$ Power rule for exponents

$4x - 8 = 6x$ If $a^x = a^y$, then $x = y$.

$-8 = 2x$ Solve for x.

$-4 = x$

The solution set is $\{-4\}$.

Objective 3 Practice Exercises

For extra help, see Examples 4–5 on pages 919–920 of your text.

Solve each equation.

4. $25^{1-t} = 5$

4. _____

5. $8^{2x+1} = 4^{4x}$

6. $\left(\dfrac{3}{4}\right)^x = \dfrac{16}{9}$

Objective 4   Use exponential functions in applications involving growth or decay.

Video Examples

Review these examples for Objective 4:	Now Try:
7. The amount of radioactive material in a sample is given by the function $A(t) = 90\left(\dfrac{1}{2}\right)^{t/18}$, where $A(t)$ is the amount present, in grams, t days after the initial measurement.	**7.** An industrial city in Ohio has found that its population is declining according to the equation $y = 70,000(2)^{-0.01x}$, where x is the time in years from 1910.

a. How many grams will be present initially?

Start with the given function. Replace t with 0.

$$A(t) = 90\left(\frac{1}{2}\right)^{t/18}$$

$$A(0) = 90\left(\frac{1}{2}\right)^{0/18}$$

$$A(0) = 90$$

Initially, there were 90 grams of the sample.

a. According to the model, what will the city's population be in the year 1910?

_____

b. How many grams will be present after 3 days? Round to the nearest hundredth.

Start with the given function. Replace t with 3.

$$A(t) = 90\left(\frac{1}{2}\right)^{t/18}$$

$$A(3) = 90\left(\frac{1}{2}\right)^{3/18}$$

$$A(3) \approx 80.18$$

After 3 days, there were about 80.18 grams in the sample.

b. According to the model, what will the city's population be in the year 2020?

_____

Objective 4 Practice Exercises

For extra help, see Examples 6–7 on pages 920–921 of your text.

Solve each problem.

7. The population of Canadian geese that spend the summer at Gemini Lake each year has been growing according to the function $f(x) = 56(2)^{0.2x}$, where x is the time in years from 1990. Find the number of geese in 2010.

7. _____

8. A sample of a radioactive substance with mass in grams decays according to the function $f(x) = 100(10)^{-0.2x}$, where x is the time in hours after the original measurement. Find the mass of the substance after 10 hours.

8. _____

9. A culture of a certain kind of bacteria grows according to $f(x) = 7750(x)^{0.75x}$, where x is the number of hours after 12 noon. Find the number of bacteria in the culture at 12 noon.

9. _____

Chapter 12 INVERSE, EXPONENTIAL, AND LOGARITHMIC FUNCTIONS

12.4   Logarithmic Functions

Learning Objectives
1   Define a logarithm.
2   Convert between exponential and logarithmic forms.
3   Solve logarithmic equations of the form $\log_a b = k$ for a, b, or k.
4   Define and graph logarithmic functions.
5   Use logarithmic functions in applications involving growth or decay.

Key Terms

Use the vocabulary terms listed below to complete each statement in exercises 1–2.

logarithm   **logarithmic equation**

1. The _____ of a positive number is the exponent indicating the power to which it is necessary to raise a given number (the base) to give the original number.

2. An equation with a logarithm in at least one term is a _____.

Objective 1   Define a logarithm.

For extra help, see page 924 of your text.

Objective 2   Convert between exponential and logarithmic forms.

Video Examples

Review these examples for Objective 2:

1.

   a. Write $5^3 = 125$ in logarithmic form.

   $\log_5 125 = 3$

   b. Write $81^{-1/4} = \dfrac{1}{3}$ in logarithmic form.

   $\log_{81}\left(\dfrac{1}{3}\right) = -\dfrac{1}{4}$

   c. Write $\log_{16} 4 = \dfrac{1}{2}$ in exponential form.

   $16^{1/2} = 4$

Now Try:

1.

   a. Write $8^2 = 64$ in logarithmic form.

   _____

   b. Write $81^{3/4} = 27$ in logarithmic form.

   _____

   c. Write $\log_{16} \dfrac{1}{4} = -\dfrac{1}{2}$ in exponential form.

   _____

d. Write $\log_2\left(\dfrac{1}{64}\right) = -6$ in exponential form.

$$2^{-6} = \dfrac{1}{64}$$

d. Write $\log_{1/2} 4 = -2$ in exponential form.

_____

Objective 2 Practice Exercises

For extra help, see Example 1 on page 924 of your text.

Write in exponential form.

1. $\log_{10} 0.001 = -3$

1. _____

2. $\log_4 \dfrac{1}{16} = -2$

2. _____

Write in logarithmic form.

3. $2^{-7} = \dfrac{1}{128}$

3. _____

Objective 3   Solve logarithmic equations of the form $\log_a b = k$ for a, b, or k.

Video Examples

Review this example for Objective 3:
2d. Solve the equation.

$$\log_{64} \sqrt[4]{8} = x$$

$$\log_{64} \sqrt[4]{8} = x$$

$\qquad 64^x = \sqrt[4]{8}$ Write in exponential form.

$\qquad \left(8^2\right)^x = 8^{1/4}$ Write with the same base.

$\qquad 8^{2x} = 8^{1/4}$ Power rule for exponents

$\qquad 2x = \dfrac{1}{4}$

$\qquad x = \dfrac{1}{8}$

The solution set is $\left\{\dfrac{1}{8}\right\}$.

Now Try:
2d. Solve the equation.

$$\log_{81} \sqrt[3]{9} = x$$

_____

Objective 3 Practice Exercises

For extra help, see Examples 2–3 on pages 925–926 of your text.

Solve each equation.

4. $x = \log_{32} 8$

4. _____

5. $\log_{1/3} r = -4$

5. _____

6. $\log_a 4 = \frac{1}{2}$

6. _____

Objective 4   Define and graph logarithmic functions.

Video Examples

Review these examples for Objective 4:

4. Graph $f(x) = \log_5 x$.

Begin by writing $y = \log_5 x$ in exponential form as $x = 5^y$. Then, create a table of values, plot the points and draw a smooth curve through them.

$x = 5^y$	y
$\frac{1}{5}$	-1
1	0
5	1
25	2

Now Try:

4. Graph $f(x) = \log_3 x$.

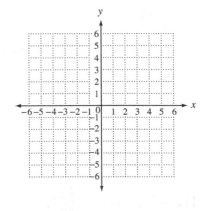

5. Graph $f(x) = \log_{1/3} x.$

Begin by writing $y = \log_{1/3} x.$ in exponential form. Then, create a table of values, plot the points and draw a smooth curve through them.

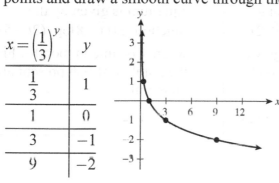

$x = \left(\dfrac{1}{3}\right)^y$	y
$\dfrac{1}{3}$	1
1	0
3	−1
9	−2

5. Graph $f(x) = \log_{1/4} x.$

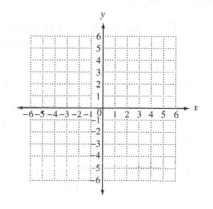

Objective 4 Practice Exercises

For extra help, see Examples 4–5 on pages 926–927 of your text.

Graph each logarithmic function.

7. $y = \log_9 x$

7.

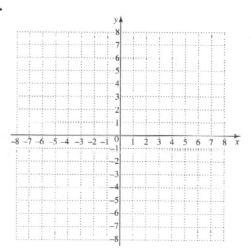

8. $y = \log_{1/2} x$

8.

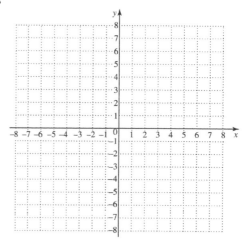

Objective 5   Use logarithmic functions in applications involving growth or decay.

Video Examples

Review this example for Objective 5:

6. A company analyst has found that total sales in thousands of dollars after a major advertising campaign are given by $S(x) = 100\log_2(x+2)$, where x is time in weeks after the campaign was introduced. Find the amount of sales two weeks after the campaign was introduced.

Two weeks after the campaign, $x = 2$, so we have

$$S(2) = 100\log_2(2+2)$$

$$S(2) = 100\log_2(2)^2$$

$$S(2) = 100\cdot 2\log_2 2$$

$$S(2) = 200\cdot 1 = 200$$

Two weeks after the campaign, sales were $200,000.

Now Try:

6. The number of fish in an aquarium is given by the function $f(t) = 8\log_5(2t+5)$, where t is time in months. Find the number of fish present after 10 months.

_____

Objective 5 Practice Exercises

For extra help, see Example 6 on page 927 of your text.

Solve each problem.

9. The population of foxes in an area t months after the foxes were introduced there is approximated by the function $F(t) = 500\log_{10}(2t+1)$. Find the number of foxes in the area when the foxes were first introduced into the area.

9. _____

10.   A population of mites in a laboratory is growing   **10.** _____
according to the function
$$p = 50\log_3(20t + 7) - 25\log_9(80t + 1),$$ where t is the
number of days after a study is begun. Find the
number of mites present 1 day after the beginning of
the study.

11.   Sales (in thousands) of a new product are   **11.** _____
approximated by
$$S = 125 + 20\log_2(30t + 4) + 30\log_4(35t - 6),$$ where
t is the number of years after the product is
introduced. Find the total sales 2 years after the
product is introduced.

Chapter 12 INVERSE, EXPONENTIAL, AND LOGARITHMIC FUNCTIONS

12.5   Properties of Logarithms

Learning Objectives
1   Use the product rule for logarithms.
2   Use the quotient rule for logarithms.
3   Use the power rule for logarithms.
4   Use properties to write alternative forms of logarithmic expressions.

Key Terms

Use the vocabulary terms listed below to complete each statement in exercises 1–4.

product rule for logarithms   **quotient rule for logarithms**

power rule for logarithms   **special properties**

1. The equations $b^{\log_b x} = x$, $x > 0$ and $\log_b b^x = x$ are referred to as

    _____ of logarithms.

2. The equation $\log_b \frac{x}{y} = \log_b x - \log_b y$ is referred to as the _____.

3. The equation $\log_b xy = \log_b x + \log_b y$ is referred to as the _____.

4. The equation $\log_b x^r = r \log_b x$ is referred to as the _____.

Objective 1   Use the product rule for logarithms.

Video Examples

Review this example for Objective 1:
1a. Use the product rule to rewrite the logarithm.

$\log_4 (6 \cdot 11)$

Use the product rule.
$\log_4 (6 \cdot 11) = \log_4 6 + \log_4 11$

Now Try:
1a. Use the product rule to rewrite the logarithm.

$\log_6 (5 \cdot 3)$

_____

Copyright © 2018 Pearson Education, Inc.

Name: _____   Date: _____

Instructor: _____   Section: _____

Objective 1 Practice Exercises

For extra help, see Example 1 on page 933 of your text.

Use the product rule to express each logarithm as a sum of logarithms.

1. $\log_7 5m$

1. _____

2. $\log_2 6xy$

2. _____

Use the product rule to express the sum as a single logarithm.

3. $\log_4 7 + \log_4 3$

3. _____

Objective 2   Use the quotient rule for logarithms.

Video Examples

Review this example for Objective 2:
2a. Use the quotient rule to rewrite the logarithm.

$$\log_4 \frac{8}{7}$$

$$\log_4 \frac{8}{7} = \log_4 8 - \log_4 7$$

Now Try:
2a. Use the quotient rule to rewrite the logarithm.

$$\log_5 \frac{4}{9}$$

_____

Objective 2 Practice Exercises

For extra help, see Example 2 on page 933 of your text.

Use the quotient rule for logarithms to express each logarithm as a difference of logarithms, or as a single number if possible.

4. $\log_2 \dfrac{5}{m}$

4. _____

5. $\log_6 \dfrac{k}{3}$

5. _____

Use the quotient rule for logarithms to express the difference as a single logarithm.

6. $\log_2 7q^4 - \log_2 5q^2$

6. _____

Objective 3   Use the power rule for logarithms.

Video Examples

Review this example for Objective 3:

3a. Use the power rule to rewrite the logarithm. Assume that $b > 0$, $x > 0$, and $b \neq 1$.

$$\log_4 3^5$$

$$\log_4 3^5 = 5 \log_4 3$$

Now Try:

3a. Use the power rule to rewrite the logarithm. Assume that $b > 0$, $x > 0$, and $b \neq 1$.

$$\log_6 4^3$$

_____

Objective 3 Practice Exercises

For extra help, see Examples 3–4 on page 935 of your text.

Use the power rule for logarithms to rewrite each logarithm or as a single number if possible.

7. $\log_m 2^7$

7. _____

8. $\log_3 \sqrt[3]{5}$

8. _____

9. $3^{\log_3 \sqrt[3]{7}}$

9. _____

Objective 4   Use properties to write alternative forms of logarithmic expressions.

Video Examples

Review these examples for Objective 4:

5. Use the properties of logarithms to rewrite each expression if possible. Assume that all variables represent positive real numbers.

a. $\log_5 6x^3$

$$\log_5 6x^3 = \log_5 6 + \log_5 x^3$$
$$= \log_5 6 + 3 \log_5 x$$

Now Try:

5. Use the properties of logarithms to rewrite each expression if possible. Assume that all variables represent positive real numbers.

a. $\log_6 36x^3$

_____

d. $3\log_b x - \left(2\log_b y + \dfrac{1}{2}\log_b z\right)$

$3\log_b x - \left(2\log_b y + \dfrac{1}{2}\log_b z\right)$

$= \log_b x^3 - (\log_b y^2 + \log_b z^{1/2})$

$= \log_b x^3 - \log_b y^2\sqrt{z}$

$= \log_b \dfrac{x^3}{y^2\sqrt{z}}$

d. $\log_b x + 4\log_b y - \log_b z$

_____

e. $2\log_3 x + \log_3(x-1) - \dfrac{1}{2}\log_3(x+1)$

$2\log_3 x + \log_3(x-1) - \dfrac{1}{2}\log_3(x+1)$

$= \log_3 x^2 + \log_3(x-1) - \log_3(x+1)^{1/2}$

$= \log_3\left(x^2(x-1)\right) - \log_3\sqrt{x+1}$

$= \log_3 \dfrac{x^3 - x^2}{\sqrt{x+1}}$

e.

$2\log_2 x + \log_2(x-1) - \dfrac{1}{3}\log_2(x^2+1)$

_____

Objective 4 Practice Exercises

For extra help, see Examples 5–6 on pages 936–937 of your text.

Use the properties of logarithms to express each logarithm as a sum or difference of logarithms, or as a single number if possible.

10. $\log_2 4p^3$

10. _____

11. $\log_7 \dfrac{8r^7}{3a^3}$

11. _____

Use the properties of logarithms to express the sum or difference of logarithms as a single logarithm, or as a single number if possible.

12. $\log_4 10y + \log_4 3y - \log_4 6y^3$

12. _____

Chapter 12 INVERSE, EXPONENTIAL, AND LOGARITHMIC FUNCTIONS

12.6 Common and Natural Logarithms

Learning Objectives
1   Evaluate common logarithms using a calculator.
2   Use common logarithms in applications.
3   Evaluate natural logarithms using a calculator.
4   Use natural logarithms in applications.

Key Terms

Use the vocabulary terms listed below to complete each statement in exercises 1–2.

common logarithm   **natural logarithm**

1. A logarithm to the base e is a _____ .

2. A logarithm to the base 10 is a _____ .

Objective 1   Evaluate common logarithms using a calculator.

Video Examples

Review these examples for Objective 1:

1. Evaluate each logarithm to four decimal places using a calculator.

   a. $\log 436.2$

   $\log 436.2 \approx 2.6397$

   c. $\log 0.125$

   $\log 0.125 \approx -0.9031$

   b. $\log 543,210$

   $\log 543,210 \approx 5.7350$

Now Try:

1. Evaluate each logarithm to four decimal places using a calculator.

   a. $\log 983.5$

   _____

   c. $\log 0.333$

   _____

   b. $\log 79,315$

   _____

Objective 1 Practice Exercises

For extra help, see Example 1 on page 940 of your text.

Use a calculator to find each logarithm. Give an approximation to four decimal places.

1. $\log 57.23$ 1. _____

2.   log 0.0914   2. _____

3.   log 87,123   3. _____

Objective 2   Use common logarithms in applications.

Video Examples

Review these examples for Objective 2:

2.   Wetlands are classified as bogs, fens, marshes, and swamps, on the basis of pH values. A pH value between 6.0 and 7.5 indicates that the wetland is a "rich fen." When the pH is between 3.0 and 6.0, the wetland is a "poor fen," and if the pH falls to 3.0 or less, it is a "bog." Suppose that the hydronium ion concentration of a sample of water from a wetland is 5.4×10^{-4}. Find the pH value for the water and determine how the wetland should be classified.

$$pH = -\log\left(5.4 \times 10^{-4}\right) \quad \text{Definition of pH}$$

$$= -\left(\log 5.4 + \log 10^{-4}\right) \quad \text{Product rule}$$

$$= -(0.7324 - 4) \quad \begin{array}{l}\text{Use a calculator to}\\\text{find log 5.4.}\end{array}$$

$$= 3.2676$$

Since the pH is between 3.0 and 6.0, the wetland is a poor fen.

3.   Find the hydronium ion concentration of a solution with pH 5.4.

$$pH = -\log\left[H_3O^+\right] \quad \text{Definition of pH}$$

$$5.4 = -\log\left[H_3O^+\right]$$

$$\log\left[H_3O^+\right] = -5.4 \quad \text{Multiply by } -1.$$

$$H_3O^+ = 10^{-5.4} \quad \begin{array}{l}\text{Write in}\\\text{exponential form.}\end{array}$$

$$\approx 4.0 \times 10^{-6} \quad \text{Use a calculator.}$$

Now Try:

2.   Suppose that the hydronium ion concentration of a sample of water from a wetland is 6.2×10^{-8}. Find the pH value for the water and determine how the wetland should be classified.

_____

3.   Find the hydronium ion concentration of a solution with pH 3.6.

_____

Objective 2 Practice Exercises

For extra help, see Examples 2–4 on pages 941–942 of your text.

Solve each problem.

4. Find the pH of a solution with the given hydronium
    ion concentration. Round the answer to the nearest
    tenth.

    a. 4.3×10^{-9} **b.** 2.8×10^{-6}

4. a._____

   b._____

5. Find the decibel level to the nearest whole number of
    the sound with intensity I of $2.5 \times 10^{13} I_0$, find the
    decibel level D to the nearest whole number.

5. _____

6. Find the hydronium ion concentration of a solution
    with the given pH value.

    a. 5.2   **b.** 1.3

6. a._____

   b._____

Objective 3   Evaluate natural logarithms using a calculator.

Video Examples

Review these examples for Objective 3:

5. Using a calculator, evaluate each logarithm to four decimal places.

    a. $\ln 436.2$

$\ln 436.2 \approx 6.0781$

    b. $\ln 54.3$

$\ln 54.3 \approx 3.9945$

    c. $\ln 0.125$

$\log 0.125 \approx -2.0794$

Now Try:

5. Using a calculator, evaluate each logarithm to four decimal places.

    a. $\ln 98$

    _____

    b. $\ln 793$

    _____

    c. $\ln 0.333$

    _____

Objective 3 Practice Exercises

For extra help, see Example 5 on page 943 of your text.

Find each natural logarithm. Give an approximation to four decimal places.

7. $\ln 76.3$

8. $\ln 0.102$

9. $\ln 50$

7. _____

8. _____

9. _____

Objective 4   Use natural logarithms in applications.

Video Examples

Review this example for Objective 4:

6. The time t in years for an investment increasing at a rate of r percent (in decimal form) to double is given by
$$t = \frac{\ln 2}{\ln(1+r)}.$$
This is called the doubling time. Find the doubling time to the nearest tenth for an investment at 4%.

$4\% = 0.04$, so $t = \dfrac{\ln 2}{\ln(1+0.04)} = \dfrac{\ln 2}{\ln 1.04} \approx 17.7$

The doubling time for the investment is about 17.7 years.

Now Try:

6. Use the formula at the left to find the doubling time to the nearest tenth for an investment at 6%.

    _____

Objective 4 Practice Exercises

For extra help, see Example 6 on page 943 of your text.

The time t in years for an amount increasing at a rate of r (in decimal form) to double (the doubling time) is given by $t = \dfrac{\ln 2}{\ln(1+r)}$. *Find the doubling time for an investment at the interest rate. Round to the nearest whole number.*

10.   3%   10. _____

The half-life of a radioactive substance is the time it takes for half of the material to decay. The amount A in pounds of substance remaining after t years is given by $\ln \dfrac{A}{C} = -\dfrac{t}{h}\ln 2$, *where C is the initial amount in pounds, and h is its half-life in years. Use the formula to solve the following problems. Round to the nearest whole number.*

11.   The half-life of radium-226 is 1620 years. How long,   11. _____
      to the nearest year, will it take for 100 pounds to
      decay to 25 pounds?

Newton's Law of Cooling describes the cooling of a warmer object to the cooler temperature of the surrounding environment. The formula can be given as $t = \dfrac{1}{k} \ln \dfrac{T_s - T_1}{T_s - T_2}$, *where t is the elapsed time, T_1 is the initial temperature measurement of the object, T_2 is the second temperature measurement of the object, and T_s is the temperature of the surrounding environment. Use this formula to solve the problem. Round to the nearest tenth.*

12.   A corpse was discovered in a motel room at midnight and its temperature was 80°F. The temperature in the room was 60°F. Assuming that the person's temperature at the time of death was 98.6° F and using $k = 0.1438$, determine t and the time of death.

12. t _____

time _____

Chapter 12 INVERSE, EXPONENTIAL, AND LOGARITHMIC FUNCTIONS

12.7 Exponential and Logarithmic Equations and Their Applications

Learning Objectives
1   Solve equations involving variables in the exponents.
2   Solve equations involving logarithms.
3   Solve applications involving compound interest.
4   Solve applications involving base e exponential growth and decay.
5   Use the change-of-base rule.

Key Terms

Use the vocabulary terms listed below to complete each statement in exercises 1–2.

compound interest   **continuous compounding**

1.   The formula for _____ is $A = Pe^{rt}$.

2.   The formula for _____ is $A = P\left(1 + \frac{r}{n}\right)^{nt}$.

Objective 1   Solve equations involving variables in the exponents.

Video Examples

Review these examples for Objective 1:

1.   Solve $4^x = 30$. Approximate the solution to three decimal places.

$$4^x = 30$$

$\log 4^x = \log 30$ If $x = y$, and $x > 0$, $y > 0$, then $\log_b x = \log_b y$.

$x \log 4 = \log 30$ Power rule

$x = \dfrac{\log 30}{\log 4}$ Divide by log 4.

$x \approx 2.453$ Use a calculator.

Check

$$4^x = 4^{2.453} \approx 30$$

The solution set is {2.453}.

Now Try:

1.   Solve $3^x = 15$. Approximate the solution to three decimal places.

_____

2. Solve $e^{0.005x} = 9$. Approximate the solution to three decimal places.

$$e^{0.005x} = 9$$

$$\ln e^{0.005x} = \ln 9 \qquad \text{If } x = y, \text{ and } x > 0, \\ y > 0, \text{ then } \ln x = \ln y.$$

$$0.005x \ln e = \ln 9 \qquad \text{Power rule}$$

$$0.005x = \ln 9 \qquad \ln e = 1$$

$$x = \frac{\ln 9}{0.005} \qquad \text{Divide by 0.005.}$$

$$x \approx 439.445 \text{ Use a calculator.}$$

The solution set is $\{439.445\}$.

2. Solve $e^{0.4x} = 15$. Approximate the solution to three decimal places.

_____

Objective 1 Practice Exercises

For extra help, see Examples 1–2 on page 946 of your text.

Solve each equation. Give solutions to three decimal places.

1. $25^{x+2} = 125^{3-x}$

1. _____

2. $4^{x-1} = 3^{2x}$

2. _____

3. $e^{0.005x} = 9$

3. _____

Objective 2   Solve equations involving logarithms.

Video Examples

Review these examples for Objective 2:

3. Solve $\log_3 (x-1)^2 = 3$. Give the exact solution.

$$\log_3 (x-1)^2 = 3$$

$$(x-1)^2 = 3^3 \qquad \text{Write in exponential form.}$$

$$(x-1)^2 = 27$$

$$x-1 = \pm\sqrt{27} \quad \begin{array}{l}\text{Take the square}\\\text{root on each side.}\end{array}$$

$$x-1 = \pm 3\sqrt{3} \quad \text{Simplify the square root.}$$

$$x = 1 \pm 3\sqrt{3} \quad \text{Add 1.}$$

Since the domain of $\log_b x$ is $(0,\ \infty)$, we disregard the negative solution, $1-3\sqrt{3}$.

Check:

$$\log_3 (x-1)^2 = 3$$

$$\log_3 \left(1+3\sqrt{3}-1\right)^2 \overset{?}{=} 3$$

$$\log_3 \left(3\sqrt{3}\right)^2 \overset{?}{=} 3$$

$$\log_3 27 \overset{?}{=} 3$$

$$3^3 \overset{?}{=} 27$$

$$27 = 27$$

The solution set is $\left\{1+3\sqrt{3}\right\}$.

4. Solve $\log_3 (5x+42) - \log_3 x = \log_3 26$.

$$\log_3 (5x+42) - \log_3 x = \log_3 26.$$

$$\log_3 \frac{5x+42}{x} = \log_3 26$$

$$\frac{5x+42}{x} = 26 \quad \begin{array}{l}\text{If } \log_b x = \log_b y\\\text{then } x = y.\end{array}$$

$$5x+42 = 26x \quad \text{Multiply by } x.$$

$$42 = 21x \quad \text{Subtract } 5x.$$

$$2 = x \quad \text{Divide by 21.}$$

Now Try:

3. Solve $\log_6 (x+1)^3 = 2$. Give the exact solution.

_____

4. Solve

$$\log_6 (2x+7) - \log_6 x = \log_6 16.$$

_____

Check:

$$\log_3 (5x+42) - \log_3 x = \log_3 26$$

$$\log_3 (5 \cdot 2 + 42) - \log_3 2 \overset{?}{=} \log_3 26$$

$$\log_3 52 - \log_3 2 \overset{?}{=} \log_3 26$$

$$\log_3 \tfrac{52}{2} \overset{?}{=} \log_3 26$$

$$\log_3 26 = \log_3 26$$

The solution set is {2}.

5. Solve $\log_2 (x+7) + \log_2 (x+3) = \log_2 77$.

$$\log_2 (x+7) + \log_2 (x+3) = \log_2 77$$

$$\log_2 [(x+7)(x+3)] = \log_2 77$$

Product rule

$$(x+7)(x+3) = 77$$

If $\log_b x = \log_b y$
then $x = y$.

$$x^2 + 10x + 21 = 77 \quad \text{Multiply.}$$

$$x^2 + 10x - 56 = 0 \quad \text{Subtract 77.}$$

$$(x-4)(x+14) = 0 \quad \text{Factor.}$$

$$x - 4 = 0 \quad \text{or} \quad x + 14 = 0$$

$$x = 4 \qquad\qquad x = -14$$

The value -14 must be rejected since it leads to the logarithm of a negative number in the original equation.
A check shows that the only solution is 4.
The solution set is {4}.

5. Solve.
$$\log_4 (4x-3) + \log_4 x = \log_4 (2x-1)$$

_____

Objective 2 Practice Exercises

For extra help, see Examples 3–5 on pages 947–948 of your text.

Solve each equation. Give exact solution.

4. $\log(-a) + \log 4 = \log(2a+5)$

4. _____

5. $\log_3(x^2 - 10) - \log_3 x = 1$ **5.** _____

6. $\ln(x+4) + \ln(x-2) = \ln 7$ **6.** _____

Objective 3   Solve applications involving compound interest.

Video Examples

Review these examples for Objective 3:

6. How much money will there be in an account at the end of 5 years if $5000 is deposited at 4% compounded monthly?

Because interest is compounded monthly, $n = 12$. The other given values are $P = 5000$, $r = 0.04$, and $t = 5$.

$$A = P\left(1 + \frac{r}{n}\right)^{nt}$$

$$A = 5000\left(1 + \frac{0.04}{12}\right)^{12 \cdot 5}$$

$$A = 5000(1.0033)^{60}$$

$$A = 6104.98$$

There will be $6104.98 in the account at the end of 5 years.

Now Try:

6. How much money will there be in an account at the end of 5 years if $10,000 is deposited at 4% compounded quarterly?

_____

7. Approximate the time it would take for money deposited in an account paying 5% interest compounded quarterly to double. Round to the nearest hundredth.

We want the number of years t for P dollars to grow to $2P$ dollars at a rate of 5% per year. In the compound interest formula, we substitute $2P$ for A, and let $r = 0.05$ and $n = 4$.

$$2P = P\left(1 + \frac{0.05}{4}\right)^{4t}$$

$$2 = 1.0125^{4t}$$

$$\log 2 = \log 1.025^{4t}$$

$$\log 2 = 4t \log 1.0125$$

$$t = \frac{\log 2}{4 \log 1.0125}$$

$$t \approx 13.95$$

It will take about 13.95 years for the investment to double.

8. Suppose that $5000 is invested at 4% interest for 3 years.

a. How much will the investment be worth if it is compounded continuously?

$$A = Pe^{rt}$$

$$A = 5000e^{0.04 \cdot 3}$$

$$A = 5000e^{0.12}$$

$$A = 5637.48$$

The investment will be worth $5637.48.

b. Approximate the amount of time it would take for the investment to double. Round to the nearest tenth.

Find the value of t that will cause A to be $2(\$5000) = \$10,000$.

7. Approximate the time it would take for money deposited in an account paying 5% interest compounded monthly to double. Round to the nearest hundredth.

_____

8. Suppose that $5000 is invested at 2% interest for 3 years.

a. How much will the investment be worth if it is compounded continuously?

_____

b. Approximate the amount of time it would take for the investment to double. Round to the nearest tenth.

_____

$$A = Pe^{rt}$$

$$10{,}000 = 5000e^{0.04t}$$

$2 = e^{0.04t}$ Divide by 5000.

$\ln 2 = \ln e^{0.04t}$ If $x = y$, then $\ln x = \ln y$.

$\ln 2 = 0.04t$ $\ln e^k = k$

$\dfrac{\ln 2}{0.04} = t$ Divide by 0.04.

$t \approx 17.3$

It will take about 17.3 years for the amount to double.

Objective 3 Practice Exercises

For extra help, see Examples 6–8 on pages 949–950 of your text.

Solve each problem.

7. How much will be in an account after 10 years if $25,000 is invested at 8% compounded quarterly? Round to the nearest cent.

7. _____

8. How much will be in an account after 5 years if $10,000 is invested at 4.5% compounded continuously? Round to the nearest cent.

8. _____

9. How long will it take an investment to double if it is placed in an account paying 9% interest compounded continuously? Round to the nearest tenth.

9. _____

Name: Date:
Instructor: Section:

Objective 4   Solve applications involving base e exponential growth and decay.

Video Examples

Review these examples for Objective 4:	Now Try:

Review these examples for Objective 4:

9. A sample of 500 g of lead-210 decays according to the function $y = y_0 e^{-0.032t}$, where t is the time in years, y is the amount of the sample at time t, and y_0 is the initial amount present at $t = 0$.

a. How much lead will be left in the sample after 20 years? Round to the nearest tenth of a gram.

Let $t = 20$ and $y_0 = 500$.

$y = 500e^{-0.032 \cdot 20} \approx 263.6$

There will be about 263.6 grams after 20 years.

b. Approximate the half-life of lead-210 to the nearest tenth.

Let $y = \frac{1}{2}(500) = 250$.

$250 = 500e^{-0.032t}$

$0.5 = e^{-0.032t}$

$\ln 0.5 = \ln e^{-0.032t}$

$\ln 0.5 = -0.032t$

$t = \frac{\ln 0.5}{-0.032} \approx 21.7$

The half-life of lead-210 is about 21.7 years.

Now Try:

9. Cesium-137, a radioactive isotope used in radiation therapy, decays according to the function $y = y_0 e^{-0.0231t}$, where t is the time in years and y_0 is the initial amount present at $t = 0$.

a. If an initial sample contains 36 mg of cesium-137, how much cesium 137 will be left in the sample after 50 years? Round to the nearest tenth.

_____

b. Approximate the half-life of cesium-137 to the nearest tenth.

_____

Name: _____   Date: _____
Instructor: _____   Section: _____

Objective 4 Practice Exercises

For extra help, see Example 9 on pages 950–951 of your text.

Solve each problem.

10. Radioactive strontium decays according to the function $y = y_0 e^{-0.0239t}$, where t is the time in years. If an initial sample contains $y_0 = 15$ g of radioactive strontium, how many grams will be present after 25 years? Round to the nearest hundredth of a gram.

10. _____

11. How long will it take the initial sample of strontium in exercise 19 to decay to half of its original amount?

11. _____

12. The concentration of a drug in a person's system decreases according to the function $C(t) = 2e^{-0.2t}$, where $C(t)$ is given in mg and t is in hours. How much of the drug will be in the person's system after one hour? Approximate answer to the nearest hundredth.

12. _____

Objective 5 Use the change-of-base rule.

Video Examples

Review this example for Objective 5:

10. Evaluate $\log_7 28$ to four decimal places.

$$\log_7 28 = \frac{\log 28}{\log 7} = 1.7124$$

Now Try:

10. Evaluate $\log_5 180$ to four decimal places.

_____

Objective 5 Practice Exercises

For extra help, see Example 10 on page 951 of your text.

Use the change-of-base rule to find each logarithm. Give approximations to four decimal places.

13. $\log_{16} 27$

13. _____

14. $\log_6 0.25$

14. _____

15. $\log_{1/2} 5$

15. _____

Chapter 13 NONLINEAR FUNCTIONS, CONIC SECTIONS, AND NONLINEAR SYSTEMS

13.1   Additional Graphs of Functions

Learning Objectives
1   Recognize the graphs of the absolute values, reciprocal, and square root functions, and graph their translations.
2   Recognize and graph step functions.

Key Terms

Use the vocabulary terms listed below to complete each statement in exercises 1–3.

asymptotes   **greatest integer function**   **step function**

1.   A _____ is a function that looks like a series of steps.

2.   The function defined by $f(x) = [\![x]\!]$ is called the _____.

3.   Lines that a graph approaches without actually touching are called

_____.

Objective 1   **Recognize the graphs of the absolute value, reciprocal, and square root functions and graph their translations.**

Video Examples

Review these examples for Objective 1:

1.   Graph $f(x) = |x + 4|$. Give the domain and range.

The graph of $f(x) = |x + 4|$ is found by shifting the graph of $y = |x|$ four units to the left.

x	y
-6	2
-5	1
-4	0
-3	1
-2	2

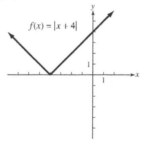

The domain is $(-\infty, \infty)$. The range is $[0, \infty)$.

Now Try:

1.   Graph $f(x) = \sqrt{x + 3}$. Give the domain and range.

_____

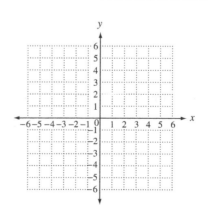

2. Graph $f(x) = \sqrt{x} - 2$. Give the domain and range.

The graph of $y = \sqrt{x} - 2$ is obtained by shifting the graph of $y = \sqrt{x}$ two units down.

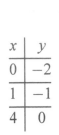

x	y
0	-2
1	-1
4	0

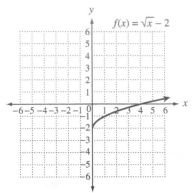

The domain is $[0, \infty)$. The range is $[-2, \infty)$.

3. Graph $f(x) = \dfrac{1}{x+2} - 1$. Give the domain and range.

The graph of $y = \dfrac{1}{x+2} - 1$ is obtained by

shifting the graph of $y = \dfrac{1}{x}$ two units to the left and then one unit down.

x	y	x	y
-6	$-\dfrac{5}{4}$	$-\dfrac{3}{2}$	1
-5	$-\dfrac{4}{3}$	-1	0
-4	$-\dfrac{3}{2}$	0	$-\dfrac{1}{2}$
-3	-2	1	$-\dfrac{2}{3}$
$-\dfrac{5}{2}$	-3	2	$-\dfrac{3}{4}$

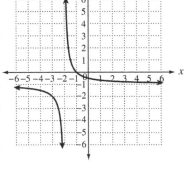

The domain is $(-\infty, -2) \cup (-2, \infty)$.

The range is $(-\infty, -1) \cup (-1, \infty)$.

2. Graph $f(x) = |x| - 1$. Give the domain and range.

_____

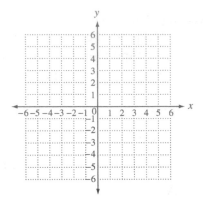

3. Graph $f(x) = \dfrac{1}{x-1} - 2$. Give the domain and range.

_____

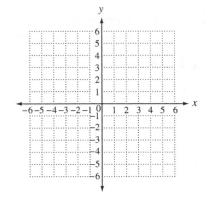

Name: _____ Date: _____

Instructor: _____ Section: _____

Objective 1 Practice Exercises

For extra help, see Examples 1–3 on pages 969–970 of your text.

Graph each function. Give the domain and range.

1. $f(x) = |x - 2| + 3$

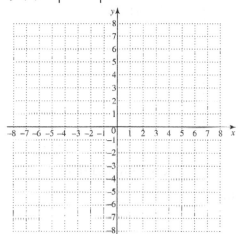

1. domain _____

range _____

2. $f(x) = \sqrt{x} + 3$

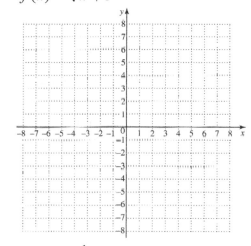

2. domain _____

range _____

3. $f(x) = \dfrac{1}{x - 3}$

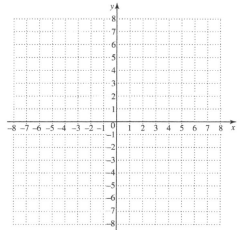

3. domain _____

range _____

Name: _____   Date: _____
Instructor: _____   Section: _____

Objective 2   Recognize and graph step functions.

Video Examples

Review these examples for Objective 2:

4. Evaluate each expression.

   d. $[\![9.76]\!]$

   $$[\![9.76]\!] = 9$$

   e. $[\![-3.5]\!]$

   $$[\![-3.5]\!] = -4$$

5. Graph $f(x) = [\![x]\!] + 2$. Give the domain and range.

   The graph of $y = [\![x]\!] + 2$ is obtained by shifting the graph of $y = [\![x]\!]$ two units up.
   If $-3 \le x < -2$, then $[\![x]\!] + 2 = -3 + 2 = -1$.
   If $-2 \le x < -1$, then $[\![x]\!] + 2 = -2 + 2 = 0$.
   If $-1 \le x < 0$, then $[\![x]\!] + 2 = -1 + 2 = 1$.
   If $0 \le x < 1$, then $[\![x]\!] + 2 = 0 + 2 = 2$, etc.

   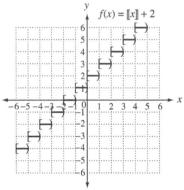

   The domain is $(-\infty, \infty)$. The range is $\{...,-2,-1,\ 0,\ 1,\ 2,...\}$ (the set of integers)

Now Try:

4. Evaluate each expression.

   d. $[\![1.76]\!]$

   _____

   e. $[\![-10.01]\!]$

   _____

5. Graph $f(x) = [\![x-2]\!]$. Give the domain and range.

   _____

   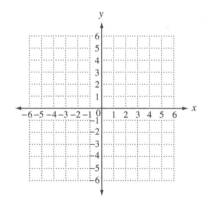

6. In 2011, U.S. first class letter postage is 46¢ for the first ounce and 18¢ for each additional ounce. Let $p(x)$ = the cost of sending a letter that weighs x ounces. Graph the function over the interval (0, 6].

For x in the interval (0, 1], $y = \$0.46$.
For x in the interval (1, 2],
   $y = \$0.46 + \$0.18 = \$0.64$.
For x in the interval (2, 3],
   $y = \$0.64 + \$0.18 = \$0.82$.
For x in the interval (3, 4],
   $y = \$0.82 + \$0.18 = \$1.00$.
For x in the interval (4, 5],
   $y = \$1.00 + \$0.18 = \$1.18$.
For x in the interval (5, 6],
   $y = \$1.18 + \$0.18 = \$1.36$.

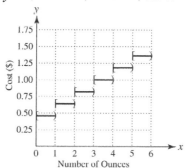

6. At the end of December, Mary's dad lent her $500 to buy an iPad. She agreed to repay the loan at $50 per month on the first of each month, starting in January. If $f(x)$ represents the amount to be repaid in month x, graph the function over its entire domain.

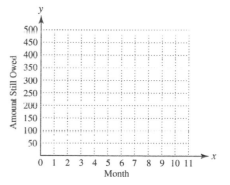

Objective 2 Practice Exercises

For extra help, see Examples 4–6 on pages 970–971 of your text.

Graph each function.

4. $f(x) = [\![x + 2]\!] - 3$

4.

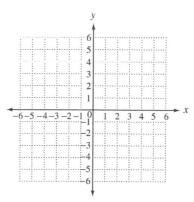

5. The cost of parking a car in an hourly parking lot is $6.00 for the first hour and $4.00 for each additional hour or fraction of an hour. Graph the function *f* that models the cost of parking a car for *x* hours over the interval (0, 6].

5.

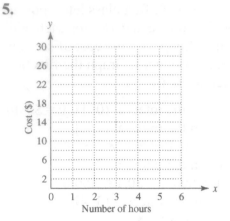

Chapter 13 NONLINEAR FUNCTIONS, CONIC SECTIONS, AND NONLINEAR SYSTEMS

13.2   Circles and Ellipses

Learning Objectives
1   Find an equation of a circle given its center and radius.
2   Determine the center and radius of a circle given its equation.
3   Recognize the equation of an ellipse.
4   Graph ellipses.

Key Terms

Use the vocabulary terms listed below to complete each statement in exercises 1–5.

conic sections   circle   center   radius   ellipse

1. A(n) _____ is the set of all points in a plane that lie a fixed distance from a fixed point.

2. A(n) _____ is the set of all points in a plane the sum of whose distances from two fixed points is constant.

3. Figures that result from the intersection of an infinite cone with a plane are called

    _____.

4. A fixed point such that every point on a circle is a fixed distance from it is the

    _____.

5. The distance from the center of a circle to a point on the circle is called the

    _____.

Objective 1   Find an equation of a circle given its center and radius.

Video Examples

Review this example for Objective 1:

3. Write an equation of a circle with center $(-2,-4)$ and radius $2\sqrt{5}$.

    Use the center-radius form.
    $$(x-h)^2 + (y-k)^2 = r^2$$
    $$[x-(-2)]^2 + [y-(-4)]^2 = \left(2\sqrt{5}\right)^2$$
    $$(x+2)^2 + (y+4)^2 = 20$$

Now Try:

3. Write an equation of a circle with center $(3,-1)$ and radius $\sqrt{6}$.

    _____

Objective 1 Practice Exercises

For extra help, see Examples 1–3 on pages 975–976 of your text.

Find the equation of a circle satisfying the given conditions.

1.   center: (3,–4); radius: 5   1. _____

2.   center: (–2,–2); radius: 3   2. _____

3.   center: (0, 3); radius: $\sqrt{2}$ 3. _____

Objective 2   Determine the center and radius of a circle given its equation.

Video Examples

Review this example for Objective 2:

4. Find the center and radius of the circle
$x^2 + y^2 - 10x + 12y + 52 = 0.$

Complete the squares on x and y.
$x^2 + y^2 - 10x + 12y + 52 = 0$

$(x^2 - 10x) + (y^2 + 12y) = -52$

$\left[\frac{1}{2}(-10)\right]^2 = 25 \quad \left[\frac{1}{2}(12)\right]^2 = 36$

$(x^2 - 10x + 25) + (y^2 + 12y + 36) = -52 + 25 + 36$

$(x - 5)^2 + (y + 6)^2 = 9$

$(x - 5)^2 + [y - (-6)]^2 = 3^2$

The graph is a circle with center at (5,–6) and radius 3.

Now Try:

4. Find the center and radius of the circle
$x^2 + y^2 - 8x - 2y + 15 = 0.$

_____

_____

Objective 2 Practice Exercises

For extra help, see Example 4 on page 977 of your text.

Find the center and radius of each circle.

4. $x^2 + y^2 + 4x + 6y - 3 = 0$ 4. center:_____

                                               radius:_____

5. $3x^2 + 3y^2 + 12y + 30x = 21$ 5. center:_____

                                               radius:_____

6. $x^2 + y^2 + 8x + 4y - 29 = 0$ 6. center:_____

                                               radius:_____

Objective 3   Recognize the equation of an ellipse.

For extra help, see pages 978–979 of your text.

543

Name: Date:

Instructor: Section:

Objective 4 Graph ellipses.

Video Examples

Review these examples for Objective 4:

5b. Graph the ellipse.

$$\frac{x^2}{9} + \frac{y^2}{49} = 1$$

Here, $a = 3$, so the x-intercepts are $(-3, 0)$ and $(3, 0)$.
Similarly, $b = 7$, so the y-intercepts are $(0, 7)$ and $(0, -7)$.
Plot the intercepts and sketch the ellipse.

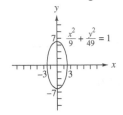

6. Graph $\dfrac{(x-1)^2}{9} + \dfrac{(y+2)^2}{4} = 1$.

The center is $(1, -2)$, $a = 3$, and $b = 2$.
The ellipse passes through the points $(4, -2)$, $(1, -4)$, $(-2, -2)$, and $(1, 0)$.

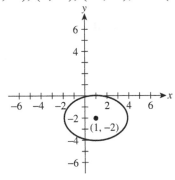

Now Try:

5b. Graph the ellipse.

$$\frac{x^2}{25} + \frac{y^2}{36} = 1$$

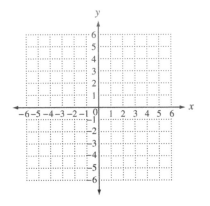

6. Graph $\dfrac{(x+1)^2}{4} + \dfrac{(y-2)^2}{9} = 1$.

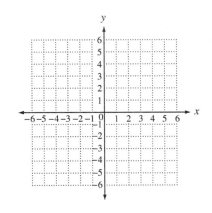

Name: _____ Date: _____

Instructor: _____ Section: _____

Objective 4 Practice Exercises

For extra help, see Examples 5–6 on pages 979–980 of your text.

Graph each ellipse.

7. $\dfrac{x^2}{16} + \dfrac{y^2}{49} = 1$

7.

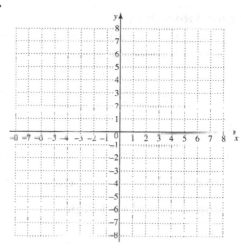

8. $\dfrac{x^2}{25} + \dfrac{y^2}{81} = 1$

8.

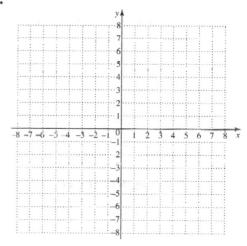

9. $\dfrac{(x+3)^2}{9} + \dfrac{(y \mid 3)^2}{4} = 1$

9.

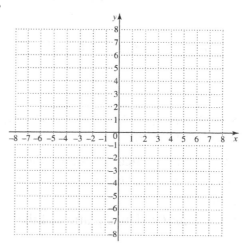

Chapter 13 NONLINEAR FUNCTIONS, CONIC SECTIONS, AND NONLINEAR SYSTEMS

13.3   Hyperbolas and Functions Defined by Radicals

Learning Objectives
1   Recognize the equation of a hyperbola.
2   Graph hyperbolas by using asymptotes.
3   Identify conic sections by their equations.
4   Graph generalized square root functions.

Key Terms

Use the vocabulary terms listed below to complete each statement in exercises 1–4.

hyperbola   **transverse axis**   **asymptotes of a hyperbola**

fundamental rectangle

1.   A(n) _____ is the set of all points in a plane such that the absolute value of the difference of the distances from two fixed points is constant.

2.   The two intersecting lines that the branches of a hyperbola approach are the_____.

3.   The asymptotes of a hyperbola are the extended diagonals of its _____.

4.   The line segment joining the two vertices of a hyperbola is called the _____.

Objective 1   Recognize the equation of a hyperbola.

For extra help, see page 985 of your text.

Objective 2   Graph hyperbolas by using asymptotes.

Video Examples

Review this example for Objective 2:	Now Try:
1. Graph $\dfrac{x^2}{25} - \dfrac{y^2}{9} = 1$.	1. Graph $\dfrac{x^2}{25} - \dfrac{y^2}{4} = 1$.

Step 1 Here $a = 5$ and $b = 3$. The x-intercepts are (5, 0) and (–5, 0).

Step 2 The vertices of the fundamental rectangle are the four pairs (5, 3), (–5, 3), (5,–3), and (–5, –3).

Steps 3 and 4 The equation of the asymptotes

are $y = \pm \dfrac{b}{a}x$, or $y = \pm \dfrac{3}{5}x$.

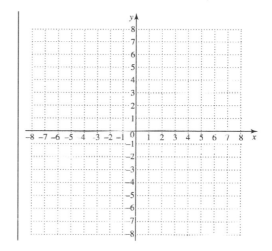

Objective 2 Practice Exercises

For extra help, see Examples 1–2 on page 986 of your text.

Graph each hyperbola.

1. $\dfrac{x^2}{9} - \dfrac{y^2}{16} = 1$

1.

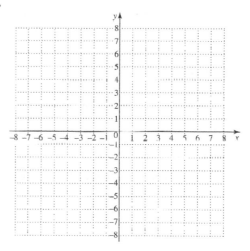

2. $\dfrac{x^2}{36} - \dfrac{y^2}{49} = 1$

2.

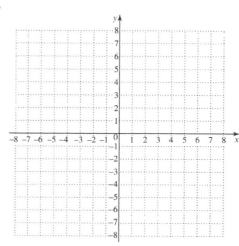

3. $\dfrac{y^2}{4} - \dfrac{x^2}{4} = 1$

3.

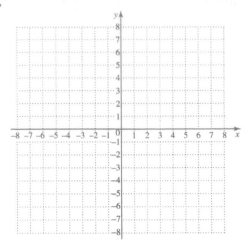

Objective 3   Identify conic sections by their equations.

Video Examples

Review these examples for Objective 3:

3. Identify the graph of each equation.

a. $25y^2 + 100 = 4x^2$

$4x^2 - 25y^2 = 100$

$\dfrac{x^2}{25} - \dfrac{y^2}{4} = 1$

Because of the subtraction symbol, the graph is a hyperbola.

b. $16x^2 + 9y = 144$

Only one of the two variables, x is squared, so this is a vertical parabola with equation

$9y = -16x^2 + 144$

$y = -\dfrac{16}{9}x^2 + 16$

c. $5x^2 = 25 - 5y^2$

$5x^2 + 5y^2 = 25$

$x^2 + y^2 = 5$

The graph of this equation is a circle with center at the origin and radius $\sqrt{5}$.

Now Try:

3. Identify the graph of each equation.

a. $3x^2 = 3y^2 + 1$

_____

b. $x^2 = 16 - y$

_____

c. $3x^2 + 3y^2 = 1$

_____

548

Objective 3 Practice Exercises

For extra help, see Example 3 on page 988 of your text.

Identify each of the following as the equation of a parabola, a circle, an ellipse, or a hyperbola.

4. $16x^2 + 16y^2 = 64$

4. _____

5. $2x^2 + 4y^2 = 8$

5. _____

6. $3x^2 - 3y = 9$

6. _____

Objective 4   Graph generalized square root functions.

Video Examples

Review these examples for Objective 4:

4.   Graph $f(x) = \sqrt{36 - x^2}$. Give the domain and range.

$$f(x) = \sqrt{36 - x^2}$$
$$y = \sqrt{36 - x^2}$$
$$y^2 = \left(\sqrt{36 - x^2}\right)^2$$
$$y^2 = 36 - x^2$$
$$x^2 + y^2 = 36$$

This is a graph of a circle with center at (0, 0), and radius 6. Since the function represents a principal square root in the original equation, the graph must be nonnegative, the upper half of the circle. The domain is [–6, 6] and

Now Try:

4.   Graph $f(x) = -\sqrt{4 - x^2}$. Give the domain and range.

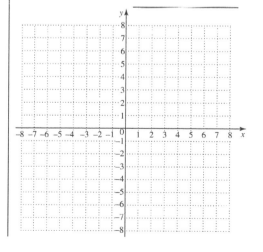

_____

the range is [0, 6].

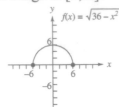

5. Graph $f(x) = -5\sqrt{1 - \dfrac{x^2}{9}}$. Give the domain and range.

$$y = -5\sqrt{1 - \frac{x^2}{9}}$$

$$\frac{y}{5} = -\sqrt{1 - \frac{x^2}{9}}$$

$$\left(\frac{y}{5}\right)^2 = \left(-\sqrt{1 - \frac{x^2}{9}}\right)^2$$

$$\frac{y^2}{25} = 1 - \frac{x^2}{9}$$

$$\frac{x^2}{9} + \frac{y^2}{25} = 1$$

This is an equation of an ellipse with x-intercepts (–3, 0) and (3, 0), and y-intercepts (0, –5) and (0, 5). Since the original equation is negative, y must be nonpositive, restricting the graph to the lower half of the ellipse. The domain is [–3, 3] and the range is [–5, 0].

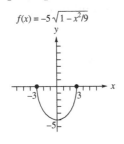

5. Graph $f(x) = \sqrt{9 - 9x^2}$. Give the domain and range.

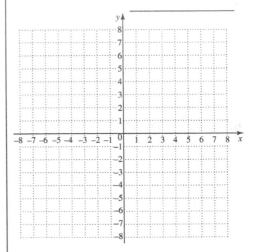

Name: _____ Date: _____
Instructor: _____ Section: _____

Objective 4 Practice Exercises

For extra help, see Examples 4–5 on page 989 of your text.

Graph each function. Give the domain and range.

7. $f(x) = \sqrt{25 - x^2}$

7.

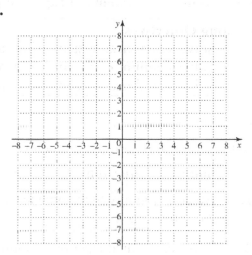

8. $f(x) = -\sqrt{9 - x^2}$

8.

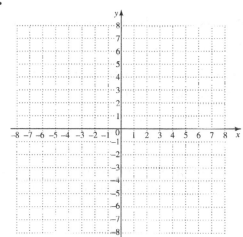

9. $f(x) = \sqrt{1 + \dfrac{x^2}{4}}$

9.

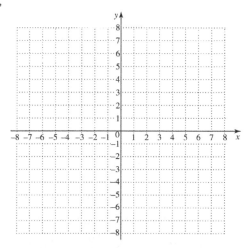

Chapter 13 NONLINEAR FUNCTIONS, CONIC SECTIONS, AND NONLINEAR SYSTEMS

13.4 Nonlinear Systems of Equations

Learning Objectives
1   Solve a nonlinear system using substitution.
2   Solve a nonlinear system with two second-degree equations using elimination.
3   Solve a nonlinear system that requires a combination of methods.

Key Terms

Use the vocabulary terms listed below to complete each statement in exercises 1–2.

nonlinear equation   **nonlinear system of equations**

1. An equation in which some terms have more than one variable or a variable of degree 2 or greater is a _____.

2. A system with at least one nonlinear equation is a _____.

Objective 1   Solve a nonlinear system using substitution.

Video Examples

Review these examples for Objective 1:

1. Solve the system.

$$x^2 + y^2 = 17 \quad (1)$$
$$2x = y + 9 \quad (2)$$

The graph of (1) is a circle and the graph of (2) is a line. Solve linear equation (2) for y.

$$2x = y + 9 \quad (2)$$
$$y = 2x - 9 \quad (3)$$

Substitute $2x - 9$ for y in equation (1).

$$x^2 + y^2 = 17$$
$$x^2 + (2x - 9)^2 = 17$$
$$x^2 + 4x^2 - 36x + 81 = 17$$
$$5x^2 - 36x + 64 = 0$$
$$(5x - 16)(x - 4) = 0$$
$$5x - 16 = 0 \quad \text{or} \quad x - 4 = 0$$
$$x = \frac{16}{5} \quad \text{or} \quad x = 4$$

Let $x = \frac{16}{5}$ in equation (3) to get $y = -\frac{13}{5}$.

Let $x = 4$ in equation (3) to get $y = -1$.

The solution set is $\left\{\left(\frac{16}{5}, -\frac{13}{5}\right), (4, -1)\right\}$.

Now Try:

1. Solve the system.

$$x^2 + y^2 = 13 \quad (1)$$
$$x - y = 1 \quad (2)$$

_____

2. Solve the system.

$$xy = -6 \quad (1)$$
$$x + y = 1 \quad (2)$$

The graph of equation (1) is a hyperbola. The graph of equation (2) is a line. Solving $xy = -6$ for x gives $x = -\dfrac{6}{y}$. Substitute $-\dfrac{6}{y}$ for x in equation (2).

$$x + y = 1$$
$$-\frac{6}{y} + y = 1$$
$$6 + y^2 - y$$
$$y^2 - y - 6 = 0$$
$$(y+2)(y-3) = 0$$
$$y + 2 = 0 \quad \text{or} \quad y - 3 = 0$$
$$y = -2 \quad \text{or} \quad y = 3$$

Substitute the results into $x = -\dfrac{6}{y}$ to obtain the corresponding values of x.

If $y = -2$, then $x = 3$.
If $y = 3$, then $x = -2$.

The solution set of the system is $\{(-2, 3),(3,-2)\}$.

2. Solve the system.

$$xy = -10 \quad (1)$$
$$2x - y = 9 \quad (2)$$

_____

Objective 1 Practice Exercises

For extra help, see Examples 1–2 on pages 993–994 of your text.

Solve each system by the substitution method.

1. $2x^2 - y^2 = -1$
       $2x + y = 7$

1. _____

2. $y = x^2 - 3x - 8$

$x = y + 3$

2. _____

3. $x = y^2 + 5y$

$3y = x$

3. _____

Objective 2   Solve a nonlinear system with two second-degree equations using elimination.

Video Examples

Review this example for Objective 2:

3. Solve the system.

$$x^2 - y^2 = 3 \quad (1)$$

$$2x^2 + y^2 = 9 \quad (2)$$

The graph of equation (1) is a hyperbola, while the graph of equation (2) is an ellipse. Solve using elimination.

$$x^2 - y^2 = 3 \quad (1)$$

$$\underline{2x^2 + y^2 = 9} \quad (2)$$

$$3x^2 = 12$$

$$x^2 = 4$$

$$x = 2 \quad \text{or} \quad x = -2$$

Now Try:

3. Solve the system.

$$x^2 + y^2 = 10 \quad (1)$$

$$2x^2 - y^2 = -7 \quad (2)$$

_____

Copyright © 2018 Pearson Education, Inc.

Use equation (2) to solve for y.

$$2x^2 + y^2 = 9 \qquad\qquad 2x^2 + y^2 = 9$$
$$2(2)^2 + y^2 = 9 \qquad 2(-2)^2 + y^2 = 9$$
$$y^2 = 1 \qquad\qquad\qquad y^2 = 1$$
$$y = -1 \quad \text{or} \quad y = 1 \quad\bigg|\quad y = -1 \quad \text{or} \quad y = 1$$

The solution set is $\{(2, 1), (2,-1), (-2, 1), (-2,-1)\}$.

Objective 2 Practice Exercises

For extra help, see Example 3 on page 995 of your text.

Solve each system using the elimination method.

4. $2x^2 + y^2 = 54$

       $x^2 - 3y^2 = 13$

4. _____

5. $2x^2 - 3y^2 = -19$

       $4x^2 + y^2 = 25$

5. _____

6. $5x^2 + y^2 = 6$

       $2x^2 - 3y^2 = -1$

6. _____

Objective 3   Solve a nonlinear system that requires a combination of methods.

Video Examples

Review this example for Objective 3:

4. Solve the system.

$$5x^2 - xy + 5y^2 = 89 \quad (1)$$
$$x^2 + y^2 = 17 \quad (2)$$

First, use the elimination method.

$$\begin{aligned} 5x^2 - xy + 5y^2 &= 89 \quad (1) \\ -5x^2 \qquad -5y^2 &= -85 \quad \text{Multiply (2) by } -5. \\ \hline -xy &= 4 \end{aligned}$$

$$y = -\frac{4}{x} \quad (3)$$

Now, substitute $y = -\dfrac{4}{x}$ in equation (2).

$$x^2 + y^2 = 17$$
$$x^2 + \left(-\frac{4}{x}\right)^2 = 17$$
$$x^2 + \frac{16}{x^2} = 17$$
$$x^4 + 16 = 17x^2$$
$$x^4 - 17x^2 + 16 = 0$$
$$(x^2 - 16)(x^2 - 1) = 0$$

$$x^2 - 16 = 0 \quad \text{or} \quad x^2 - 1 = 0$$
$$x^2 = 16 \quad \text{or} \qquad x^2 = 1$$

$x = -4$ or $x = 4$ $\quad x = -1$ or $x = 1$

Substituting these four values of x into equation (3), gives the corresponding values of x.

If $x = -4$, then $y = -\dfrac{4}{-4} = 1.$

If $x = 4$, then $y = -\dfrac{4}{4} = -1.$

If $x = -1$, then $y = -\dfrac{4}{-1} = 4.$

If $x = 1$, then $y = -\dfrac{4}{1} = -4.$

Solution set: $\{(1,-4), (-1, 4), (4,-1), (-4, 1)\}.$

Now Try:

4. Solve the system.

$$4x^2 - 2xy + 4y^2 = 64 \quad (1)$$
$$x^2 + y^2 = 13 \quad (2)$$

_____

Name: Date:
Instructor: Section:

Objective 3 Practice Exercises

For extra help, see Example 4 on pages 996–997 of your text.

Solve each system.

7. $x^2 + xy + y^2 = 43$

    $x^2 + 2xy + y^2 = 49$

7. _____

8. $x^2 + xy - y^2 = 5$

    $-x^2 + xy + y^2 = -1$

8. _____

9. $2x^2 + 3xy - 2y^2 = 50$

    $x^2 - 4xy - y^2 = -41$

9. _____

Chapter 13 NONLINEAR FUNCTIONS, CONIC SECTIONS, AND NONLINEAR SYSTEMS

13.5 Second-Degree Inequalities and Systems of Inequalities

Learning Objectives
1   Graph second-degree inequalities.
2   Graph the solution set of a system of inequalities.

Key Terms

Use the vocabulary terms listed below to complete each statement in exercises 1–2.

second-degree inequality   **system of inequalities**

1. A _____ consists of two or more inequalities to be solved at the same time.

2. A(n)_____ is an inequality with at least one variable of degree 2 and no variable with degree greater than 2.

Objective 1 Graph second-degree inequalities.

Video Examples

Review these examples for Objective 1:

2. Graph $y < -x^2 + 3$.

The boundary, $y = -x^2 + 3$, is a parabola that opens down with vertex (0, 3).
Use (0, 0) as a test point.

$$y < -x^2 + 3$$
$$\overset{?}{0 < -0^2 + 3}$$
$$0 < 3 \quad \text{True}$$

Because the final inequality is a true statement, the points in the region containing (0, 0) satisfy the inequality. The parabola is drawn as a dashed curve since the points on the parabola itself do not satisfy the inequality and the region inside (or below) the parabola is shaded.

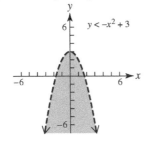

Now Try:

2. Graph $y \geq x^2 - 4$.

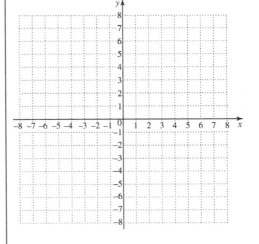

3. Graph $16x^2 < 9y^2 + 144$.

Rewrite the inequality.

$$16x^2 - 9y^2 < 144$$

$$\frac{x^2}{9} - \frac{y^2}{16} < 1$$

The boundary, drawn as a dashed curve, is the

following hyperbola. $\frac{x^2}{9} - \frac{y^2}{16} = 1$

Since the graph is a horizontal hyperbola, the desired region will be either between the branches or the regions to the right of the right branch and to the left of the left branch. Using the test point (0, 0) into the original inequality leads to $0 < 1$, a true statement. So the region between the branches containing (0, 0) is shaded.

$16x^2 < 9y^2 + 144$

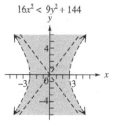

3. Graph $9y^2 - 36x^2 > 144$.

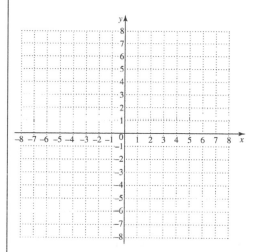

Objective 1 Practice Exercises

For extra help, see Examples 1–3 on pages 1000–1001 of your text.

Graph each inequality.

1. $x \le 2y^2 + 8y + 9$

1.

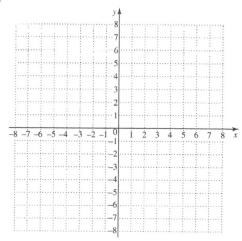

2. $x^2 + 9y^2 > 36$

2.

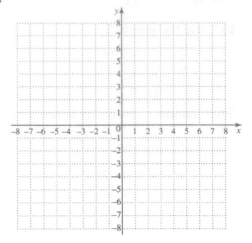

3. $9x^2 - y^2 < 36$

3.

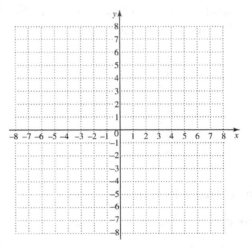

Name: Date:
Instructor: Section:

Objective 2   Graph the solution set of a system of inequalities.

Video Examples

Review these examples for Objective 2:

5. Graph the solution set of the system.

$$x^2 + y^2 \leq 16$$

$$y > x$$

Begin by graphing $x^2 + y^2 \leq 16$. The boundary line is a circle centered at the origin with radius 4. The test point $(0, 0)$ leads to a true statement, so we shade inside the circle.

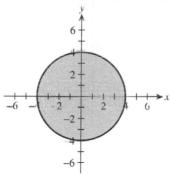

The boundary of the solution set of $y > x$ is a dashed line passing through $(0, 0)$. Using the test point $(0, 1)$ leads to a true statement, so shade above the line.

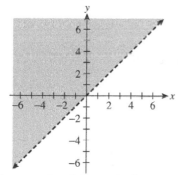

The graph of the solution set of the system is the intersection of the graphs of the two inequalities.

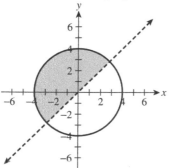

Now Try:

5. Graph the solution set of the system.

$$4y + x^2 < 0$$

$$x \geq 0$$

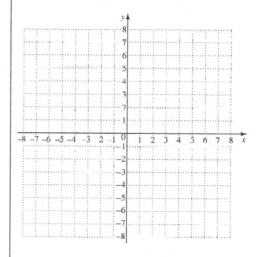

7. Graph the solution set of the system.

$$25x^2 + 9y^2 < 225$$
$$y \le -x^2 + 4$$
$$y < -x$$

The graph of $25x^2 + 9y^2 < 225$ is a dashed ellipse with $a = 3$ and $y = 5$. To satisfy the inequality, a point must lie inside the ellipse.

The graph of $y \le -x^2 + 4$ is a parabola with vertex (0, 4) opening downward. The inequality includes the points on the boundary along with the points inside the parabola. The graph of $y < -x$ includes all points below the line $y = -x$. Therefore, the graph of the system is the shaded region, which lies inside the ellipse and the parabola, and below the line.

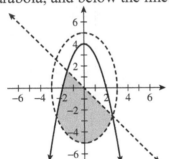

7. Graph the solution set of the system.

$$y \ge (x+2)^2 - 5$$
$$2x + y < -5$$
$$(x+3)^2 + (y-1)^2 < 9$$

Objective 2 Practice Exercises

For extra help, see Examples 4–7 on pages 1001–1003 of your text.

Graph each system of inequalities.

4. $-x + y > 2$
        $3x + y > 6$

4.

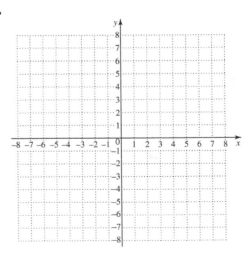

Copyright © 2018 Pearson Education, Inc.

5. $x^2 + y^2 \leq 25$

       $3x - 5y > -15$

5.

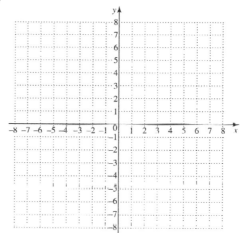

6. $x^2 + 4y^2 \leq 36$

       $-5 < x < 2$

         $y \geq 0$

6.

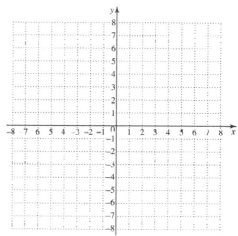

Chapter R PREALGEBRA REVIEW

R.1   Fractions

Key Terms

1. equivalent fractions   2. improper fraction   3. numerator

4. proper fraction   5. denominator   6. composite number

7. prime factorization   8. prime number   9. lowest terms

Objective 1

Practice Exercises

1. prime   3. neither

Objective 2

Now Try

2a. $5 \cdot 11$ 2b. $2 \cdot 3 \cdot 5 \cdot 7$

Practice Exercises

5. $2 \cdot 2 \cdot 2 \cdot 2 \cdot 2 \cdot 2 \cdot 2 \cdot 2$

Objective 3

Now Try

3c. $\dfrac{3}{4}$

Practice Exercises

7. $\dfrac{7}{25}$ 9. $\dfrac{33}{73}$

Objective 4

Now Try

4. $14\dfrac{4}{5}$ 5. $\dfrac{86}{7}$

Practice Exercises

11. $\dfrac{122}{9}$

Objective 5

Now Try

7b. $\dfrac{16}{21}$ 7c. $\dfrac{1}{10}$

Practice Exercises

13. $\dfrac{15}{2}$ or $7\dfrac{1}{2}$ 15. $\dfrac{45}{4}$ or $11\dfrac{1}{4}$

Objective 6

Now Try

9a. $\dfrac{23}{24}$ 10b. $\dfrac{41}{42}$

Practice Exercises

17. $\dfrac{125}{12}$ or $10\dfrac{5}{12}$

Objective 7
 Now Try

12. $14\dfrac{2}{9}$ pies

 Practice Exercises

19. $220.50 $3\dfrac{1}{4}$ yards   21. $34\dfrac{5}{8}$ yards

Objective 8
 Now Try
13c. 104 workers

 Practice Exercises
23. About 780 workers

R.2   Decimals and Percents

Key Terms
 1. decimals   2. place value   3. percent

Objective 1
 Now Try
 1c. $\dfrac{37,058}{10,000}$

 Practice Exercises
 1. $\dfrac{7}{1000}$ 3. $\dfrac{300,005}{10,000}$

Objective 2
 Now Try
 2c. 7.024

 Practice Exercises
 5. 755.098

Objective 3
 Now Try
 3a. 251.116   4a. 32.4   5a. 5130.2

 5b. 0.0986

 Practice Exercises
 7. 2.3424   9. 0.4292

Objective 4
  Now Try
6a. 0.35   6b. $3.\overline{5}$; 3.556

  Practice Exercises
11. $0.\overline{4}$ or 0.444

Objective 5
  Now Try
9c. 0.0043   9d. 91%

  Practice Exercises
13. 3.62   15. 0.84%

Objective 6
  Now Try
10a. $\dfrac{13}{25}$

  Practice Exercises
17. $\dfrac{7}{200}$

Objective 7
  Now Try
12. $14, $42 saved

  Practice Exercises
19. $810   21. 20%

Chapter 1 THE REAL NUMBER SYSTEM

1.1   Exponents, Order of Operations, and Inequality

Key Terms
1. exponential expression
2. base
3. inequality
4. exponent

Objective 1
Now Try
1a. 81

Practice Exercises
1. 27
3. 0.16

Objective 2
Now Try
2a. 85

Practice Exercises
5. 45

Objective 3
Now Try
3a. 215

Practice Exercises
7. −8
9. 48

Objective 4
Now Try
4c. true

Practice Exercises
11. false

Objective 5
Now Try
5d. $5 > 3$

Practice Exercises
13. $7 = 13 - 6$
15. $20 \geq 2 \cdot 7$

Objective 6
Now Try
6a. $11 < 15$

Practice Exercises
17. $8 \leq 12$

1.2   Variables, Expressions, and Equations

Key Terms

1. equation
2. variable
3. algebraic expression
4. solution
5. constant

Objective 1
  Now Try
1d. 252, 567   2a. 64   2b. 20

  Practice Exercises
1. 8   3. $\dfrac{28}{13}$

Objective 2
  Now Try
3c. $12 - x$ 3f. $5(x - 6)$

  Practice Exercises
5. $8x - 11$

Objective 3
  Now Try
4a. yes   4b. no

  Practice Exercises
7. no   9. yes

Objective 4
  Now Try
5c. $3x - 7 = 12$

  Practice Exercises
11. $6(5 + x) = 19$

Objective 5
  Now Try
6a. equation   6b. expression

  Practice Exercises
13. expression   15. equation

1.3   Real Numbers and the Number Line

Key Terms

1. whole numbers
2. additive inverse
3. integers
4. natural numbers
5. absolute value
6. number line
7. irrational number
8. coordinate
9. negative number
10. positive number
11. real numbers
12. set-builder notation
13. signed numbers
14. rational number

Answers

Objective 1
Now Try
1a. -282

2.

3a. 7

3b. 0, 7

3c. $-10, 0, 7$

3d. $-10, -\dfrac{5}{8}, 0, 0.\overline{4}, 5\dfrac{1}{2}, 7, 9.9$

3e. $\sqrt{5}$

3f. all of the numbers

Practice Exercises
1. -75 pounds

3.

Objective 2
Now Try
4. true

Practice Exercises
5. false

Objective 3
Practice Exercises
7. 25

9. -4.5

Objective 4
Now Try
5b. 10

5c. 10

5e. -10

Practice Exercises
11. 1.22

1.4 Adding Real Numbers

Key Terms
1. sum

2. addends

Objective 1
Now Try
1a. 6

1b. -5

2a. -12

2b. -31

2c. -40

Practice Exercises
1. -18

3. $-5\dfrac{5}{8}$

Objective 2
Now Try
4a. -12

4b. -7

Practice Exercises

5. $-\dfrac{1}{6}$

Objective 3
  Now Try
5b. 15

  Practice Exercises
7. –3   9. –6.8

Objective 4
  Now Try
6a. $-10 + 11 + 2; 3$ 6b. $(-9 + 15) + 6; 12$

  Practice Exercises
11. $-10 + [20 + (-4)]; 6$

1.5   Subtracting Real Numbers

Key Terms
  1. minuend   2. subtrahend   3. difference

Objective 1
  Now Try
  1. 2

  Practice Exercises
  1. 3   3. –7

Objective 2
  Now Try
2b. –4   2c. –38   2d. 2

2e. $\dfrac{61}{45}$

  Practice Exercises
  5. 4.4

Objective 3
  Now Try
3a. –5   3b. $\dfrac{19}{2}$

  Practice Exercises
  7. 18   9. $-\dfrac{23}{18}$

Objective 4
  Now Try
4a. $-17 - 9; -26$ 4b. $[25 + (-6)] - 8; 11$ 5. 5464 ft

Practice Exercises

11. $(-4+12)-9; \; -1$

1.6 Multiplying and Dividing Real Numbers

Key Terms

1. quotient   2. reciprocals   3. product

4. dividend   5. divisor

Objective 1
  Now Try

1a. −56   1c. −45.08

  Practice Exercises
  1. −28   3. −13.12

Objective 2
  Now Try

2a. 30   2b. 135

  Practice Exercises

5. $\dfrac{4}{5}$

Objective 3
  Now Try

3a. 3   3b. −2   3c. 3

3d. $\dfrac{4}{11}$

  Practice Exercises
  7. 6   9. undefined

Objective 4
  Now Try

4d. $-\dfrac{5}{4}$

  Practice Exercises

11. $-\dfrac{11}{2}$

Objective 5
  Now Try

5a. −432   5b. 103

  Practice Exercises

13. 7   15. $\dfrac{19}{4}$

Objective 6
  Now Try
  6a. $16[5+(-7)]$; -32 6d. $0.08[18-(-4)]$; 1.76
  Practice Exercises
  17. $85-\dfrac{3}{10}[50-(-10)]$; 67

Objective 7
  Now Try
  8d. $\dfrac{36}{x}=-4$
  Practice Exercises
  19. $\dfrac{2}{3}x=-7$ 21. $\dfrac{x}{-4}=1$

1.7   Properties of Real Numbers

Key Terms
  1. identity element for addition
  2. identity element for multiplication

Objective 1
  Now Try
  1a. -12 1b. 2
  Practice Exercises
  1. 4   3. $(4+z)$

Objective 2
  Now Try
  2a. 4   2b. $[(-3)\cdot 4]$ 4a.   107
  4b. 12,600
  Practice Exercises
  5. $[(-4+3y)]$

Objective 3
  Now Try
  6a. $\dfrac{7}{9}$
  Practice Exercises
  7. 4   9. $\dfrac{6}{7}$

Objective 4
  Now Try
  7b. -8 7d. $\dfrac{5}{8}$

Practice Exercises
11. 0; identity

Objective 5
Now Try
8f. 375   9c. $4x+5y-z$

Practice Exercises
13. $2an-4bn+6cn$ 15. $2k-7$

1.8 Simplifying Expressions

Key Terms
1. numerical coefficient   2. term

3. like terms

Objective 1
Now Try
1c. $71+18x$ 1d. $11-7x$

Practice Exercises
1. $8x+27$ 3. $10x+3$

Objective 2
Practice Exercises
5. $\dfrac{7}{9}$

Objective 3
Practice Exercises
7. like   9. unlike

Objective 4
Now Try
2c. $19x$ 3a. $49y+15$

Practice Exercises
11. $2x-14$

Objective 5
Now Try
4. $11+10x+8x+4x$; $11+22x$

Practice Exercises
13. $6x+12+4x=10x+12$
15. $4(2x-6x)+6(x+9)=-10x+54$

Chapter 2 EQUATIONS, INEQUALITIES, AND APPLICATIONS

2.1   The Addition Property of Equality

Key Terms
1. equivalent equations
2. linear equation
3. solution set

Objective 1
Practice Exercises
1. no
3. yes

Objective 2
Now Try
1. $\{21\}$
3. $\{-19\}$
4. $\{7\}$

Practice Exercises
5. $\left\{\dfrac{1}{2}\right\}$

Objective 3
Now Try
7. $\{29\}$
8. $\{8\}$

Practice Exercises
7. $\{7\}$
9. $\{7.2\}$

2.2   The Multiplication Property of Equality

Key Terms
1. multiplication property of equality
2. addition property of equality

Objective 1
Now Try
1. $\{14\}$
6. $\{39\}$
4. $\{24\}$
5. $\{36\}$

Practice Exercises
1. $\{-17\}$
3. $\{6.4\}$

Objective 2
Now Try
7. $\{4\}$

Practice Exercises
5. $\{-5\}$

2.3   More on Solving Linear Equations

Key Terms
1. contradiction   2. conditional equation   3. identity

Objective 1
Now Try
3. $\{4\}$

Practice Exercises
1. $\left\{\dfrac{5}{2}\right\}$ 3. $\left\{-\dfrac{1}{5}\right\}$

Objective 2
Now Try
6. {all real numbers}   7. $\varnothing$

Practice Exercises
5. {all real numbers}

Objective 3
Now Try
8. $\{-18\}$ 10. $\{2\}$

Practice Exercises
7. $\{2\}$ 9. $\{10\}$

Objective 4
Now Try
11a. $67 - t$

Practice Exercises
11. $\dfrac{17}{p}$

2.4   An Introduction to Applications of Linear Equations

Key Terms
1. supplementary angles   2. complementary angles

3. right angle   4. straight angle   5. consecutive integers

Objective 1
Practice Exercises
1. Read the problem; assign a variable to represent the unknown; write an equation; solve the equation; state the answer; check the answer.

Objective 2
Now Try
1. 16

Practice Exercises
3. $-2(4-x) = 24$; 16

Objective 3
Now Try
2. 52

Practice Exercises
5. $x + (x + 5910) = 34,730$; Mt. Rainier: 14,410 ft; Mt. McKinley: 20,320 feet
7. $x + (5 + 3x) + 4x = 29$; Mark: 3 laps; Pablo: 14 laps; Faustino: 12 laps

Objective 4
Now Try
5. 352 and 353   6. $-2, 0, 2, 4$

Practice Exercises
9. 27, 28

Objective 5
Now Try
7. 18°

Practice Exercises
11. 133°   13. 27°

2.5   Formulas and Additional Applications from Geometry

Key Terms
1. vertical angles   2. formula   3. perimeter

4. area

Objective 1
Now Try
1a. $W = 5.5$

Practice Exercises
1. $a = 36$ 3. $h = 12$

Objective 2
Now Try
2. 12 ft   3. 15 ft, 20 ft, 30 ft

Practice Exercises
5. 1.5 years

Objective 3
Now Try
5b. 54°, 126°

Answers

Practice Exercises
7. 35°, 35°   9. 129°, 51°

Objective 4
Now Try
6. $t = \dfrac{d}{r}$ 7. $a = P - b - c$ 9b. $y = 6x + 5$

Practice Exercises
11. $n = \dfrac{S}{180} + 2$ or $n = \dfrac{S + 360}{180}$

2.6 Ratio, Proportion, and Percent

Key Terms
1. proportion   2. ratio   3. terms

4. cross products

Objective 1
Now Try
1a. $\dfrac{11}{17}$ 1b. $\dfrac{4}{15}$ 2. 24-ounce jar, $0.054 per oz

Practice Exercises
1. $\dfrac{8}{3}$ 3. 45-count box

Objective 2
Now Try
4. $\left\{-\dfrac{3}{4}\right\}$

Practice Exercises
5. $\left\{\dfrac{10}{3}\right\}$

Objective 3
Now Try
5. $259.20

Practice Exercises
7. 15 inches   9. $135

Objective 4
Now Try
6a. 140   6b. 950

Practice Exercises
11. 2%

2.7 Solving Linear Inequalities

Key Terms

1. three-part inequality
2. interval
3. lincar inequality
4. inequalities
5. interval notation

Objective 1
Now Try

1.

Practice Exercises

1. $(3, \infty)$;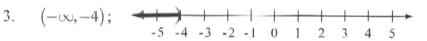

3. $(-\infty, -4)$;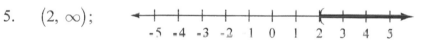

Objective 2
Now Try

3. $[-3, \infty)$

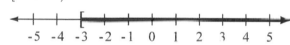

Practice Exercises

5. $(2, \infty)$;

Objective 3
Now Try

4a. $(-\infty, -5]$

4b. $(-\infty, -4)$

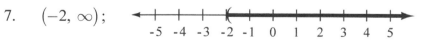

Practice Exercises

7. $(-2, \infty)$;

9. $(-\infty, 4]$

Objective 4
  Now Try
  5. $[2, \infty)$

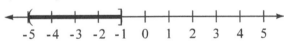

  Practice Exercises

11. $(-\infty, 5]$;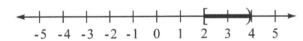

Objective 5
  Now Try
  8. 10 feet

  Practice Exercises
13. 89   15. all numbers greater than 5

Objective 6
  Now Try
  9. $(-5, -1]$

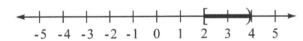

10a. $[2, 4)$

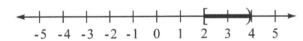

  Practice Exercises

17. $[-5, -3)$;

Chapter 3 GRAPHS OF LINEAR EQUATIONS AND INEQUALITIES IN TWO VARIABLES

3.1 Linear Equations and Rectangular Coordinates

Key Terms
1. line graph
2. linear equation in two variables
3. coordinates
4. x-axis
5. y-axis
6. ordered pair
7. rectangular (Cartesian) coordinate system
8. quadrants
9. origin
10. plot
11. scatter diagram
12. table of values
13. plane

Objective 1
Now Try
1a. 2010-2011, 2012-2013, 2014-2015
1b. 2011-2012
1c. 300 degrees

Practice Exercises
1. 2013-2014
3. 1600 and 1000 degrees; 600 degrees

Objective 2
Practice Exercises
5. $\left(0, \dfrac{1}{3}\right)$

Objective 3
Now Try
2a. yes
2b. no

Practice Exercises
7. no, not a solution
9. yes, a solution

Objective 4
Now Try
3a. (5, 13)

Practice Exercises
11. (a) $(-4, -5)$; (b) $(2, 7)$; (c) $\left(-\dfrac{3}{2}, 0\right)$; (d) $(-2, -1)$; (e) $(-5, -7)$

Objective 5
Now Try
4a.

x	y
1	-4
5	12
2	0
3	4

$(1, -4), (5, 12), (2, 0), (3, 4)$

Practice Exercises

13. $(-4, 4), (0, 4), (6, 4)$

Objective 6
Now Try

5.
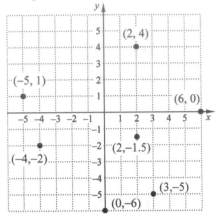

6a.

x (months)	y (balance due)
2	10,153.48
5	10,565.20
11	11,388.64

6b.
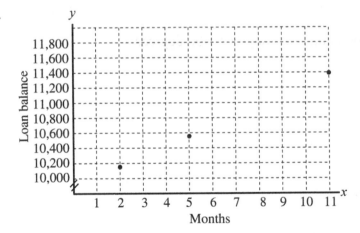

Practice Exercises

15.
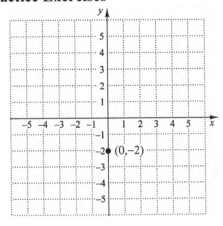

17.
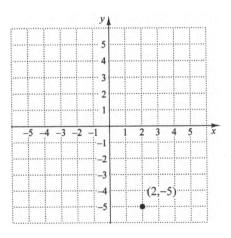

3.2 Graphing Linear Equations in Two Variables

Key Terms
1. y-intercept
2. x-intercept
3. graphing
4. graph

Objective 1
Now Try

2.

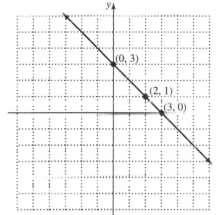

Practice Exercises

1. $(0, -2), \left(\dfrac{2}{3},\ 0\right), (2,\ 4)$

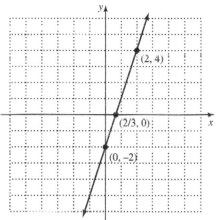

3. $\left(0, -\dfrac{1}{2}\right), (1,\ 0), (-3,\ 2)$

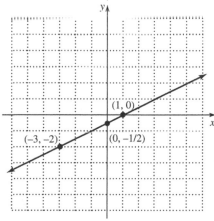

Answers

Objective 2
Now Try
3.

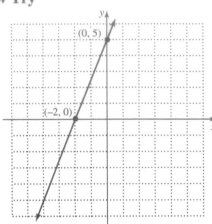

Practice Exercises
5.

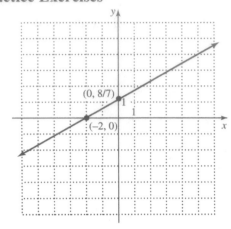

Objective 3
Now Try
5.

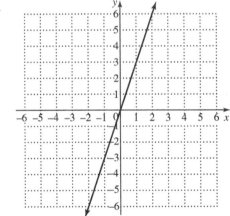

Practice Exercises
7. $-3x - 2y = 0$

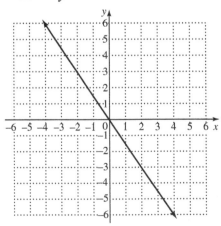

9. $y = 2x$

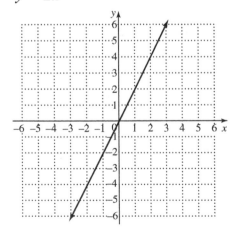

Objective 4
Now Try

6.

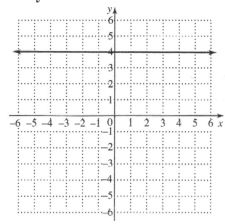

7.

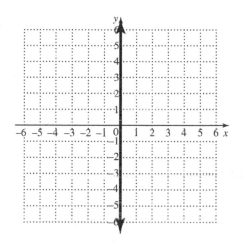

Practice Exercises

11. $y + 3 = 0$

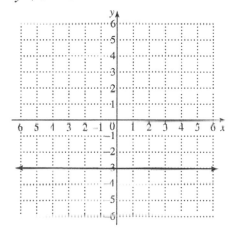

Objective 5
Now Try

8a. 0 calculators, $45
5000 calculators, $42
20,000 calculators, $33
45,000 calculators, $18

8b. (0, $45), (5, $42), (20, $33), (45, $18)

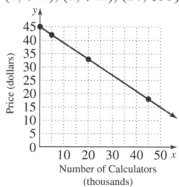

8c. 30,000 calculators, $27

Practice Exercises

13. 2013, 325;
     2014, 367;
     2015, 409;
     2016, 451

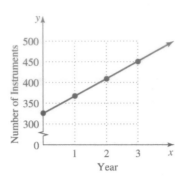

3.3   The Slope of a Line

Key Terms

1. perpendicular lines   2. slope   3. rise

4. parallel lines   5. run

Objective 1

Now Try

1. $\frac{4}{1}$, or 4   2a. $-\frac{16}{9}$ 2b. $\frac{9}{17}$

3. 0   4. undefined slope

Practice Exercises

1. –2   3. 0

Objective 2

Now Try

5c. $-\frac{1}{8}$ 5a. $\frac{7}{4}$

Practice Exercises

5. $\frac{2}{3}$

Objective 3

Now Try

6b. parallel

Practice Exercises

7. 1; 1; parallel   9. –3; $\frac{1}{3}$; perpendicular

Objective 4

Now Try

7. an increase of 5 employees/yr

Practice Exercises

11. 4500 people/yr

3.4 Slope-Intercept Form of a Linear Equation

Key Terms
1. slope-intercept form   2. standard form   3. *y*-intercept

4. slope

Objective 1
Now Try
1a. slope: –12; *y*-intercept: (0, 6)   1c. slope: 23; *y*-intercept (0, 0)

Practice Exercises

1. slope: $\frac{3}{2}$; *y*-intercept: $\left(0, -\frac{2}{3}\right)$ 3. $y = -3x + 3$

Objective 2
Now Try

2b. 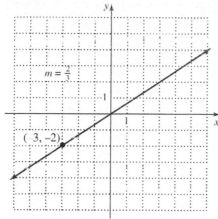   3.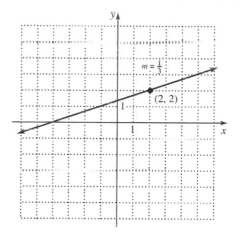

Practice Exercises

5.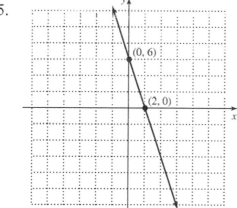

Objective 3
Now Try

4a. $y = \frac{3}{4}x - 6$ 4b. $y = 6x + 10$

Practice Exercises
7. $y = -2x - 11$

Answers

Objective 4
  Now Try

5a.
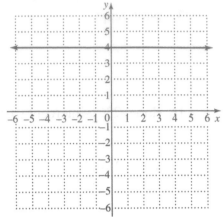

Practice Exercises

9.
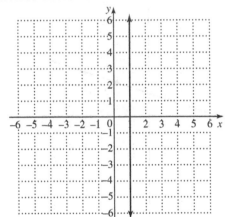

11. $x = 3$

Objective 5
  Now Try

7a. $y = 8x + 15$ 7b. (5, 55)

  Practice Exercises

13. (a) $y = 1.25x + 25$; (b) (0, 25), (5, 31.25), (10, 37.50)

3.5   Point-Slope Form of a Linear Equation and Modeling

Key Terms

1. point-slope form   2. standard form   3. slope-intercept form

Objective 1

Now Try

1b. $y = -\dfrac{2}{3}x + 15$

Practice Exercises

1. $y = \dfrac{1}{4}x - 9$ 3. $y = 2x - 8$

Objective 2

Now Try

2. $y = -\dfrac{3}{4}x + \dfrac{81}{4}$; $3x + 4y = 81$

Practice Exercises

5. $3x - 2y = 0$

Objective 3

Now Try

3a. $y = -\dfrac{3}{4}x + \dfrac{7}{2}$ 3b. $y = \dfrac{4}{3}x + 16$

Practice Exercises

7. $2x + 3y = 9$ 9. $3x + y = -28$

Objective 4

Now Try

4.

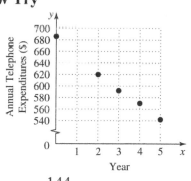

$$y = -\dfrac{144}{5}x + 686$$

Practice Exercises

11.

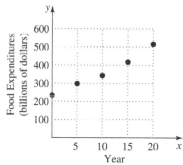

$$y = \dfrac{141}{10}x + 233$$

Answers

3.6 Graphing Linear Inequalities in Two Variables

Key Terms
1. boundary line   2. linear inequality in two variables

Objective 1
Now Try

2.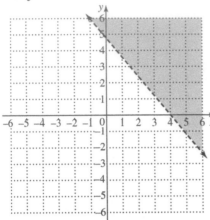

3.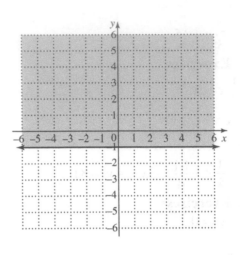

Practice Exercises

1.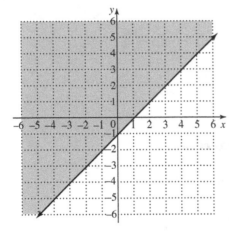

3.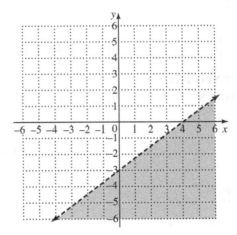

Objective 2
Now Try   **Practice Exercises**

4.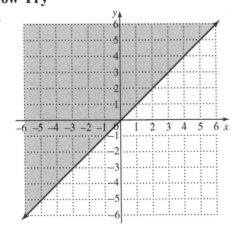

5.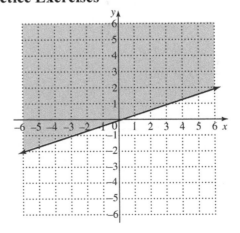

Chapter 4 SYSTEMS OF LINEAR EQUATIONS AND INEQUALITIES

4.1 Solving Systems of Linear Equations by Graphing

Key Terms
1. independent equations
2. consistent system
3. solution set of a system
4. solution of a system
5. dependent equations
6. inconsistent system
7. system of linear equations

Objective 1
Now Try
1b. no

Practice Exercises
1. no   3. no

Objective 2
Now Try

2.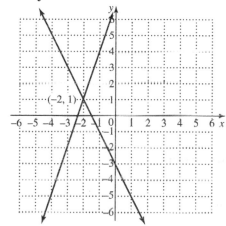

Practice Exercises

5.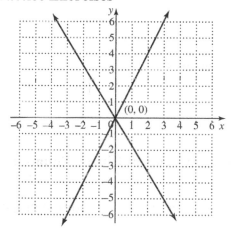

Objective 3
Now Try
3a. $\{(x, y)| \ 4x - 2y = 8\}$ 3b. $\varnothing$

Practice Exercises
7. no solution

Objective 4
Now Try
4c. neither parallel nor same line; exactly one solution

Practice Exercises
9. (a) neither   (b) intersecting lines   (c) one solution

11. (a) dependent   (b) one line   (c) infinitely many solutions

Answers

4.2 Solving Systems of Linear Equations by Substitution

Key Terms
1. ordered pair
2. substitution
3. dependent system
4. inconsistent system

Objective 1
Now Try
1. $\{(1, 6)\}$
2. $\{(9, -4)\}$
3. $\{(8, -2)\}$

Practice Exercises
1. $(2, 4)$
3. $(4, -9)$

Objective 2
Now Try
4. $\varnothing$
5. $\{(x, y)|\ 5x + 4y = 20\}$

Practice Exercises
5. $\{(x, y)\,|\,-x + 2y = 6\}$

Objective 3
Now Try
6. $\{(-2, 5)\}$

Practice Exercises
7. $(-9, -11)$
9. $\{(x, y)\,|\,0.6x + 0.8y = 1\}$

4.3 Solving Systems of Linear Equations by Elimination

Key Terms
1. elimination method
2. addition property of equality
3. substitution

Objective 1
Now Try
1. $\{(8, 3)\}$

Practice Exercises
1. $(8, 3)$
3. $(5, 0)$

Objective 2
Now Try
4. $\{(-4, 9)\}$
3. $\{(1, -3)\}$

Practice Exercises
5. $\left(\dfrac{1}{2}, 1\right)$

Objective 3
Now Try

5. $\left\{\left(\dfrac{9}{17}, \dfrac{11}{17}\right)\right\}$

Practice Exercises

7. $(2, -4)$

9. $(3, -2)$

Objective 4
Now Try

6a. $\varnothing$

6b. $\left\{(x, y) \mid 9x - 7y = 5\right\}$

Practice Exercises

11. $\left\{(x, y) \mid -x - 2y = 3\right\}$

4.4 Applications of Linear Systems

Key Terms

1. $d = rt$

2. system of linear equations

Objective 1
Now Try

1. 56 cm, 26 cm

Practice Exercises

1. 32, 18

3. length: 14 ft; width: 11 ft

Objective 2
Now Try

2. 1500 general admission tickets; 750 reserved seats

Practice Exercises

5. 30 $5 bills; 60 $10 bills

Objective 3
Now Try

3. 32 oz of 65% solution; 48 oz of 85% solution

Practice Exercises

7. $6 coffee: 100 lbs; $12 coffee: 50 lb

9. $1.60 candy: 20 lb; $2.50 candy: 10 lb

Objective 4
Now Try

4. Enid: 44 mph; Jerry: 16 mph

Practice Exercises

11. plane A: 400 mph; plane B: 360 mph

4.5   Graphing Linear Inequalities in Two Variables

Key Terms
1. solution set of a system of linear inequalities   2. system of linear inequalities

Objective 1
 Now Try

1.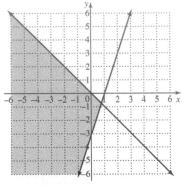

2.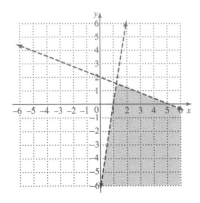

3.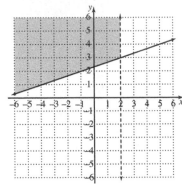

Practice Exercises

1.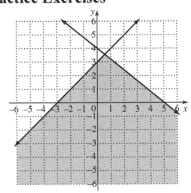

3.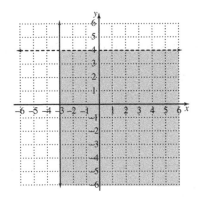

Chapter 5 EXPONENTS AND POLYNOMIALS

5.1 The Product Rule and Power Rules for Exponents

Key Terms

1. power   2. exponential expression   3. base

Objective 1

Now Try

1. 4^5; 1024   2a. base: 2; exponent: 6; value: 64

2b. base:2; exponent: 6; value: −64

2c. base: −2; exponent: 6; value 64

Practice Exercises

1. $\left(\dfrac{1}{3}\right)^5$ 3. −6561; base: 3; exponent: 8

Objective 2

Now Try

3a. 9^{13} 3d. m^{27} 4. $18x^{11}$

3f. 108

Practice Exercises

5. $8c^{15}$

Objective 3

Now Try

5a. 4^{15} 5c. x^{30}

Practice Exercises

7. 7^{12} 9. $(-3)^{21}$

Objective 4

Now Try

6a. $64a^3b^3$

Practice Exercises

11. $-0.008a^{12}b^3$

Objective 5

Now Try

7b. $\dfrac{1}{1024}$

Practice Exercises

13. $-\dfrac{8x^3}{125}$

15. $-\dfrac{128a^7}{b^{14}}$

Objective 6

Now Try

8a. $\dfrac{5^5}{2^3}$, or $\dfrac{3125}{8}$

8d. $-x^{27}y^{13}$

Practice Exercises

17. $32a^9 b^{14} c^5$

Objective 7

Now Try

9a. $28x^5$

Practice Exercises

19. $36x^5$

21. $28q^{11}$

5.2   Integer Exponents and the Quotient Rule

Key Terms

1. power rule for exponents

2. base; exponent

3. product rule for exponents

Objective 1

Now Try

1a. 1   1b. 1   1c. -1 1d. 1   1e. 88

1f. 1

Practice Exercises

1. -1 3. 0

Objective 2

Now Try

2a. $\dfrac{1}{64}$ 2d. $\dfrac{49}{36}$ 2f. $\dfrac{1}{8}$ 2h. x^8 3b. $\dfrac{y}{x^7}$

3c. $\dfrac{qr^5}{4p^3}$ 3d. $\dfrac{625b^4}{a^4}$

Practice Exercises

5. $\dfrac{1}{m^{18}n^9}$

Objective 3
Now Try

4b. $\dfrac{1}{16}$ 4d. z^{10} 4e. $\dfrac{a^3}{25}$

Practice Exercises

7. $\dfrac{k^4 m^5}{2}$ 9. $\dfrac{p^8}{3^5 m^3}$ or $\dfrac{p^8}{243 m^3}$

Objective 4
Now Try

5d. $\dfrac{243}{32 p^{20}}$ 5e. $\dfrac{x^{12} y^{15}}{8000}$

Practice Exercises

11. $a^{16} b^{22}$

5.3   An Application of Exponents: Scientific Notation

Key Terms

1. scientific notation   2. power rule   3. quotient rule

Objective 1
Now Try

1b. 4.771×10^{10} 1c. 4.63×10^{-2}

Practice Exercises

1. 2.3651×10^4 3. -2.208×10^{-4}

Objective 2
Now Try

2b. $27{,}960{,}000$ 2d. -0.000164

Practice Exercises

5. 0.0064

Objective 3
Now Try

3a. 2.7×10^8, or $270{,}000{,}000$ 3b. 3×10^{-8}, or 0.00000003

4. 5.45×10^3 kg/m^3

Practice Exercises

7. 2.53×10^2 9. 4.86×10^{19} atoms

5.4   Adding and Subtracting Polynomials

Key Terms
1. degree of a term
2. descending powers
3. term
4. trinomial
5. polynomial
6. monomial
7. degree of a polynomial
8. binomial
9. like terms

Objective 1
Now Try

1a. $-2, -1, 3; \ 3$

1b. $-8, 1; \ 2$

Practice Exercises

1. $1; \ 1$

3. $\dfrac{2}{5}, -3, 1; \ 3$

Objective 2
Now Try

2e. $15m^3 + 29m^2$

2b. $-8x^7$

Practice Exercises

5. $2.6z^8 - 0.9z^7$

Objective 3
Now Try

3a. $8x^3 + 4x^2 + 6$; degree 3; trinomial

3d. $4x^5$; degree 5; monomial

Practice Exercises

7. $n^8 - n^2$; degree 8; binomial

9. $5c^5 + 3c^4 - 10c^2$; degree 5; trinomial

Objective 4
Now Try

4a. 1285

4b. 1357

Practice Exercises

11. a. 71; b. -19

Objective 5
Now Try

5a. $5x^3 - 5x^2 + 4x$

6b. $7x^3 + 8x^2 - 14x - 1$

Practice Exercises

13. $5m^3 - 2m^2 - 4m + 4$

15. $3r^3 + 7r^2 - 5r - 2$

Objective 6
Now Try

7b. $-11x^3 - 7x + 1$

Practice Exercises

17. $8b^4 - 4b^3 - 2b^2 - b - 2$

Objective 7
  Now Try

10b. $x^2 y + 4xy$

  Practice Exercises

19. $-3a^6 + a^4 b - 3b^2$ 21. $-7x^2 y + 3xy + 5xy^2$

5.5   Multiplying Polynomials

Key Terms
  1. inner product   2. FOIL   3. outer product

Objective 1
  Now Try

1a. $-72p^4$

  Practice Exercises

  1. $10m^4$ 3. $-27a^4 b^3$

Objective 2
  Now Try

2a. $32x^4 + 64x^3$

  Practice Exercises

  5. $6m + 14m^3 + 6m^4$

Objective 3
  Now Try

  3. $4x^7 - 2x^5 + 37x^4 - 18x^2 + 9x$
  4. $28x^4 - 33x^3 + 51x^2 + 17x - 15$

  Practice Exercises

  7. $x^3 + 27$ 9. $6x^4 + 11x^3 - 9x^2 - 4x$

Objective 4
  Now Try

  5. $x^2 + 3x - 54$ 6. $16xy + 72y - 14x - 63$
7b. $45p^2 + 11pq - 4q^2$

  Practice Exercises

11. $3 + 10a + 8a^2$

5.6   Special Products

Key Terms

1. binomial   2. conjugate

Objective 1

Now Try

1. $x^2 + 12x + 36$ 2b. $64a^2 - 48ab + 9b^2$ 2c. $4a^2 + 36ak + 81k^2$

Practice Exercises

1. $49 + 14x + x^2$ 3. $16y^2 - 5.6y + 0.49$

Objective 2

Now Try

3a. $x^2 - 81$ 4c. $x^2 - \dfrac{9}{25}$ 4b. $121x^2 - y^2$

4d. $4p^5 - 144p$

Practice Exercises

5. $64k^2 - 25p^2$

Objective 3

Now Try

5a. $x^3 + 18x^2 + 108x + 216$

5b. $81x^4 - 540x^3 + 1350x^2 - 1500x + 625$

Practice Exercises

7. $a^3 - 9a^2 + 27a - 27$

9. $256s^4 + 768s^3t + 864s^2t^2 + 432st^3 + 81t^4$

5.7   Dividing a Polynomial by a Monomial

Key Terms

1. dividend   2. quotient   3. divisor

Objective 1

Now Try

1. $x^2 - 3x$ 2. $3n^2 - 4n - \dfrac{2}{n}$ 3. $\dfrac{7z^4}{2} - 4z^3 - \dfrac{5}{z} - \dfrac{3}{z^2}$

4. $-2a^3b^2 - 4a^2b + 3$

Practice Exercises

1. $2a^3 - 3a$ 3. $-13m^2 + 4m - \dfrac{5}{m^2}$

5.8   Dividing a Polynomial by a Polynomial

Key Terms

1. divisor   2. quotient   3. dividend

Objective 1

  Now Try

  2. $2x^2 - 2x - 2 + \dfrac{-5}{5x - 1}$ 3. $x^2 + 10x + 100$

  4. $3x^2 + 5x + 5 + \dfrac{8x + 29}{x^2 - 4}$ 5. $x^2 + \dfrac{1}{8}x + \dfrac{5}{8} + \dfrac{6}{8x - 8}$

  Practice Exercises

  1. $-3x + 4$ 3. $a^2 + 1$

Objective 2

  Now Try

  6. $L = 2r^2 - r + 5$ units

  Practice Exercises

  5. $3t + 4$ units

Chapter 6 FACTORING AND APPLICATIONS

6.1　Greatest Common Factors; Factor by Grouping

Key Terms

1. factoring
2. factored form
3. greatest common factor
4. factor

Objective 1
　Now Try
1b.　8　　　　　　　　1c. 1

　Practice Exercises
1.　28　　　　　　　3.　6

Objective 2
　Now Try
2a.　$6x^4$　　　　　　2b. q^3

　Practice Exercises
5.　$k^2m^4n^4$

Objective 3
　Now Try
3a.　$8x\left(x^4+3\right)$　　　3b. $4y^2\left(5y^2-3y+1\right)$　　　5a. $(y+8)(y+4)$

　Practice Exercises
　7.　$10x\left(2x+4xy-7y^2\right)$　9.　$-13x^8\left(-2+x^4-4x^2\right)$

Objective 4
　Now Try
6a.　$(2+y)(x+7)$　　　6c. $(x-7)(4x+5y)$

　Practice Exercises
11.　$(2x+y)(x-7y)$

6.2 Factoring Trinomials

Key Terms

1. factoring   2. greatest common factor   3. prime polynomial

Objective 1

Now Try

2. $(y-7)(y-5)$ 4. $(a-17)(a+2)$ 5a. prime

6. $(p-7q)(p+2q)$

Practice Exercises

1. prime   3. $(x-11)(x+3)$

Objective 2

Now Try

7. $7x^4(x-5)(x-2)$

Practice Exercises

5. $2ab(a-3b)(a-2b)$

6.3 Factoring Trinomials by Grouping

Key Terms

1. coefficient   2. trinomial

Objective 1

Now Try

2a. $(7x-5)(2x+1)$ 2b. $(3m-7)(m+2)$ 2c. $(5x+3y)(2x-y)$

3. $3x^3(5x-3)(2x+7)$

Practice Exercises

1. $(4b+3)(2b+3)$ 3. $(5c-7t)(2c-3t)$

6.4 Factoring Trinomials Using the FOIL Method

Key Terms

1. inner product   2. FOIL   3. outer product

Objective 1

Now Try

1. $(3x+2)(x+5)$ 2. $(3x+1)(5x+7)$ 3. $(4x-1)(5x-2)$

4. $(4x+7)(2x-3)$ 5. $(6x-5y)(4x+3y)$

Practice Exercises

1. $(2q+1)(4q+3)$ 3. $2(2c-d)(c+4d)$

6.5   Special Factoring Techniques

Key Terms

  1. difference   2. perfect square trinomial

Objective 1

  Now Try

1a. $(z+6)(z-6)$ 1b. prime   1c. prime

3a. $10(3x+7)(3x-7)$ 3c. $\left(p^2+16\right)(p+4)(p-4)$

  Practice Exercises

  1. $(x-7)(x+7)$ 3. prime

Objective 2

  Now Try

  4. $(p+8)^2$ 5a. $(x-12)^2$ 5b. $(8m+3)^2$

5d. $5x(2x+5)^2$

  Practice Exercises

  5. $(3j+2)^2$

Objective 3

  Now Try

6a. $(t-6)(t^2+6t+36)$ 6b. $(3k-y)(9k^2+3ky+y^2$

  Practice Exercises

  7. $(2a-5b)(4a^2+10ab+25b^2)$ 9. $3(x-4)(x^2+4x+16)$

Objective 4

  Now Try

7a. $(6x+1)(36x^2-6x+1)$ 7b. $6(x+2y)(x^2-2xy+4y^2)$

  Practice Exercises

11. $8(a+2b)(a^2-2ab+4b^2)$

6.6   A General Approach to Factoring

Key Terms

  1. factoring by grouping   2. FOIL

Objective 1

  Now Try

1c. $(y+z)(7x-5)$ 2d. prime   2b. $(3t-4w)(9t^2+12tw+16w^2)$

3d. $(4k+1)(k-2)$ 4a. $(c-3)(m+x)$

  Practice Exercises

  1. $-6x(2x+1)$ 3. $2(x+1)^2$ 5. $2(4x-y)(16x^2+4xy+y^2)$

  7. $(2x-3y)^2$ 9. $(2w-5x)(7w+3x)$ 11. $(a-3b+5)(a-3b-5)$

6.7 Solving Quadratic Equations using the Zero-Factor Property

Key Terms
1. standard form   2. quadratic equation

Objective 1
Now Try

1a. $\left\{-12, \dfrac{7}{4}\right\}$ 2a. $\{-3, 8\}$ 3. $\left\{\dfrac{4}{5}, 3\right\}$

4c. $\left\{-\dfrac{1}{4}, 6\right\}$ 4b. $\{0, 11\}$

Practice Exercises

1. $\left\{-\dfrac{5}{2}, 4\right\}$ 3. $\left\{-4, \dfrac{3}{5}\right\}$

Objective 2
Now Try

6. $\{-5, 0, 5\}$

Practice Exercises

5. $\{-9, 0, 1\}$

6.8 Applications of Quadratic Equations

Key Terms
1. legs   2. hypotenuse

Objective 1
Now Try
1. width: 3 m, length: 5 m

Practice Exercises
1. width: 8 in., length: 24 in.   3. height: 4 ft, width: 6 ft

Objective 2
Now Try
3. 8, 10   2. $-4, -3$ or 3, 4

Practice Exercises
5. 6, 8

Objective 3
Now Try
4. 16 ft

Practice Exercises
7. 45 m, 60 m, 75 m   9. 20 mi

Objective 4
Now Try
5. 1 sec

Practice Exercises
11. 40 items or 110 items

Chapter 7 FACTORING EXPRESSIONS AND APPLICATIONS

7.1   The Fundamental Property of Rational Expressions

Key Terms
1. rational expression   2. lowest terms

Objective 1
Now Try

1a. $\dfrac{3}{2}$

1b. $\dfrac{7}{5}$

1c. undefined

1d. 0

Practice Exercises

1. a. $-\dfrac{11}{9}$; b. -4

3. a. $-\dfrac{1}{6}$; b. $-\dfrac{7}{2}$

Objective 2
Now Try

2a. $y \neq \dfrac{1}{7}$

2b. $m \neq 5,\ m \neq -4$

2c. never undefined

Practice Exercises

5. none

Objective 3
Now Try

3b. $\dfrac{3}{k^3}$

4a. $\dfrac{7}{9}$

4c. $\dfrac{m+6}{2m+3}$

5. -1

Practice Exercises

7. $\dfrac{-5b}{8c}$

9. $\dfrac{9(x+3)}{2}$

Objective 4
Now Try

7. $\dfrac{-(10x-7)}{4x-3},\ \dfrac{-10x+7}{4x-3},\ \dfrac{10x-7}{-(4x-3)},\ \dfrac{10x-7}{-4x+3}$

Practice Exercises

11. $\dfrac{-(2p-1)}{-(1-4p)};\ \dfrac{-2p+1}{-1+4p};\ \dfrac{-(2p-1)}{-1+4p};\ \dfrac{-2p+1}{-(1-4p)}$

7.2　Multiplying and Dividing Rational Expressions

Key Terms

1. reciprocal　　　2. rational expression　　　3. lowest terms

Objective 1

Now Try

1b. $\dfrac{4}{3x}$　　　2. $\dfrac{s^2}{6(r-s)}$

Practice Exercises

1. $\dfrac{10m^3n}{3}$　　　3. $\dfrac{x+4}{2(x-4)}$

Objective 2

Now Try

4a. $\dfrac{10b}{11a^3}$　　　4b. $\dfrac{w^2+6w-7}{w^2-6w}$

Practice Exercises

5. $\dfrac{1}{4r^2+2r+3}$

Objective 3

Now Try

6. $\dfrac{2}{25}$　　　7. $\dfrac{8x}{(x-3)^2}$　　　8. $\dfrac{-(m-8)}{5m(m+9)}$

Practice Exercises

7. $-\dfrac{1}{2}$　　　9. $\dfrac{-3(6k-1)}{3k-1}$

7.3　Least Common Denominators

Key Terms

1. equivalent expressions　　　2. least common denominator

Objective 1

Now Try

2. $120a^4$　　　3a. $9w(w-2)$　　　3b. $(b+4)(b+1)(b-4)^2$

Practice Exercises

1. $108b^4$　　　3. $w(w+3)(w-3)(w-2)$

Objective 2
Now Try

4a. $\dfrac{65}{30}$

5a. $\dfrac{76}{24c-20}$

5b. $\dfrac{3(z+2)}{z(z-7)(z+2)}$ or $\dfrac{3z+6}{z^3-5z-14z}$

Practice Exercises
5. $30r$

7.4 Adding and Subtracting Rational Expressions

Key Terms
1. greatest common factor

2. least common multiple

Objective 1
Now Try
1b. $2x$

Practice Exercises

1. $\dfrac{4}{w^2}$

3. $\dfrac{1}{x-2}$

Objective 2
Now Try

2a. $\dfrac{137}{315}$

3. $\dfrac{4}{x-10}$

4. $\dfrac{7x^2+11x+8}{(x+2)(x+1)(x-4)}$

Practice Exercises

5. $\dfrac{7z^2-z-6}{(z+2)(z-2)^2}$

Objective 3
Now Try

8. 7

9. $\dfrac{8x^2+37x+15}{(x+5)(x-5)^2}$

10. $\dfrac{p+25}{(p+1)(p-1)(p-5)}$

Practice Exercises

7. $\dfrac{8z}{(z-2)(z+2)}$ or $\dfrac{8z}{z^2-4}$

9. $\dfrac{2m^2-m+2}{(m-2)(m+2)^2}$

7.5 Complex Fractions

Key Terms
1. complex fraction   2. LCD

Objective 1

Objective 2
  Now Try

1b. $\dfrac{90}{x}$ 2. $\dfrac{bc^2}{a^2}$ 3. $\dfrac{-9x+65}{2x-5}$

Practice Exercises

1. $\dfrac{7m^2}{2n^3}$ 3. $\dfrac{9s+12}{6s^2+2s}$ or $\dfrac{3(3s+4)}{2s(3s+1)}$

Objective 3
  Now Try

4b. $\dfrac{12}{x}$ 5. $\dfrac{5n-9}{3(6n+5)}$

Practice Exercises

5. $\dfrac{(x-2)^2}{x(x+2)}$

Objective 4
  Now Try

7. $\dfrac{1}{xy-1}$

Practice Exercises

11. $\dfrac{r}{s}$

7.6 Solving Equations with Rational Expressions

Key Terms
1. extraneous solution   2. proposed solution

Objective 1
  Now Try

1a. expression; $\dfrac{1}{2}x$ 1b. equation; $\{14\}$

Practice Exercises

1. equation; $\{-2\}$ 3. expression; $\dfrac{41x}{15}$

Answers

Objective 2
Now Try
7. $\{-4\}$ 6. $\{0\}$ 5. $\{6\}$
Practice Exercises
5. $\{-4, 16\}$

Objective 3
Now Try
9. $b = aq + c$
Practice Exercises

7. $f = \dfrac{d_0 d_1}{d_0 + d_1}$ 9. $q = \dfrac{2pf - Ab}{Ab}$ or $\dfrac{2pf}{Ab} - 1$

7.7   Applications of Rational Expressions

Key Terms
1. numerator   2. denominator   3. reciprocal

Objective 1
Now Try
1. 3

Practice Exercises

1. $-\dfrac{2}{3}$ or 1   3. $\dfrac{3}{5}$

Objective 2
Now Try
4. 3 miles per hour

Practice Exercises
5. 24 miles per hour

Objective 3
Now Try

5. $\dfrac{2}{5}$ hour

Practice Exercises

7. $2\dfrac{2}{5}$ hr   9. 72 hr

Chapter 8 EQUATIONS, INEQUALITIES, GRAPHS, AND SYSTEMS REVISITED

8.1 Review of Solving Linear Equations and Inequalities in One Variable

Key Terms

1. equivalent equations 2. linear (first-degree) equation in one variable

3. solution set 4. contradiction 5. conditional equation

6. identity 7. solution

8. linear inequality in one variable

Objective 1
  Now Try
1a. equation   1b. expression

  Practice Exercises
1. equation

Objective 2
  Now Try
3. {−3}   5. {3}   7a. {9}; conditional

7b. ∅; contradiction   7c. {all real numbers}; identity

  Practice Exercises
3. {8}   5. identity; {all real numbers}

Objective 3

Objective 4
  Now Try
10. (−∞, 2];

  Practice Exercises
7. [1, ∞);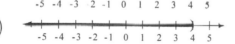

9. (−∞, 4)

Objective 5
  Now Try
11. [−4, 2]

  Practice Exercises
11. [−5, −3);

8.2   Set Operations and Compound Inequalities

Key Terms
1. union   2. compound inequality   3. intersection

Objective 1

Objective 2
Now Try
1. {30, 50}
Practice Exercises
1. $\{0,2,4\}$ 3. $\{1,3,5\}$

Objective 3
Now Try

3. $(-\infty,-3)$

2.   [12, 16]

4. $\varnothing$

Practice Exercises
5. $[-4,3)$

Objective 4
Now Try
5. {20, 30, 40, 50, 70}
Practice Exercises
7. $\{0,1,2,3,4,5\}$ 9. $\{0,1,2,3,4,5\}$

Objective 5
Now Try

6. $(-\infty, 2) \cup [5, \infty)$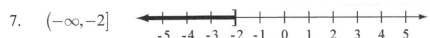

8. $(-\infty, \infty)$

7. $(-\infty,-2]$

Practice Exercises
11. $(-\infty,\infty)$

8.3   Absolute Value Equations and Inequalities

Key Terms
1.   absolute value equation   2.   absolute value inequality

Objective 1
Practice Exercises

1.    3.

Objective 2
Now Try

1. $\left\{-3, \dfrac{7}{5}\right\}$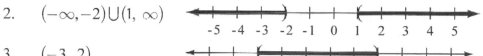

Practice Exercises

5. $\{3\}$

Objective 3
Now Try

2. $(-\infty, -2) \cup (1, \infty)$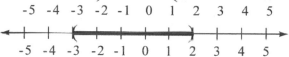

3. $(-3, 2)$

Practice Exercises

7. $(-\infty, -6) \cup (10, \infty)$

9. $\left(-4, \dfrac{16}{5}\right)$

Objective 4
Now Try

5. $\{-4, 10\}$

Practice Exercises

11. $\{-42, 50\}$

Objective 5
Now Try

7. $\{-5, -1\}$

Practice Exercises

13. $\left\{-\dfrac{3}{2}, 2\right\}$ 15. $\left\{-5, -\dfrac{3}{4}\right\}$

Objective 6
Now Try

8a. $\varnothing$ 8b. $\{-5\}$ 9a. $(-\infty, \infty)$

9b. $\varnothing$ 9c. $\{4\}$

Practice Exercises

17. $(-\infty, \infty)$

Objective 7
  Now Try
  10. between 16.055 and 17.745 oz, inclusive

  Practice Exercises
  19. between 16.224 and 17.576 oz, inclusive
  21. between 16.393 and 17.407 oz, inclusive

8.4   Review of Graphing Linear Equations in Two Variables; Slope

Key Terms
  1. slope   2. rise   3. run
  4. y-intercept   5. linear equation in two variables
  6. coordinate   7. origin   8. x-intercept
  9. x-axis   10. quadrant   11. ordered pair
  12. y-axis   13. components   14. plot
  15. rectangular (Cartesian) coordinate system
  16. graph of an equation   17. linear equation in two variables

Objective 1
  Practice Exercises
  1.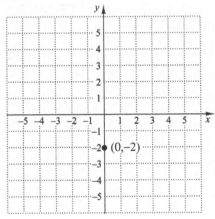

  3.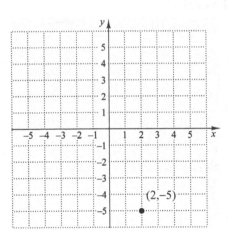

Objective 2
  Now Try
  1a. $(0,-2)$ 1b. $(5, 0)$ 1c. $(-5,-4)$
  1d. $(10, 2)$

  Practice Exercises
  4. (a) $(2, 0)$; (b) $\left(4,-\dfrac{5}{2}\right)$; (c) $\left(-\dfrac{2}{5}, 3\right)$; (d) $\left(0, \dfrac{5}{2}\right)$; (e) $\left(\dfrac{2}{5}, 2\right)$

  5. $(2, 0)$, $(0,-7)$, $\left(3, \dfrac{7}{2}\right)$, $(4, 7)$

  6. $(-4, 4)$, $(0, 4)$, $(6, 4)$, $(-12, 4)$

Objective 3
 Now Try
2.

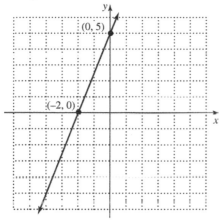

Practice Exercises
7.

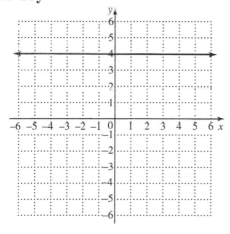

Objective 4
 Now Try
4a.

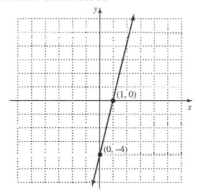

4b.

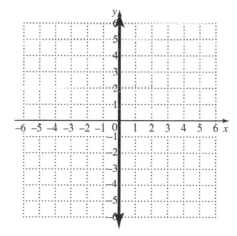

Practice Exercises

9. $x - 1 = 0$

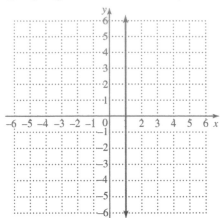

Objective 5
Now Try

5. $-\dfrac{16}{9}$

Practice Exercises

11. -2

13. $-\dfrac{1}{14}$

Objective 6
Now Try

8.

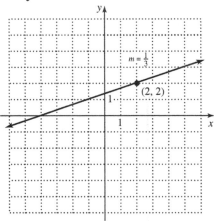

Practice Exercises

15.

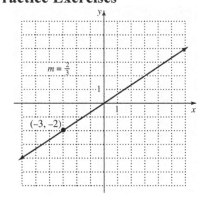

8.5   Review of Solving Linear Equations and Inequalities

Key Terms
1. independent equations
2. consistent system
3. system of equations
4. Solution set of a system
5. dependent equations
6. inconsistent system
7. linear system

Objective 1
Now Try
1.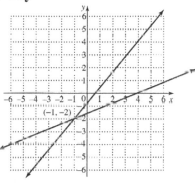

2. $\{(2, 3)\}$

5. $\{(-1, 3)\}$

Practice Exercises
1.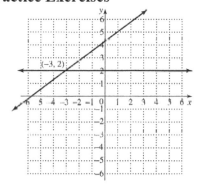

3. $\{(-3, -1)\}$

Objective 2
Now Try
6. dependent; $\{(x, y) \mid 4x + 3y = 12\}$
Practice Exercises
5. $\{(x, y) \mid x + 2y = 4\}$

8.6 Systems of Linear Equations in Three Variables; Applications

Key Terms

1. ordered triple   2. dependent system   3. inconsistent system

Objective 1

Now Try

Practice Exercises

1. The planes intersect in one point.

Objective 2

Now Try

1. $\{(-3, 1, 2)\}$

Practice Exercises

3. $\{(-1, 2, 1)\}$ 5. $\{(4, -4, 1)\}$

Objective 3

Now Try

2. $\{(2, -5, 3)\}$

Practice Exercises

7. $\{(4, 2, -1)\}$

Objective 4

Now Try

4. $\{(x, y, z) \mid x - 5y + 2z = 0\}$ 3. $\emptyset$

Practice Exercises

9. $\emptyset$; inconsistent system

11. $\{(x, y, z) \mid -x + 5y - 2z = 3\}$; dependent equations

Objective 5

Now Try

7. 20 lb of $4 candy; 30 lb of $6 candy; 50 lb of $10 candy

Practice Exercises

13. $8 coffee: 15 lb; $10 coffee: 12 lb; $15 coffee: 23 lb

Chapter 9 RELATIONS AND FUNCTIONS

9.1   Introduction to Relations and Functions

Key Terms

1. range
2. relation
3. dependent variable
4. domain
5. function
6. independent variable

Objective 1

Now Try

2a. function
2c. not a function

Practice Exercises

1. not a function
3. not a function

Objective 2

Now Try

3a. domain: $\{13\}$; range: $\{-2, -1, 4\}$; not a function

4c. domain: $(-\infty, \infty)$; range: $(-\infty, \infty)$

4d. domain: $[-4, \infty)$; range: $(-\infty, \infty)$

Practice Exercises

5. function; domain: $\{A, B, C, D, E\}$; range: $\{V, W, X, Z\}$

Objective 3

Now Try

5b. function
6a. function; domain: $(-\infty, \infty)$

Practice Exercises

7. function
9. function; $(-\infty, -6) \cup (-6, \infty)$

9.2   Function Notation and Linear Functions

Key Terms

1. linear function
2. function notation
3. constant function

Objective 1

Now Try

1a. $f(3) = 17$
3. $g(a-1) = 4a - 11$
4c. $f(-6) = 21$

6b. $f(x) = -\dfrac{2}{3}x + \dfrac{7}{3}$, $f(-3) = \dfrac{13}{3}$

Practice Exercises

1. (a) -13; (b) -7; (c) $-3x - 7$ 3. (a) 9; (b) 9; (c) 9

Answers

Objective 2
Now Try

7a. domain: $(-\infty, \infty)$
   range: $(-\infty, \infty)$

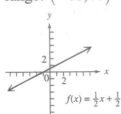

$f(x) = \frac{1}{2}x + \frac{1}{2}$

7b. domain: $(-\infty, \infty)$
   range: $\{-3\}$

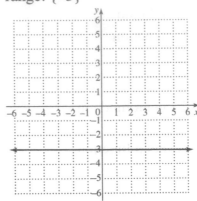

Practice Exercises
5.

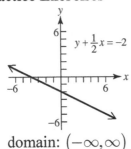

$y + \frac{1}{2}x = -2$

domain: $(-\infty, \infty)$
range: $(-\infty, \infty)$

9.3 Polynomial Functions, Operations, and Graphs

Key Terms
1. squaring function
2. polynomial function of degree n
3. cubing function
4. identity function

Objective 1
Now Try
1a. 1   1b. −44

Practice Exercises
1. (a) −7; (b) −17   3. (a) 28; (b) 198

Objective 2
Now Try
2a. $3x^2 - 6x + 21$ 2b. $9x^2 - 8x + 3$ 4. $42x^3 + 47x^2 + 10x;\ -741$

Practice Exercises
5. (a) $-15x^3 - 10x$; (b) 25

Objective 3
Now Try
6a.

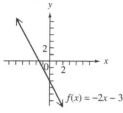

domain: $(-\infty, \infty)$;
range: $(-\infty, \infty)$

6b.

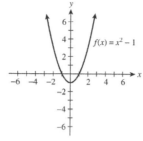

domain: $(-\infty, \infty)$;
range: $[-1, \infty)$

6c.

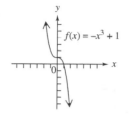

domain: $(-\infty, \infty)$;
range: $(-\infty, \infty)$

Practice Exercises
7.

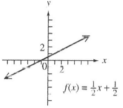

domain: $(-\infty, \infty)$;
range: $(-\infty, \infty)$

9.

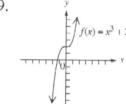

domain: $(-\infty, \infty)$;
range: $(-\infty, \infty)$

9.4   Variation
Key Terms
1. constant of variation

2. varies inversely

3. varies directly

Objective 1

Objective 2
Now Try
2. 125 psi   3. 153.86 sq cm

Practice Exercises
1. $k = 0.25$; $y = 0.25x$ 3. 100 newtons

Objective 3
Now Try
4. 640 cycles per second
Practice Exercises
5. 40 lb

Objective 4
  Now Try
  6. 1280 psi

  Practice Exercises
  7. 96   9. 750°

Objective 5
  Now Try
  7. About 63.5 cm^3

  Practice Exercises
  11. 9 hr

Chapter 10 ROOTS, RADICALS, AND ROOT FUNCTIONS

10.1 Radical Expressions and Graphs

Key Terms

1. radicand
2. perfect square
3. index (order)
4. square root
5. radical expression
6. principal square root
7. irrational number
8. perfect cube
9. radical
10. cube root

Objective 1

Now Try

1. $9, -9$
2a. 13
2d. $-\dfrac{3}{7}$
3b. 37
3c. $n^2 + 5$

Practice Exercises

1. $25, -25$
3. $\dfrac{30}{7}$

Objective 2

Now Try

4a. irrational
4b. rational
4c. not a real number

Practice Exercises

5. not a real number

Objective 3

Now Try

5c. 7
5h. -5
6c. none
6a. 5

Practice Exercises

7. -3
9. -1

Objective 4

Now Try

7a.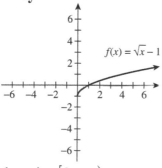

domain: $[0, \infty)$

range : $[-1, \infty)$

7b.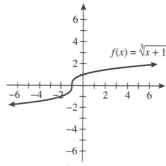

domain: $(-\infty, \infty)$

range: $(-\infty, \infty)$

Practice Exercises

11.

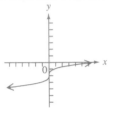

domain: $(-\infty, \infty)$

range: $(-\infty, \infty)$

Objective 5
Now Try

9c. -2

9e. w^{10}

Practice Exercises

13. 9

15. $-x^4$

Objective 6
Now Try

10a. 3.742

Practice Exercises

17. -31.464

10.2   Rational Exponents

Key Terms
1. power rule for exponents
2. product rule for exponents
3. quotient rule for exponents

Objective 1
Now Try

1b. 11

1e. -3

1d. not a real number

Practice Exercises

1. -5

3. -15

Objective 2
Now Try

2a. 9

2c. -216

3a. $\dfrac{1}{4}$

Practice Exercises

5. 7776

Objective 3
Now Try

4a. $\sqrt[3]{23}$

4d. $\left(\sqrt[3]{2x}\right)^4 - 3\left(\sqrt[5]{x}\right)^2$

5c. $10^2 = 100$

Practice Exercises

7. $4\left(\sqrt[5]{y}\right)^2 + \sqrt[5]{5x}$

9. $a^{1/4}$, or $\sqrt[4]{a}$

Objective 4
Now Try

6a. $5^{5/2}$

6d. $\dfrac{x^2}{y^{1/4}}$

7a. $x^{7/6}$

7c. $x^{1/4}$

Practice Exercises

11. $a^{5/6}$

10.3 Simplifying Radical Expressions

Key Terms

1. hypotenuse

2. index; radicand

3. legs

Objective 1
Now Try

1a. $\sqrt{14}$

1c. $\sqrt{33mn}$

2a. $\sqrt[3]{21}$

2b. $\sqrt[3]{35xy}$

2c. $\sqrt[5]{8w^4}$

2d. cannot be simplified using the product rule directly

Practice Exercises

1. $\sqrt{42tx}$

3. cannot be simplified using the product rule directly

Objective 2
Now Try

3a. $\dfrac{6}{7}$

3b. $\dfrac{\sqrt{13}}{9}$

3c. $-\dfrac{7}{5}$

3e. $-\dfrac{a^2}{5}$

3f. $\dfrac{\sqrt[4]{m}}{3}$

Practice Exercises

5. $\dfrac{\sqrt[5]{7x}}{2}$

Objective 3
Now Try

4a. $2\sqrt{21}$

4b. $9\sqrt{2}$

4c. cannot be simplified

4d. $4\sqrt[3]{4}$

4e. $-2\sqrt[5]{16}$

5a. $10y\sqrt{y}$

5b. $4m^2r^4\sqrt{3mr}$

5c. $-2n^2t\sqrt[3]{4nt^2}$

5d. $-3y^2\sqrt[4]{5x^3y}$

6a. $\sqrt[4]{11^3}$, or $\sqrt[4]{1331}$

6c. $z\sqrt[4]{z}$

Answers

Practice Exercises

7. $\sqrt[3]{x^2}$ 9. $5ab^2\sqrt[3]{10a^2b}$

Objective 4
 Now Try

7. $\sqrt[6]{63}$

 Practice Exercises

11. $\sqrt[8]{28}$

Objective 5
 Now Try

8. $12\sqrt{2}$

 Practice Exercises

13. 26 15. $6\sqrt{2}$

Objective 6
 Now Try

9. $\sqrt{34}$

 Practice Exercises

17. 5

10.4 Adding and Subtracting Radical Expressions

Key Terms

1. unlike radicals 2. like radicals

Objective 1
 Now Try

1c. $-\sqrt{6}$ 1d. $13\sqrt{2x}$ 1e. cannot be simplified

2a. $-3\sqrt[3]{2}$ 2b. $5z\sqrt[4]{2y^2z}$ 3a. $\dfrac{2\sqrt{5}}{3}$

3b. $\dfrac{17}{w^2}$

 Practice Exercises

1. $12\sqrt{x}$ 3. $\dfrac{8y^2\sqrt[3]{y}}{15}$

10.5 Multiplying and Dividing Radical Expressions

Key Terms

1. conjugate
2. rationalizing the denominator

Objective 1

Now Try

2a. $\sqrt{10} - \sqrt{55} + \sqrt{6} - \sqrt{33}$

Practice Exercises

1. $\sqrt{10} - 4\sqrt{5} + 2\sqrt{3} - 4\sqrt{6}$

3. $4 - \sqrt[3]{25}$

Objective 2

Now Try

3b. $\dfrac{3\sqrt{42}}{7}$

4a. $-\dfrac{3\sqrt{10}}{8}$

5a. $\dfrac{\sqrt[3]{10}}{5}$

Practice Exercises

5. $\dfrac{y\sqrt{21b}}{6b}$

Objective 3

Now Try

6a. $-4(\sqrt{3} - 2)$

6c. $\dfrac{\sqrt{15} + \sqrt{6} + \sqrt{10} + 2}{3}$

Practice Exercises

7. $-4\sqrt{3} + 8$

9. $\dfrac{\sqrt{3} + 2\sqrt{6} + \sqrt{2} + 4}{7}$

Objective 4

Now Try

7b. $\dfrac{1 - \sqrt{2x}}{2}$

Practice Exercises

11. $\dfrac{2 - 9\sqrt{2}}{3}$

10.6 Solving Equations with Radicals

Key Terms
1. extraneous solution
2. radical equation
3. proposed solution

Objective 1
Now Try
1. $\{10\}$
2. $\varnothing$

Practice Exercises
1. $\{11\}$
3. $\{17\}$

Objective 2
Now Try
3. $\{5\}$
5. $\{4\}$

Practice Exercises
5. $\{8\}$

Objective 3
Now Try
6. $\{2\}$

Practice Exercises
7. $\{-31\}$
9. $\dfrac{3}{2}$

Objective 4
Now Try
7. $h = \dfrac{3v}{\pi r^2}$

Practice Exercises
11. $C = \dfrac{1}{4\pi^2 f^2 L}$

10.7 Complex Numbers

Key Terms

1. complex number   2. complex conjugate   3. imaginary part

4. real part   5. standard form (of a complex number)

6. pure imaginary number   7. nonreal complex number

Objective 1

Now Try

1a. $4i$ 2b. $-\sqrt{30}$ 3a. 5

Practice Exercises

1. $-9i\sqrt{2}$ 3. $6i$

Objective 2

Objective 3

Now Try

4a. $10 - 9i$ 5b. $-2i$

Practice Exercises

5. $-7 + 2i$

Objective 4

Now Try

6a. $-14 + 8i$

Practice Exercises

7. 29 9. $-14 + 10i$

Objective 5

Now Try

7a. $\dfrac{18}{29} + \dfrac{13i}{29}$

Practice Exercises

11. $\dfrac{37}{97} + \dfrac{38}{97}i$

Objective 6

Now Try

8a. 1

Practice Exercises

13. -1 15. $-i$

Chapter 11 QUADRATIC EQUATIONS, INEQUALITIES, AND FUNCTIONS

11.1 Solving Quadratic Equations by the Square Root Property

Key Terms

1. quadratic equation   2. zero-factor property

Objective 1

Now Try

1a. $\left\{-\dfrac{7}{3},\ 4\right\}$

Practice Exercises

1. $\left\{-\dfrac{1}{3},\ \dfrac{2}{5}\right\}$ 3. $\{-2, 2\}$

Objective 2

Now Try

2b. $\left\{2\sqrt{10}, -2\sqrt{10}\right\}$, or $\left\{\pm 2\sqrt{10}\right\}$ 2c. $\left\{3\sqrt{2}, -3\sqrt{2}\right\}$, or $\left\{\pm 3\sqrt{2}\right\}$

Practice Exercises

5. $\left\{-7\sqrt{2},\ 7\sqrt{2}\right\}$

Objective 3

Now Try

5. $\left\{\dfrac{-5-4\sqrt{2}}{2},\ \dfrac{-5+4\sqrt{2}}{2}\right\}$ 4b. $\left\{-5-\sqrt{14}, -5+\sqrt{14}\right\}$

Practice Exercises

7. $\{-6, 2\}$ 9. $\left\{\dfrac{-4-4\sqrt{2}}{3},\ \dfrac{-4+4\sqrt{2}}{3}\right\}$

Objective 4

Now Try

6a. $\left\{-4i\sqrt{2},\ 4i\sqrt{2}\right\}$ 6b. $\{-2-7i,\ -2+7i\}$

Practice Exercises

11. $\{-1-6i,\ -1+6i\}$

11.2   Solving Quadratic Equations by Completing the Square

Key Terms

1. perfect square trinomial

2. square root property

3. completing the square

Objective 1

Now Try

3. $\left\{\dfrac{11-\sqrt{89}}{2}, \dfrac{11+\sqrt{89}}{2}\right\}$

2. $\left\{5-\sqrt{35}, 5+\sqrt{35}\right\}$

Practice Exercises

1. $\left\{-4-2\sqrt{3}, -4+2\sqrt{3}\right\}$

3. $\{1, 8\}$

Objective 2

Now Try

4. $\left\{-\dfrac{4}{3}, \dfrac{2}{3}\right\}$

5. $\left\{\dfrac{-3-\sqrt{11}}{2}, \dfrac{-3+\sqrt{11}}{2}\right\}$

Practice Exercises

5. $\left\{-1-\dfrac{1}{2}i, -1+\dfrac{1}{2}i\right\}$

Objective 3

Now Try

7. $\left\{-2-\sqrt{2}, -2+\sqrt{2}\right\}$

Practice Exercises

7. $\left\{\dfrac{1-\sqrt{3}}{2}, \dfrac{1+\sqrt{3}}{2}\right\}$

9. $\{-1, 2\}$

Answers

11.3 Solving Quadratic Equations by the Quadratic Formula

Key Terms
1. discriminant
2. quadratic formula

Objective 1

Objective 2
  Now Try
  3. $\left\{\dfrac{1-\sqrt{7}}{2}, \dfrac{1+\sqrt{7}}{2}\right\}$
  4. $\{1-2i, 1+2i\}$

  Practice Exercises
  1. $\{2, 4\}$
  3. $\{5-3i, 5+3i\}$

Objective 3
  Now Try
  5c. 0; one rational solution; factoring

  5b. 80; two irrational solutions; quadratic formula

  5a. 81; two rational solutions; factoring

  5d. –48; two nonreal complex solutions; quadratic formula

  Practice Exercises
  5. C

11.4 Equations Quadratic in Form

Key Terms
1. standard form
2. quadratic in form

Objective 1
  Now Try
  1. $\left\{-7, \dfrac{5}{4}\right\}$
  Practice Exercises
  1. $\left\{-\dfrac{35}{4}, -3\right\}$
  3. $\left\{-7, -\dfrac{7}{2}\right\}$

Objective 2
  Now Try
  2. 2 mph
  3. Tom: 12 hours; Huck: 24 hours

  Practice Exercises
  5. 550 mph

Objective 3
  Now Try
  4a. $\{2\}$

  Practice Exercises
  7. $\{2, 5\}$ 9. $\{9\}$

Objective 4
  Now Try
  6b. $\left\{-1, -\dfrac{3}{4}, \dfrac{3}{4}, 1\right\}$

  Practice Exercises
  11. $\left\{-27, -3\sqrt{3}, 3\sqrt{3}, 27\right\}$

11.5   Formulas and Further Applications

Key Terms
  1. quadratic function   2. Pythagorean theorem

Objective 1
  Now Try
  1a. $t = \pm\dfrac{\sqrt{mxF}}{F}$ 2. $q = \dfrac{-k \pm k\sqrt{5}}{2p}$

  Practice Exercises
  1. $d = \dfrac{k^2 l^2}{F^2}$ 3. $a = \dfrac{-c \pm c\sqrt{2}}{b}$

Objective 2
  Now Try
  3. south: 72 mi; east 54 mi

  Practice Exercises
  5. 10 ft

Objective 3
  Now Try
  4. 2.5 in.

  Practice Exercises
  7. 2.5 ft   9. 4 ft

Objective 4
  Now Try
  5. 4.1 sec

  Practice Exercises
  11. 1.2 sec

Answers

11.6 Graphs of Quadratic Functions

Key Terms
1. axis
2. vertex
3. quadratic function
4. parabola

Objective 1

Objective 2
Now Try
1. vertex: (0,–1), axis: $x = 0$;
domain: $(-\infty, \infty)$; range: $[-1, \infty)$

2. Vertex: (3, 0); axis: $x = 3$
domain: $(-\infty, \infty)$; range: $[0, \infty)$

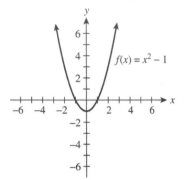

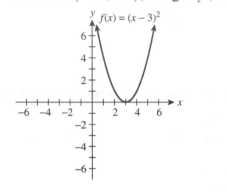

3. vertex: (–2,–1); axis: $x = -2$
domain: $(-\infty, \infty)$; range: $[-1, \infty)$

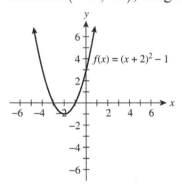

Practice Exercises
1.

Vertex: (0, −4)
Axis: $x = 0$
Domain: $(-\infty, \infty)$
Range: $[-4, \infty)$

3.

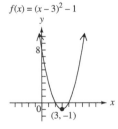

Vertex: (3, −1)
Axis: $x = 3$
Domain: $(-\infty, \infty)$
Range: $[-1, \infty)$

Objective 3
  Now Try
  4. vertex: (0, 0); axis: $x = 0$
      domain: $(-\infty, \infty)$; range: $(-\infty, 0]$

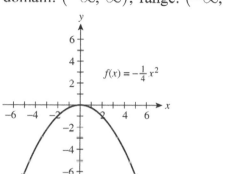

  5. vertex: (1, 1); axis: $x = 1$
      domain: $(-\infty, \infty)$; range: $[1, \infty)$

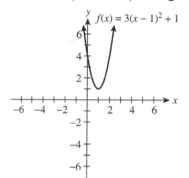

Practice Exercises
  5. down; narrower; vertex: $(-1, 0)$; domain: $(-\infty, \infty)$; range: $(-\infty, 0]$

Objective 4
  Now Try
  6. $y = 0.01x^2 + 0.77x + 8.36$

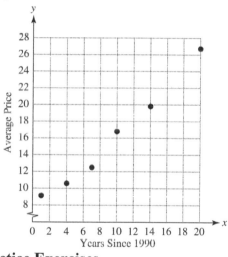

Practice Exercises
  7. linear; positive

  9.

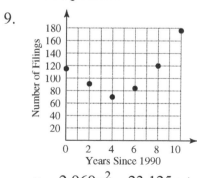

  $y = 2.969x^2 - 23.125x + 115$

11.7 More about Parabolas and Their Applications

Key Terms
1. discriminant   2. vertex

Objective 1
Now Try

2. $(1, 1)$ 3. $\left(\dfrac{3}{2}, \dfrac{1}{2}\right)$

Practice Exercises
1. $(1, 3)$ 3. $(-6, 0)$

Objective 2
Now Try

4. vertex: $(-2, 1)$, axis: $x = -2$;
   domain: $(-\infty, \infty)$; range: $[1, \infty)$

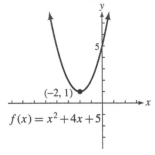

$f(x) = x^2 + 4x + 5$

Practice Exercises
5.

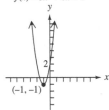

$f(x) = 3x^2 + 6x + 2$

Vertex: $(-1, -1)$
Axis: $x = -1$
Domain: $(-\infty, \infty)$
Range: $[-1, \infty)$

Objective 3
Now Try
5b. 0; 1 x-intercept

Practice Exercises
7. 0   9. 1

Objective 4
Now Try
7. maximum height: 286 feet after 1.5 seconds

Practice Exercises

11. 24 (a square)

Objective 5

Now Try

8. vertex: $(1, 2)$, axis: $y = 2$;

domain: $[1, \infty)$ range: $(-\infty, \infty)$;

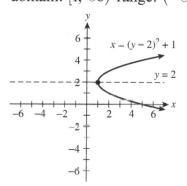

Practice Exercises

13.

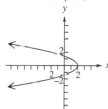

Vertex: $(2, 0)$
Axis: $y = 0$
Domain: $(-\infty, 2]$
Range: $(-\infty, \infty)$

15.

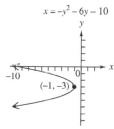

Vertex: $(-1, -3)$
Axis: $y = -3$
Domain:
$(-\infty, -1]$
Range: $(-\infty, \infty)$

11.8   Polynomial and Rational Inequalities

Key Terms
1. rational inequality   2. quadratic inequality

Objective 1
Now Try
2. $(-\infty - 4) \cup (-3, \infty)$

3. $(-1, 2)$

Practice Exercises
1. $\left(-3, \frac{1}{2}\right)$

3. $\varnothing$

Objective 2
Now Try
5. $\left(-\infty, -\frac{3}{2}\right] \cup \left[-\frac{1}{3}, \frac{1}{2}\right]$

Practice Exercises
5. $(-\infty, -5] \cup [-3, 1]$

Objective 3
Now Try
7. $(-\infty, 3) \cup [8, \infty)$

Practice Exercises
7. $(-\infty, 1) \cup [8, \infty)$ 9. $[-2, 3)$

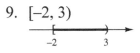

Chapter 12 INVERSE, EXPONENTIAL, AND LOGARITHMIC FUNCTIONS

12.1 Composition of Functions

Key Terms
1. composition   2. composite function

Objective 1
Now Try

3b. $(f \circ g)(x) = -3x^2 + 12$ 3a. 0

3d. $(g \circ f)(x) = (-3x - 3)^2 - 5$
$$= 9x^2 + 18x + 4$$

Practice Exercises

1. a. 25; b. 97; c. $8x^2 - 7$

12.2 Inverse Functions

Key Terms
1. one-to-one function   2. inverse of a function

Objective 1
Now Try

1b. One-to-one; $G^{-1} = \{(2, 3), (-2, -3), (-4, 5), (6, 0)\}$

Practice Exercises

1. One-to-one; $\{(-1, -3), (2, -2), (3, -1), (4, 0)\}$

3. One-to-one; $\{(0, 0), (-1, 1), (2, -2), (-3, 4), (-5, 3)\}$

Objective 2
Now Try

2b. Not one-to-one   2a. One-to-one

Practice Exercises

5. Not one-to-one

Objective 3
Now Try

3a. $f^{-1}(x) = \frac{1}{4}x + \frac{1}{4}$ 3c. $f^{-1}(x) = \sqrt[3]{\frac{x + 3}{2}}$

Practice Exercises

7. $f^{-1}(x) = \frac{x + 5}{2}$ 9. Not one-to-one

Answers

Objective 4
Now Try
4b.

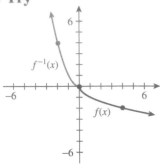

Practice Exercises
11. Not one-to-one

12.3   Exponential Functions

Key Terms
1. inverse   2. exponential equation

Objective 1

Objective 2
Now Try
1.

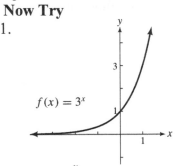

2.

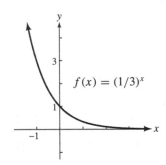

3.

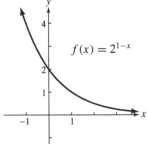

Practice Exercises
1.

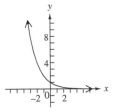

3.

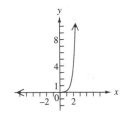

Objective 3
  Now Try

4. $\left\{\dfrac{3}{2}\right\}$

5a. $\{-2\}$

  Practice Exercises

5. $\left\{\dfrac{3}{2}\right\}$

Objective 4
  Now Try

7a. 70,000

7b. about 32,656

  Practice Exercises

7. 896 geese

9. 7750 bacteria

12.4 Logarithmic Functions

Key Terms
  1. logarithm

  2. logarithmic equation

Objective 1

Objective 2
  Now Try

1a. $\log_8 64 = 2$

1b. $\log_{81} 27 = \dfrac{3}{4}$

1c. $16^{-1/2} = \dfrac{1}{4}$

1d. $\left(\dfrac{1}{2}\right)^{-2} = 4$

  Practice Exercises

1. $10^{-3} = 0.001$

3. $\log_2 \dfrac{1}{128} = -7$

Objective 3
  Now Try

2d. $\left\{\dfrac{1}{6}\right\}$

  Practice Exercises
5. $\{81\}$

Answers

Objective 4

Now Try

4.

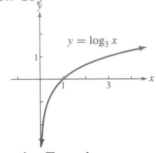

5.

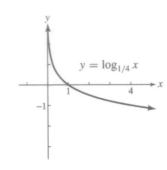

Practice Exercises

7.

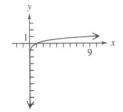

Objective 5

Now Try

6. 16 fish

Practice Exercises

9. 500 foxes

11. 335,000 items

12.5 Properties of Logarithms

Key Terms

1. special properties

2. quotient rule for logarithms

3. product rule for logarithms

4. power rule for logarithms

Objective 1

Now Try

1a. $\log_6 5 + \log_6 3$

Practice Exercises

1. $\log_7 5 + \log_7 m$

Objective 2

Now Try

2a. $\log_5 4 - \log_5 9$

Practice Exercises

5. $\log_6 k - \log_6 3$

Objective 3
  Now Try
3a. $3\log_6 4$
  Practice Exercises
7. $7\log_m 2$ 9. $\sqrt[3]{7}$

Objective 4
  Now Try

5a. $2+3\log_6 x$ 5d. $\log_b \dfrac{xy^4}{z}$ 5e. $\log_2 \dfrac{x^2(x-1)}{\sqrt[3]{x^2+1}}$

  Practice Exercises
11. $\log_7 8 + 7\log_7 r - \log_7 3 - 3\log_7 a$

12.6   Common and Natural Logarithms

Key Terms
  1. natural logarithm   2. common logarithm

Objective 1
  Now Try
1a.   2.9928   1c. −0.4776   1b. 4.8994
  Practice Exercises
1.   1.7576   3.   4.9401

Objective 2
  Now Try
2.   pH = 7.2076; rich fen   3. 2.5×10^{-4}
  Practice Exercises
5.   134 dB

Objective 3
  Now Try
5a.   4.5850   5b. 6.6758   5c. −1.0996
  Practice Exercises
7.   4.3347   9.   3.9120

Objective 4
  Now Try
6.   11.9 years
  Practice Exercises
11.   3240 years

12.7   Exponential and Logarithmic Equations and Their Applications

Key Terms
1. continuous compounding
2. compound interest

Objective 1
Now Try
1. 2.465
2. 6.770

Practice Exercises
1. {1}
3. {439.445}

Objective 2
Now Try

3. $-1+\sqrt[3]{36}$
4. $\left\{\dfrac{1}{2}\right\}$
5. {1}

Practice Exercises
5. {5}

Objective 3
Now Try
6. $12,201.90
7. 13.89 years
8a. 5309.18

8b. 34.7 years

Practice Exercises
7. $55,200.99
9. 7.7 years

Objective 4
Now Try
9a. 11.3 mg
9b. 30.0 years

Practice Exercises
11. 29 years

Objective 5
Now Try
10. 3.2266

Practice Exercises
13. 1.1887
15. −2.3219

Chapter 13 NONLINEAR FUNCTIONS, CONIC SECTIONS, AND NONLINEAR SYSTEMS

13.1   Additional Graphs of Functions

Key Terms
1.   step function

2.   greatest integer function

3.   asymptotes

Objective 1
Now Try
1.

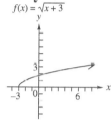

Domain: $[-3, \infty)$

Range: $[0, \infty)$

2.

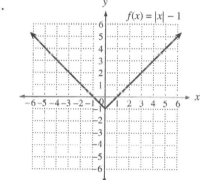

Domain: $(-\infty, \infty)$

Range: $[-1, \infty)$

3.

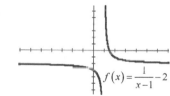

Domain: $(-\infty, 1) \cup (1, \infty)$

Range: $(-\infty, -2) \cup (-2, \infty)$

Practice Exercises
1.

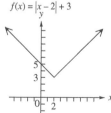

Domain: $(-\infty, \infty)$

Range: $[3, \infty)$

3.

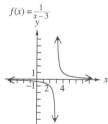

Domain: $(-\infty, 3) \cup (3, \infty)$

Range: $(-\infty, 0) \cup (0, \infty)$

Answers

Objective 2
Now Try

4d. 1

4e. −11

5.

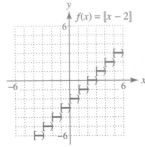

The domain is $(-\infty, \infty)$.

The range is $\{\dots,-2,-1, 0, 1, 2, \dots\}$ or the set of integers.

6.

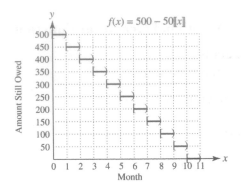

Practice Exercises

5.

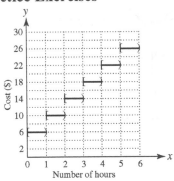

13.2   Circles and Ellipses

Key Terms

1. circle   2. ellipse   3. conic sections

4. center   5. radius

Objective 1

Now Try

3. $(x-3)^2 + (y+1)^2 = 6$

Practice Exercises

1. $(x-3)^2 + (y+4)^2 = 25$ 3. $x^2 + (y-3)^2 = 2$

Objective 2

Now Try

4. center: $(4, 1)$; radius: $\sqrt{2}$

Practice Exercises

5. center: $(-5, -2)$; radius: 6

Objective 3

Objective 4

Now Try

5b. $\dfrac{x^2}{25} + \dfrac{y^2}{36} = 1$

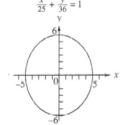

6.

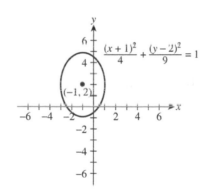

Practice Exercises

7. $\dfrac{x^2}{16} + \dfrac{y^2}{49} = 1$

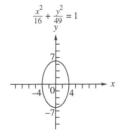

9.

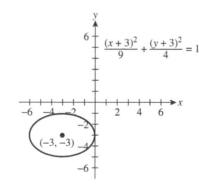

13.3 Hyperbolas and Functions Defined by Radicals

Key Terms

1. hyperbola   2. asymptotes of a hyperbola   3. fundamental rectangle

4. transverse axis

Objective 1

Objective 2
  Now Try
  1. $\dfrac{x^2}{25} - \dfrac{y^2}{4} = 1$

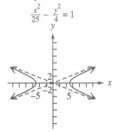

  Practice Exercises
  1. $\dfrac{x^2}{9} - \dfrac{y^2}{16} = 1$ 3. $\dfrac{y^2}{4} - \dfrac{x^2}{4} = 1$

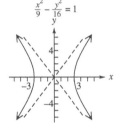 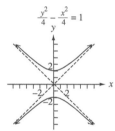

Objective 3
  Now Try
  3a. hyperbola   3b. parabola   3c. circle

  Practice Exercises
  5. ellipse

Objective 4
  Now Try
  4. $f(x) = -\sqrt{4 - x^2}$ 5. $f(x) = \sqrt{9 - 9x^2}$

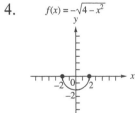 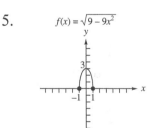

  domain: $[-2, 2]$ domain: $[-1, 1]$
  range: $[-2, 0]$ range: $[0, 3]$

Practice Exercises

7.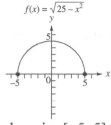

$f(x) = \sqrt{25 - x^2}$

domain: $[-5, 5]$
range: $[0, 5]$

9.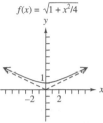

$f(x) = \sqrt{1 + x^2/4}$

domain:
$(-\infty, \infty)$

range: $[1, \infty)$

13.4   Nonlinear Systems of Equations

Key Terms
1. nonlinear equation   2. nonlinear system of equations

Objective 1
Now Try

1. $\{(-2, -3), (3, 2)\}$ 2. $\left\{\left(\dfrac{5}{2}, -4\right), (2, -5)\right\}$

Practice Exercises

1. $\{(12, -17), (2, 3)\}$ 3. $\{(0, 0), (-6, -2)\}$

Objective 2
Now Try

3. $\{(1, 3), (1, -3), (-1, 3), (-1, -3)\}$

Practice Exercises

5. $\{(2, 3), (2, -3), (-2, 3), (-2, -3)\}$

Objective 3
Now Try

4. $\{(2, -3), (-2, 3), (3, -2), (-3, 2)\}$

Practice Exercises

7. $\{(6, 1), (1, 6), (-6, -1), (-1, -6)\}$

9. $\{(4, 3), (-4, -3), (3i, -4i), (-3i, 4i)\}$

13.5 Second-Degree Inequalities and Systems of Inequalities

Key Terms

1. system of inequalities

2. second-degree inequality

Objective 1

Now Try

2. $y \geq x^2 - 4$

3.

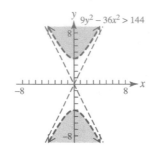

$9y^2 - 36x^2 > 144$

Practice Exercises

1. $x \leq 2y^2 + 8y + 9$

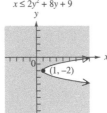

$(1, -2)$

3.

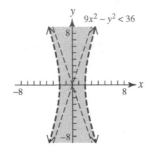

$9x^2 - y^2 < 36$

Objective 2

Now Try

5. $4y + x^2 < 0$
$x \geq 0$

7.

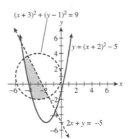

$(x + 3)^2 + (y - 1)^2 = 9$

$y = (x + 2)^2 - 5$

$2x + y = -5$

Practice Exercises

5. $x^2 + y^2 \leq 25$
$3x - 5y > -15$

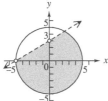